FAO
FORESTRY
PAPER
65

Forest legislation in selected African countries

Based on the review and analysis
of forest legislation in 11 member countries
of the African Timber Organization

by
Franz Schmithüsen
FAO Consultant

FOOD
AND
AGRICULTURE
ORGANIZATION
OF THE
UNITED NATIONS
Rome, 1986

M-30
ISBN 92-5-102322-0

PREFACE

In the recent past it has become clear that, for most developing countries, sustained growth and development can be achieved only through mobilizing the energies of rural people and the resources of the rural areas. This increasing focus on rural development has heightened understanding among governments of the importance of forests and forest outputs to development of the rural sector. As a result strategies are being developed which add a social dimension to the traditional production and protection objectives of forestry.

At the same time, many of these countries are in transition from an era of forest exploitation to one of sustainable forest management and production. This change carries with it the need to strengthen all elements of the forestry sector, from public administration to manpower development, resource planning, revenue enhancement, environmental conservation and forest products processing and marketing.

As these policies of forestry for rural development and modernization of the forestry sector are adopted, forestry laws and regulations must be reviewed and updated to implement them. The past few years have seen a number of countries introduce new and comprehensive forestry laws. Many others are in the process of preparing such laws, or will soon be doing so. It is to assist in these efforts that this publication has been issued.

The document was originally prepared to support efforts being undertaken by member countries of the African Timber Organization (ATO) to modernize and harmonize their forestry legislation, and was also used as the background document for a seminar on forestry legislation for forestry officials of ATO member governments. Subsequently, the author was engaged to expand the analysis and review sections of the report in order to make it useful as well to forestry officials and others in any country where revision of forestry laws is contemplated. We believe this aim has been accomplished.

Although the focus of the document remains on the member countries of ATO, the examples, analysis and general commentary should be of interest and value to anyone dealing in the subject of forestry legislation. For this reason, the report is being published as an FAO Forestry Paper and will be given wide circulation.

The document is a joint product of the FAO Legal Office and the Forestry Department, and the support provided by the legal staff is gratefully acknowledged. Thanks are due also to the member governments and the secretariat of the African Timber Organization whose cooperation made preparation of the report possible.

M. A. Flores Rodas
Assistant Director-General
Forestry Department

FOREWORD

Forest legislation is one of the important institutional elements for the development and expansion of the forest sector. It provides the structural framework within which national forest policies are set and in turn reflects or should reflect their objectives and priorities. It is also an indispensable instrument for the implementation of those policies.

The timber resources of the tropical forest zone in Africa have gained during the last two decades an increasing importance as a source of raw material both for national markets and for international trade in high quality tropical logs and processed forest products. The role of forests and forestry for a balanced and integrated development of rural areas is of concern to many governments and efforts are being undertaken in order to foster forestry for local community development. The member countries of the African Timber Organization (ATO) are aware of the economic potential of the forest resources but also of the danger of their uncontrolled depletion. Important policy and legislative measures have been taken in order to ensure a more rational and long-term utilization of the forest potential and to obtain a greater contribution from forestry activities to the socio-economic advancement of the region.

This publication gives evidence of the dynamic evolution that has taken place in the field of forest law as well as of the multitude of problems that have been subject to specific regulations.

This document reviews the situation of the forest legislation in eleven ATO member states on the basis of the available information and documentation. It summarizes the structure and principal content of the selected texts, in order to facilitate a general understanding of their development and of the specific problems that are of concern in each particular case. For a more detailed knowledge one will, of course, have to refer to the full text of the laws and regulations presented.

The document has been prepared as a source of information for government officials, in particular in national forest administration, as well as for inter-governmental and international agencies concerned with the regional aspects of forestry development. It should also be of interest as background material in higher forest education and for further comparative research on forest policy and forest law. It is hoped that the document will contribute to:

(i) an increased understanding of the importance of forest legislation for determining and implementing national policies and management systems for an improved utilization of forest resources;

(ii) an appreciation of the numerous changes that have already taken place in the field of legislation and of the need to envisage further modifications and improvements;

(iii) a more intensive regional exchange of views on forest policy objectives and their subsequent incorporation into forest law at national level, as well as to a closer cooperation among the ATO member states in the respective fields.

The coverage of the various countries is necessarily selective, depending on the availability of material and on the relevancy of the texts in a broader regional context. It would be highly desirable if the collection of laws and regulations could be continued and updated in the future by supplementary volumes, making available additional and new material in the field of forest legislation.

TABLE OF CONTENTS

INTRODUCTION

Among the most immediate tasks of the public authority responsible for the overall aspects of forest resources utilization are the preparation of sectorial development programmes, the establishment of an appropriate legislative framework, and ensuring the implementation in practice of the applicable provisions and regulations. The formulation of a realistic concept of how to use the available forest areas and the determination of precise norms for their management and conservation thus form part of the conditions that are indispensable for a sustained expansion of forestry.

It is the role of the government's forest policy to determine the sectorial objectives within the broad socio-economic framework of the country. It is the role of forest legislation to translate the requirements of such a policy into certain obligatory provisions.

Forest legislation, generally speaking, comprises the specific laws referring to the sector and their implementing regulations. One has also to include the regulations that concern the granting of timber harvesting contracts, leases and licences.

In a considerable number of countries, there exists in addition specialized legislation on wildlife and national parks management, hunting control and nature protection. These subjects are generally related to forest resources management and conservation, and are in fact included in certain countries in the forest law itself.

Moreover, there exists a whole range of other laws and regulations, the provisions of which may be relevant to the forestry sector. As examples, one may cite the laws and regulations on land-use planning and land tenure, legislation refering to public administration and state organizations, and certain elements of the fiscal and investment laws.

Modern and effective national forest legislation has to consider the particular socio-economic conditions prevailing at national, regional and local levels. Its content must not be inconsistent with the country's fundamental social and political concepts and options; on the contrary the applicable legislation should adequately reflect the principal economic development objectives of the government.

Conflicts which may occur, for instance, between the application of long-term forest management principles and the requirements of a rapid expansion of timber harvesting, or between a policy of reservation of permanent forest areas and the existence of customary land-use rights of the local population, have to be settled within the determination of national priorities and objectives and not merely by the provisions of a repressive legislation.

If such aspects are not considered in advance, the stipulations of the forest law risk being ineffective or even becoming inapplicable. One has to admit that many legislative provisions adopted in previous laws suffered precisely from these shortcomings. This refers in particular to regulations concerning land tenure and customary rights, land-use planning on forest areas and soils, and timber allocation procedures to small entrepreneurs and rural communities.

PART I

TENDENCIES AND TRENDS IN THE FOREST LEGISLATION OF ATO MEMBER COUNTRIES

1. FOREST LAWS AS PART OF THE NATURAL RESOURCES LEGISLATION

1.1 From Timber Harvesting Regulations to Integrated Land Management

Forest legislation deals with an important segment of natural resources and represents for that reason a substantive element of natural resources legislation in the more general sense. The recent forest laws adopted in the ATO member countries are gradually moving away from the narrow aspects of timber harvesting and felling controls and becoming more orientated towards planning and management of forest land. In addition, they tend to incorporate provisions that refer directly to environmental and nature conservation issues.

1.2 Acknowledgment of Resource Management Objectives

Practically all forest laws presently applicable in the ATO member countries contain some reference to the multiple role and uses of forests for environmental protection and raw material production. This reference is generally made either in the initial statement of the laws on its principal objectives and/or in the definitions of forest land and of protection forests. The following examples may be quoted:

- Cameroon: The purpose of the law is defined as to assure the conservation, exploitation and development of forest, wildlife and fishery resources (For. Law 1981, Sec. 1). Forest includes land capable of producing wood or other non-agricultural produce; of providing habitat for wildlife; and of providing an indirect effect on soil, climate or water regime (Sec. 3).

- Congo: The purpose of protection forests is the conservation or reestablishment of forest stands, flora, fauna, soils and water regime (For. Law 1974, Sec. 3, 2°). The use of forest resources has to be organized in such a manner as to avoid the destruction of forest domain and to assure its permanence, expansion and rational utilization (Sec. 23).

- Equatorial Guinea: Forests represent a natural resource that is to be conserved for the benefit of the present and future generations and to be used rationally and in a sustained manner (For. Law 1981, Instructory Statement). Forest benefits include the use of timber and other forest products, as well as soil and water protection and its role as habitat for wildlife (Sec. 2).

- Gabon: The purpose of the law is to promote the rational utilization of the forest domain, of wildlife and fishery, in order to increase the contribution of the forest sector to the economic, social, cultural and scientific development of the country (For. Law 1982, Sec. 1, Sec. 3). Rational utilization implies, in particular, an improved knowledge of the available resources, a better equilibrium between use and renewal of the resource, and where necessary the reconstitution of the resource in order to guarantee its sustained availability (Sec. 2 lines 1,2 and 3). Forests include land capable of producing timber and other forest products, protecting wildlife, and exercising direct and indirect effects on soil, climate and the water regime (Sec. 9).

- Ghana: One of the reasons for constituting forest reserves is to maintain or reestablish forest vegetation in order to safeguard water supply, to assist the well-being of forest and agricultural crops grown on the land or in the vicinity thereof, and to secure the supply of forest produce to the inhabitants of villages (For. Law, Sec. 4, N° 4). Integrated resources management is of concern to the Forestry Commission, established in 1982. It is, among other tasks, responsible for reviewing national practices relating to forests and forestry resources, formulating recommendations of national policy on forests and the exploitation of forest resources including game and wildlife, and for ensuring that forests are maintained and protected as an economic resource (Law 42/1982, Sec. 34, N° a, c).

- Ivory Coast: Protection zones may be constituted and include land with important soil erosion control and water management functions (For. Law 1965, Sec. 2). The permanent forest domain shall be used for timber production and guarantee the ecological equilibrium whereas the rural forest domain constitutes a land reserve for agricultural development (Decree 78/231, Sec. 2).

- Liberia: Forests are an important part of the country's natural resources, may contribute to the economic and social welfare, and are to be devoted to their most productive use for the permanent good of the whole people. The conservation and utilization shall be made wisely and in accordance with such restrictions as will ensure perpetual benefits from this heritage (For. Law 1953, Introductory Part).

- Tanzania: Forests include all lands capable of producing wood or other products, of exercising an influence on the climate or on the water regime, or providing shelter for live-stock and wildlife (Draft Law 1982, Sec. 2).

- Zaire: Forests include lands capable of producing timber and forest products, or exercising an indirect effect on the climate, water regime and soil (Decree 1949, Sec. 1).

1.3 Forest Land-Use Policies

New legislation or amendements provide for a more systematic and rational apportioning of forests among the principal land-uses in accordance with the various needs of national development. This refers in particular to the necessity that exists in many countries to provide more land for the expansion of permanent agriculture in order to cope with an expanding rural population, and to maintain at the same time a certain proportion of the remaining forests for continous timber production, environmental protection and communal forest development.

1.4 Long-Term Management Practices

Timber harvesting in most tropical forest zones is, generally speaking, still in an exploitation stage. Only the best trees of a small number of species are removed when new forest zones are opened up to logging. The future production of the moist tropical forests, especially the harvest of the second logging cycle, depends mainly on utilization of additional tree species not removed during previous logging; on the increment of commercial trees of lower diameter classes remaining in the forest after the first exploitation; and on natural regeneration, that occurs without additional silvicultural intervention. For this type of extensive forestry, the prescription of minimum diameter limits, the prohibition of returning to the over-logged area before the second logging cycle, and the determination of some form of annual allowable cut are the most immediate and practical requirements that may be imposed on the operators.

The application of this type of cutting regime represents a first step towards a long-term management strategy. In a subsequent phase, when a national land-use policy has determined which areas should be demarcated and managed permanently under tree cover, the annual production volume must be related to sustained yield principles based on annual increment of the remaining stands, on natural regeneration and/or reforestation.

1.5 Sustained Use of Forest Resources

Practically all of the ATO member countries have made efforts to improve their forest legislation so that it could provide the necessary instruments for introducing long-term management practices and support more consistently a rational utilization of the forest resources. One of the following sections reviews the presently applicable provisions related to management and it becomes evident that quite a number of basic relevant concepts have gradually been introduced in the laws and regulations.

On the other hand, one has to admit that the need to adjust the annual harvesting volumes to the long-term production potential of the forests has not yet found full acceptance in the various laws and regulations. It will thus be important to formulate in future revisions of the legislation more specific operational provisions that implement the generally acknowledged principle of a sustained use of natural resources.

1.6 Wildlife, National Parks and Nature Protection as Part of the Forest Legislation

The growing awareness of most of the ATO member countries of the importance of wildlife management and nature conservation has lead to the adoption of legislative provisions related more specifically to the environmental function of forests and forest land. Some forest laws (e.g. in Cameroon, Equatorial Guinea, Gabon and Liberia) incorporate the respective regulation as applicable to forests and other land. This type of legislation offers a broader framework for the utilization and management of natural resources, based on a multiple use concept. The new forest laws of Cameroon and Gabon include, for instance, several provisions which appear of particular interest in this context, such as the possibility of establishing national parks, fully protected nature reserves, protected wildlife areas and buffer zones around such areas, game reserves and game sanctuaries. It is also possible to create special game ranches, in which wildlife management and hunting are subject to particular development and promotionary measures.

1.7 Specialized Wildlife and National Parks Legislation

A second approach is to deal with wildlife protection and hunting control, national park management and nature conservation separately and to adopt specialized laws and regulations. Obviously, there exists a close interlinkage between forest laws that determine principally timber production and forest resources development, and this specialized legislation. Among the ATO member countries, those that have adopted or are considering adopting a separate wildlife and national park legislation included Congo, Ivory Coast, Ghana, Nigeria, Tanzania and Zaire 1/.

1/ The presented law collection is principally concerned with the use and management of forests and forest industry development. For this reason the subject of wildlife and national parks cannot be reviewed in detail. For a general discussion of these issues one may refer to the recent FAO study by du Saussay (1981): "La législation sur la faune et les aires protégées en Afrique".

1.8 Provisions on Inland-Fishery Development

In most French-speaking countries of West and Central Africa, fishery development in rivers and fresh-water ponds has traditionally been associated with the forest sector and is generally under the responsibility of the same national administration 1/. For this reason, the forest laws of Gabon and Cameroon include special regulations on this subject. The provisions of the latter country appear to be particularly comprehensive and management oriented.

1.9 Complementary Legislation

Among the laws that are of particular relevance to forest resources development the following should be mentioned:

- Legislation on land-use planning affects heavily the scope and effectiveness of forest legislation. The provisions of forest laws on reservation, demarcation and management of permanent production and protection forests as well as the removal of harvestable timber from land to be turned over to agricultural development, require a close coordination with the respective provisions on national and regional land-use planning and practices.

- Land tenure legislation may be of importance in particular in connection with the constitution of state, communal and farm forests.

- Rural development legislation and in particular agricultural legislation has an important bearing on integrating forestry into rural development. It is therefore essential to make forestry and rural development legislation compatible, in order to make both more effective.

- Soil conservation and land rehabilitation laws are complementary to forest legislation. The protection of critical water-sheds and the rehabilitation of degraded land depends usually on the protection of the remaining forest cover or on reforestation for protection purposes. Specific soil conservation legislation has been enacted, for instance, in Nigeria.

2. CONSTITUTION OF THE FOREST DOMAIN AND CATEGORIES OF FOREST LAND

2.1 Permanent Forest Land

The legislation of all ATO member countries provides for the possibility of setting up permanent forest reserves or national forests. The forest laws of Gabon and Cameroon stipulate respectively that at least 40%, and 20% of the national territory is to be reserved as permanent forest land. Forest laws generally include detailed provisions regulating applicable procedures for forest reservation and in particular for dealing with customary rights. The regulations of the anglophone countries are more elaborated in this respect and ensure a larger degree of participation of the local population during the inquiry on constituting permanent forest tenure.

1/ Marine fishery development is generally subject to a separate specialized fisheries legislation. See FAO (1982): "Compendium de la législation sur la pêche - Région du COPACE".

Timber harvesting in forest reserves is usually subject to special restrictions. In some countries forest utilization contracts may only be granted on forest land not classified as forest reserves.

2.2 Classification of Permanent Forest Land According to Principal Uses

Permanent forest land may be classified according to certain functional uses such as production, protection and recreation forests, and reforestation areas. The forest laws of several countries provide for the reservation of forests and other land for various other purposes such as the constitution of national parks; partially and fully protected nature reserves; wildlife sanctuaries and wildlife management areas.

2.3 Changes in Permanent Forest Tenure

The reservation of permanent forest land is a dynamic process. Due to increase of population and changes of the socio-economic conditions in rural areas, certain modifications in the pattern of the established permanent forest land tenure may be necessary. The forest legislation of all countries provides for the possibility to dereserve permanent forest areas under certain conditions. Changes of permanent forest tenure are generally to be made by law or decree.

The forest laws of Nigeria and Tanzania provide for considerable flexibility in establishing and managing the country's permanent forest estate. Ivory Coast has recently adopted regulatory measures which allow for a completely new classification of the previously established forest reserves.

2.4 Protection Forest Land

Forests which have not been constituted under permanent tenure are classified in most forest laws as protected forest land. In Cameroon the term "communal forests" and in Ivory Coast the expression "rural forest domain" are used for such land. Part of the protected forests may eventually be put under permanent tenure whereas other areas may be required for agricultural development.

The provisions of the forest legislation which refer to protected forests are generally concerned with the protection of certain tree species, and certain categories of forest produce used by the local population. Most laws also provide for the enforcement of minimum cutting limits on protected forest land.

2.5 Salvage Logging

The existing regulations do not appear sufficiently comprehensive and flexible to deal with the problems of forest land which has not yet been constituted under permanent tenure.

- Ivory Coast has recently adopted regulations on the management of the rural forest domain, which are of considerable interest in this context. On forest land that is to be converted to agricultural use, all utilizable timber has to be harvested prior to the start of agricultural development programmes.

Provisions that facilitate systematic salvage logging operations prior to changes in land-use, also exist, for instance, in Cameroon, Ghana and Liberia.

2.6 Summary of Categories of Forest Land as Established by the Prevailing Forest Laws

The tenurial status of forest land and its functional sub-division may vary considerably from country to country. On the whole, it follows, however, the previously indicated pattern. The following summarizes the categories of forest land as determined by the basic forest legislation of the ATO member countries. Wildlife areas and national parks are only mentioned if refered to in the respective forest laws.

- Cameroon: The permanent state forest domain may comprise integral nature reserves, national parks, sanctuaries for certain wild animals or plant species, game reserves, production forests, protection forests, recreation forests, forest plantations, zoological and botanical gardens, and game ranches. No particular subdivision is foreseen for local council forests and private forests. The communal forests are land not yet classified under permanent forest tenure; their status may be compared to that of the protected forests, which exist in several other ATO member countries.

- Central African Republic: The permanent state forest domain includes state forests, fauna and flora reserves, integral nature reserves, and national parks. Forests owned by local councils and public agencies such as communal and rural forests are also under permanent tenure. The remaining areas are classified as the customary forest domain; its status corresponds largely to that of protected forest land.

- Congo: All forest land belongs to the forest domain of the state. Part of the forests may be classified under permanent tenure as production and protection forests, national parks, nature reserves, integral nature reserves, and forests for local community development. Forests which have not been subject to a special classification procedure are considered as protected forest land.

- Equatorial Guinea: Forests are only classified as state, communal and private forests. Special provisions on permanent forest tenure and different categories of functional uses have not been included in the forest law.

- Gabon: The permanent state forest domain comprises permanent production forests, reforestation areas, national parks on forest land, protection forests, botanical gardens and sanctuaries for certain species, and wildlife management areas. State forests which have not been constituted under permanent forest tenure are considered as protected forests. The forest law does not contain provisions on communal and private forest tenure.

- Ghana: The forest law refers only to forest reserves which as a matter of principle may comprise lands which are the property of the government, tribal or stool lands, and private lands. In practice forest reserves are generally established on tribal and stool lands. Forest land which has not been classified under permanent tenure as a forest reserve may be declared a protected area in virtue of Sec. 12 of Decree N° 273/74. Such declaration implies the requirement of a licence for farming in such areas, the probihition to damage trees and timber, and the restriction of the use of fire.

- Ivory Coast: The permanent state forest domain comprises the classified forest reserves and the protection and reforestation areas. New regulations refer to this category of forest land as the permanent forest domain of the state. The remainder of the state forests have the status of protected forests and are now called the rural forest domain of the state. The forest law also acknowledges the status of registered and immatriculated communal and private forests but no distinction is made with regard to permanent or temporary tenure.

- Liberia: The forest law provides for the creation of government forest reserves, which are to be managed as the permanent forest estate. There may also be created native authority forest reserves with a temporary tenurial status; such reserves are protected as potential government forest reserves. There exists also the status of communal forests for smaller forest areas, and of national parks, both under permanent tenure. The law does not include provisions on private forest land but more recent regulations are specifically concerned with private forests.

- Nigeria: The forest legislation generally provides for the constitution of government forest reserves and native authority or local government council forest reserves. Both categories have the status of permanent forest tenure. Other forest land may be declared to become a government protected forest, or a native authority or local government council forest. This declaration may precede the formal inquiry on constituting a forest reserve. There exists also a provision to create communal forests on the request of any native community. No special provisions of the forest law refer to private lands.

- Tanzania: The forest law distinguishes between unreserved forest land, provisional state forest reserves, state forest reserves, and local authority forest reserves under permanent tenure. Local authority forest reserves may include forest reserves of district councils, city and municipal councils, and local and village councils. Forests on private land are only mentioned in the context of dedication convenants and official resource assessment.

- Zaire: The Decree on the Forest Regime provides for the constitution of classified forests (forest reserves) on state forest land, land to be reforested for protection reasons, and forest land owned by traditional communities. Forest lands that have not been declared under permanent tenure are considered as protected forests. Private forest tenure is acknowledged in principle and certain limitations with regard to the clearing of the forest cover on such land have been made.

3. CUSTOMARY USAGE RIGHTS AND OWNERSHIP OF FOREST LAND

3.1 Importance of Customary Usage Rights

Customary usage rights are covered in considerable detail by the forest legislation of all ATO member countries. They are an essential element of the prevailing forest laws which acknowledges the interdependence between rural people and their forest environment. The practice of usage rights is an important contribution for securing part of the livelihood of farmers and villagers. These rights generally comprise the following uses:

- the use of trees, shrubs, dead wood and branches for firewood;

- the felling of trees for construction timber and fence posts;

- the collection of forest produce from trees such as bark, latex, gum, resin fruits and nuts;

- the collection of forest produce other than trees such as medicinal or edible plants, fibres of tree climbers, copal, honey, stones, laterite and limestones;
- the practice of hunting and fishing;

- the practice of grazing in open forests and the use of branches and leaves for fodder and forage;

- the clearance of forests for subsistance agriculture;

- rights of way and water usage rights.

3.2 Regulation Concerning Usage Rights

Usage rights may not be practised to the same degree on the different categories of forest land. On unreserved forest land the usage rights may generally be exercised freely in accordance with traditions and customs. Regulations are only made in order to limit certain uses or methods of use that would endanger the sustained practice of the rights. If, however, certain zones have been classified and demarcated as part of the permanent forest estate, the rights of the local population are set out in more detail and likely to be restricted in quantity and quality. Forest clearing for subsistance agriculture, for instance, may be admitted on unreserved forest land, whereas it is generally prohibited in legally constituted permanent forests. Grazing is usually admitted on unreserved land, but excluded or severely limited in forest reserves. Customary uses may also be completely prohibited or abolished in certain state forests either by the general regulations or on the basis of local arrangements that provide compensation to the villagers.

3.3 Administrative Authorization and Fees

The practice of usage rights within the legally admitted scope is, in principle, free of charge and may be done without any formal authorization. In many instances the applicable regulations provide, however, that the practice of these rights is subject to special permits and to the payment of fees. Such an approach is quite cumbersome for the villagers as well as for the local administration, and it is doubtful that this type of provision is generally required and justified. There may be cases which call for administrative regulation of usage rights through the introduction of permits if, for instance, the uses tend to be excessive and devastating. But on the whole it would be more practical to provide in the legislation only the possibility of issuing permits if required under certain circumstances. It would then be the task of the local institutions to determine whether such administrative measures should be introduced or not.

3.4 Management of Usage Rights

From a review of the presently applicable provisions that regulate the various forestry uses of the local population, the impression emerges that the usage rights are considered as an inevitable burden and strain on the existing forest resource, that is tolerated by the regulations and the responsible administrations. The social dimension of usage rights as an important part of the daily subsistence of rural people, determining their interest in the existing forests, is not much taken into account in the existing laws. Regulations that provide for the registration of local and customany uses as established rights of certain communities and local groups of people are rarely to be found. If an inquiry and registration is to take place, it is usually in connection with the constitution of state forest reserves and with the likely objective to limit or abolish the practice of such uses. Nor do there exist legal provisions and administrative procedures that would facilitate an adjustment of customary uses to the long-term potential of the resource.

In order to improve this situation, it will be necessary to strengthen the positive aspects of forest legislation with regard to forest usage rights. Forest laws should provide a firm support for an active management of forests and forest lands in order to satisfy local needs and demands. They should encourage and facilitate local participation and establish responsibilities of rural communities in decision making on forest zones in which usage rights are of importance. The regulations of Nigeria and Tanzania, and to a somewhat lesser degree those of Ghana, offer positive examples for such an approach.

3.5 Impact of Industrial Timber Harvesting and Commercial Use of Forest Products

The issue of forest use by local communities - be it in the form of traditional usage rights, under specific authorizations or only tolerated in practice - has also to be seen in the context of modern industrial timber harvesting and of other commercial forest uses. These operations are generally authorized by permits and utilization contracts, granted in accordance with the forest law. Nevertheless they may seriously interfere with the needs of local people.

If new forest zones are opened up to large scale exploitation, hunting may become immediately more difficult or impossible. Trees that have some specific local uses may be removed by the contractors or be destroyed during the felling operations. Timber that could be used as raw material for local handicraft and village industries may be harvested by commercial firms. Products like nuts and resin, that could be collected and processed by villagers in order to make a living, are exploited by people from outside the community on a commercial scale.

3.6 Protection of Usage Rights in Connection with the Granting of Forest Concessions and Logging Permits

The laws, regulations and individual agreements of practically all countries provide that the exercise of the usage rights by the local population must not be disturbed by the operations made under forest concessions. Certain tree species, categories of forest produce or determined forest areas in the vicinity of villages may be excluded from commercial logging operations.

3.7 Maintaining an Adequate Forest Environment for the Practice of Usage Rights

Conflicts with the interests of the local population may, however, occur even if the official permits and contracts stipulate that local usage rights have to be respected by the contractors. What does it mean to the local community if the hunting rights of villagers are respected in principle, but if the game disappears after commercial timber harvesting followed by forest clearing? The inherent deficiency of this situation results again from the fact that local forest uses may be tolerated but not established as rights in their own, specifically determined and, where necessary, quantified. The difficulties also arise, because large scale timber harvesting or the commercial use of other forest products may change the forest environment to such an extent that the practice of customary uses may become marginal or impossible.

One can thus conclude that the compatibility of industrial timber harvesting and other large scale commercial uses on the one side, and the protection of the existing customary uses of trees and forests on the other hand, merit considerable attention. Those of the ATO member countries that have granted or are in the process of granting large forest concessions and timber utilization contracts need to review the respective regulations and contractual arrangements. The objective of such a review should be that customary usages rights should not only be admitted but that their practice should be facilitated and sustained.

3.8 Abolishment of Usage Rights and Compensation

It may be necessary to limit and even abolish local usage practices, that are not compatible with the principal objectives of forest management or are detrimental to the forest and vegetation cover itself. Such restrictions may be justified and essential in order to protect the resource in the long-term interest of the local community itself, and for maintaining a stable environment in the country. A typical example is the prohibition of land clearing and the restriction of grazing in forest on steep hills with important protective functions. In order that such limitations and restrictions are to be respected, they must be explained to the population. If formally acknowledged usage rights are concerned, it is necessary to negotiate the envisaged restrictions with the concerned group of people and to offer some kind of compensation. One important type of compensation that merits being considered with much attention in this context is the designation of certain forest areas for constitution as community or communal forests.

3.9 Possession and Ownership of Forest Land

Possession and ownership of forest land comprise in principle state forest tenure, communal forest tenure and private forest tenure. The possession and ownership of forests may also not yet be determined in a definite manner or may eventually change. This refers usually to forest land which has not been the object of formal reservation procedures. In several of the ATO member countries only a limited proportion of forests has been declared either state or communal forest reserves. The remaining forest land is generally under a rather flexible form of state tenure, but part of it can eventually be constituted as communal forest land.

3.10 State Forest Tenure

State tenure of forest land exists in all ATO member countries. In Cameroon, Congo, Equatorial Guinea, Gabon, Ivory Coast, Liberia and Zaire, state tenure of forest reserves and protected forests is the predominant or even exclusive form of forest ownership. In these countries, therefore, forest laws are mainly concerned with the utilization and management of state forests.

3.11 Communal Forest Tenure

The term communal forest tenure comprises in this context all forms of possession and full-right ownership of forest land by formally constituted groups of people or local communities. This definition includes firstly possession and ownership of forests and forest lands by community governments, such as towns, villages and local authorities. It includes secondly forests and forest land that are possessed or owned by recognized non-governmental groups of people, such as tribal and customary organizations.

The forest laws of all ATO member countries - with the exception of Gabon - include some provisions on communal forest tenure. In Congo, the forests belong to the state but a new provision has been inserted in the law which allows for the declaration of forests for local community development.

3.12 Various Forms of Communal Forests

Communal forests may be constituted under various tenurial arrangements. The most advanced form of communal forest land is full ownership on the basis of registered land titles. "Full ownership" is used here to mean ownership that entitles the community to the use of and the benefits from their forest land. It includes - at least as a matter of principle - the possibility of granting harvesting licences and permits, to grant land leases and to exchange or sell the land itself. The use of fully owned communal forests is, however, subject to numerous rules and limitations as determined by the forest and/or the general administrative legislation. Full communal ownership on the basis of registered land titles is fairly rare in the ATO member countries.

A more common type of communal forest tenure may be called "restricted ownership", where customary title is formally recognized, but practical powers to alienate, manage and exploit the forest are narrowly circumscribed. The ownership of the land may be recognized, but not the timber rights, for example. Management and timber harvesting authority in such cases may be vested in the state forestry agency. Revenues from the forest may or may not be fully paid to the customary owners.

In practical terms, restricted ownership may not be distinguishable from customary usage rights without ownership. Usage rights may include the right to cut timber, pasture livestock and clear and cultivate land, but they are frequently limited to removing fallen timber and non-timber forest products. Customary rights-holders may or may not receive a share of the fees from commercial exploitation.

The nature of the usage and property rights that are recognized, the procedures provided for their recognition and the limitations placed on their exercise all vary greatly. The number of resulting combinations is large and can only be appreciated in detail through knowledge of the forestry and land legislation of the country, including the customary laws.

3.13 Regulations on Managing Communal Forest Land

There must be a clear understanding of the major purposes for constituting communal forests and the principal uses and benefits that can accrue to the community. There must also be an understanding that the potential of these forests is limited and that an adjustment has to be made between the local demand and the long-term yield of the resource.

The forest laws of Ghana, Nigeria and Tanzania include specific provisions on the management and development of communal forest land. These provisions generally refer to the use of trees and forest products by the local communities, to reforestation, yield regulations and management plans, payments from a revolving fund and technical assistance by the forest service. The most effective and pragmatic framework for the utilization of communal forests are rules and management prescriptions that have been agreed upon by the community, subject to approval by the competent governmental services. This possibility exists, for instance, in Nigeria and Tanzania where local rules, regulations and by-laws may be made on the use and management of communal forests.

The legislation of the other ATO member countries is generally little concerned with management procedures on communal forest land.

3.14 Private Forest Land

Private forest tenure is recognized in the forest laws of Cameroon, Equatorial Guinea, Ivory Coast, Liberia, Tanzania and Zaire. Private forests are, however, of a very limited importance in these countries. This refers in particular to the natural tropical forests which are almost exclusively under public ownership. Private forest land is generally limited to small individual holdings which result from plantations by farmers and villagers. The provisions of the forest legislation usually concentrate on the issue of permits and authorizations for timber harvesting. Little or nothing is said on the management and development of private forests. In Liberia, where some larger private forest holdings exist, special regulations have recently been adopted, which provide for measures of assistance by the Forest Development Authority.

4. FOREST RESOURCES ALLOCATION METHODS

4.1 Importance of Forest Utilization Contracts, Harvesting Licences, and Logging Permits

In the early stages of forestry development the state is frequently not in a position to organize the utilization of large areas of tropical forests by direct intervention of the national forest service in the form of state operated timber extraction. Other timber allocation systems that imply the involvement of private companies and/or specialized semi-autonomous public agencies are generally used in order to organize timber harvesting operations. In all ATO member countries forest utilization contracts, timber harvesting licences, and logging permits play a dominant role in the utilization of tropical forest resources.

4.2 Improvements in Forest Resources Allocation Procedures

Considerable efforts have been undertaken during the last two decades by several of the ATO member countries to improve the formal procedures for resources allocation as well as the substantive content of timber harvesting arrangements on public forest land. The negative experience with previous timber leases and licences as well as the rapidly increasing value of high quality tropical timber have made it necessary to introduce and implement more coherent allocation systems and to improve the supporting legislative and/or contractual provisions. Legislation and agreements for long-term utilization contracts that appear of a particular interest, exist, for instance, in Congo, Gabon and Liberia.

4.3 Regulations on Granting and Operating Forest Utilization Contracts

Particular emphasis has been put on the elaboration of an appropriate regulation with regard to:

- the relation between the envisaged resource commitment in terms of contract duration and size of granted area, and the proposed level of investment and local processing;

- the preparation and subsequent implementation of utilization contracts in accordance with forest management regulations;

- the determination of provisions that ensure specific contributions to the development of social and economic infrastructures in the forest areas and to the training of national staff;

- the encouragement of local timber processing or restriction on log exports, and the increasing participation of the national sector in the expansion of the industry;

- the establishment of a performance control system for the companies to which utilization contracts have been granted;

- the introduction of forest revenue and export tax assessment procedures that enable the granting governments to capture a greater proportion of the resource rent from the committed harvestable timber stock.

4.4 Contract Duration

A comparison of contract duration in various countries shows:

- that unrealistically long contracts, which have sometimes existed in the past, are disappearing. A duration of around 20 years seems to become the upper limit for long-term contracts;

- that with an increasing differentiation in contract tenure the individual duration of concessions becomes more realistically related to a determined investment level;

- that an effective timber allocation policy requires short-term and medium-term arrangements as well as long-term contracts in order to achieve a balanced development in timber harvesting and processing;
  and
- that an allocation system, principally based on medium- and long-term arrangements may need additional measures of support (e.g. formation of cooperatives, credit facilities) for national operators, who may find it otherwise difficult to expand into integrated timber harvesting and processing.

4.5 Provisions on Maximum Contract Duration in ATO Member Countries

In Cameroon renewable timber harvesting licences with a 5 years duration are issued. In the Central African Republic large concessions usually have been granted for a duration of 15 years. Congo has introduced provisions that permit the granting of wood transformation contracts with a negotiable duration (e.g. 14 years); and of timber harvesting contracts for periods not exceeding 7 years. In Equatorial Guinea, the maximum duration of forest concessions is 20 years.

The Gabonese regulations of 1968 provided for the granting of industrial permits with a duration of up to 30 years. The new Forest Law does not determine a maximum period but establishes that the type of permits and licences to be issued as well as the granting procedures are to be determined by regulations.

In Ivory Coast, different contract durations have already been introduced during the period 1965/1968. 5-year permits were granted to logging companies; 10-year permits to companies operating a sawmill and 15-year permits to integrated wood-processing units. Expiring permits are generally renewed for 5-year periods. In Liberia, the duration of long-term contracts is fixed by agreement, generally in the range of 20 years. New regulations of Zaire provide for a maximum duration of contracts of 20 years.

In Ghana, Nigeria and Tanzania, the forest legislation does not include similar provisions; the duration of permits and licences is ususally fixed on a case-by-case basis by agreement.

4.6 Size of Granted Areas

As a matter of principle the size of a granted concession area should depend on such considerations as:

- the effective investment level of the envisaged logging and/or timber processing operation;

- the intended contract duration;

- the actual and potential utilization possibilities of the forest areas to be granted; and
- the prescription of management plans that regulate maximum annual cutting volumes.

Several countries have followed this approach in negotiating new forestry projects with the aim of adjusting the concession area to be granted to the annual raw material requirements of the proposed forest industry and to the annual allowable cut as determined by the management plan. The forest laws of Cameroon and Gabon provide in addition for a maximum concession area (200 000 ha), that may be attributed to one single operator or to a company and its affiliates.

The Forest Law of Equatorial Guinea determines in detail the linkage between the size of the granted area and contract duration. Areas with less than 25 000 ha are issued for 5 years; areas between 25 000 and 50 000 ha for 5 years with renewal; areas with more than 50 000 ha for 10 years; and areas beyond 80 000 ha for 10 years with a renewal for another period of 5 to 10 years.

4.7 Raw Material Supply Contracts Based on Annual Production Volumes

Congo and Zaire have replaced the granting of determined concession areas by introducing the concept of raw material supply contracts, which guarantee a determined annual production volume of commercial species for a certain period of time. The advantage of this system is principally that it leaves a greater flexibility in the management of forest land to the public forest administration. On the other hand, this approach will only work if detailed inventory information and technically sound forest management plans are available.

4.8 Forest Inventories, Management Plans and Annual Allowable Cut

The legislation and forest utilization contracts of most ATO member countries include provisions on the preparation of forest inventories and management plans. The preparation of forest management plans prior to timber harvesting is, however, in some countries, only obligatory in forest reserves.

The introduction of an annual allowable cut that is determined for each forest utilization contract and regulates the timber harvesting volumes of the operators is a necessary element for industry planning and long-term resources management. Annual logging volumes should be calculated for each important commercial species or for groups of species. During the initial period, no restrictions will be necessary for species which are of limited commercial interest. Considering that utilization standards are improving (harvesting of lower-grade material and additional species), the annual allowable cut should be adjusted through periodic revisions of the management plan. Forest land in the vicinity of expanding zones of shifting cultivation or officially earmarked for conversion to other land-uses (salvage logging areas), should also be considered when determining the annual production volumes.

4.9 Specific Provisions in ATO Member Countries

The new Forest Law of Cameroon (1981) stipulates that timber harvesting in all areas must be preceded by forest inventories. Management plans are obligatory for all forests classified as state forest reserves. These provisions offer a possibility to plan the advancement of logging and to prescribe annual harvesting volumes within certain regional units. Congo in its 1974 Forestry Code has introduced detailed provisions for the establishment of forest management units as well as for the preparation of forest inventories and management plans, which determine maximum annual harvesting volumes for the principal commercial species. Equatorial Guinea requires that the concessionaire submit a project proposal which includes detailed inventory information.

The Forest Law of Gabon stipulates that the opening of new logging zones must be preceded by inventories. The inventory information and a logging plan have to be submitted by the operators. Management plans are obligatory for all formally constituted state forest reserves. In Liberia, the preparation of a management plan is part of the timber harvesting agreement. In Zaire, recently introduced provisions on the granting of raw material supply contracts include the requirement of the submission of a logging plan for the initial five years by the prospective operator. Management plans that determine an annual allowable cut are current practice in Ghana, Nigeria and Tanzania The Forest Law of the latter country includes a specific provision on the preparation of working or management plans in the classified state forest reserves.

4.10 Minimum Cutting Limits

The forest legislation of most countries provides minimum diameter or girth limits for the cutting of species of commercial interest. Minimum cutting limits vary between 50-80 cm diameter depending upon the growth pattern of a particular species. The point of measurement is fixed above buttresses, at breast height (DBH), or at 4 metres from the ground. The determination of minimum diameters or girth limits for forest exploitations is a first management regulation, which prevents trees of smaller dimensions being cut before they reach full maturity. Minimum cutting limits should remain enforced, until the forests can be treated under more elaborate management and silvicultural systems.

4.11 Reforestation

In Liberia, the concession agreements formerly included a formal obligation to reforest such part of the concession area as required for a continuous wood supply from the granted forests. In general, this provision has not proved to be successful. Concessionaires have either failed to establish forest plantations or did not have the experience to carry out efficient operations. New regulations now provide for plantation establishment by the Forest Development Authority, and for the concessionaires to pay a silvicultural fee per m^3 removed timber. A similar approach has been adopted in Ivory Coast, Cameroon, Congo and Gabon, where reforestation is carried out by specialized governmental agencies, which finance their operations from reforestation and management taxes levied on timber extraction. In Ghana, Nigeria and Tanzania, no reforestation obligations are imposed on the logging companies; all silvicultural work and reforestation are carried out by the Government Forest Service. Zaire has recently introduced a policy which leaves the responsibility for reforestation to the logging operators.

4.12 Status and Use of Forest Roads

The regulations which are applicable in the ATO member countries generally provide that the concessionnaires have the right to construct forest roads within their concession area. They may also construct roads through the area of other concessionnaires at their own expense or use the logging roads of such concessionnaires, provided that they pay an appropriate share of construction and maintenance costs. In some cases, the construction of forest roads outside of the concession area is subject to a special administrative authorization. Forest roads, constructed by the concessionaires, are usually considered as general traffic roads and may be used by the public. Access may be restricted if it interferes seriously with current logging operations, or if logging activities present a hazard to public use of the road.

4.13 Planning and Construction of Forest Roads

Forest legislation or contractual arrangements include little on the planning of the basic network of major logging roads and on minimum construction standards (road width, minimum curb radius, maximum slopes) of such roads. More attention should be given in order to integrate major forest concession roads into the national road network. The location of logging roads of possible public interest should be identified and minimum construction standards, compatible with those of general traffic roads, be determined prior to concession granting.

4.14 Involvement of the Local Population in the Process of Granting Utilization Contracts

A large proportion of the granted utilization contracts are on forest land controlled by the state. The local population is usually not involved during the process of granting or renewal of such contracts. The new regulations of Cameroon include detailed provisions requiring that all forestry projects, which imply the granting of large forest areas must be presented to the local representative from the area concerned. The project is submitted to a special committee in which regional and local governmental officers, the presidents of village councils, local chiefs and representatives, and the industrial operator participate. The local representatives are informed and can make suggestions on and objections against the project. This procedure should contribute to protecting the interest of rural people - and in particular the practice of their usage rights - more effectively. It could be of interest to other countries as well.

4.15 Infrastructural Improvements

Legislation and/or agreements of several countries have in the past included a provision that the concessionaires had to carry out certain infrastructural improvements (e.g. construction of dispensaries, maintenance of communal roads, supply of water and electric energy) in favour of the local population. Such provisions were, however, frequently ill-defined and subject to disputes. For this reason, Ivory Coast has modified its legislation and abolished the clauses on the execution of works of public interest (travaux d'intérêt général). Instead, the concessionaires are required to pay a certain sum per granted concession area, which is to be used for improving the living conditions of local communities. The Forest Law of Cameroon establishes the payment of a Territorial Fee to be paid into an Inter-Communal Development Fund. It also provides for the levy of a Financial Contribution to Socio-Economic Infrastructural Developments, which is to be received by the local councils in the territory of which the logging zones are situated. In Congo, the Forest Law provides that the obligations related to infrastructural improvements have to be determined in the list of requirements (cahier des charges particuliers), which is to be attached to each utilization contract.

5. TIMBER PROCESSING AND PROMOTION OF LESSER-USED SPECIES

5.1 Principle of Increased Timber Processing and National Participation

It is the intention of all ATO member countries to build up and expand their forest industries, to increase the percentage of locally processed timber, and to promote greater national participation in the forest sector as a whole. Several measures, including the supporting legislation, have been introduced during the last years, which should speed up this process. The forest laws of several countries include a general reference to these objectives. In particular the Gabonese Forest Law establishes that its principal objectives include a more rational use of the timber resources and an increased control of timber harvesting; the promotion of forest industries with a larger proportion of processing; and a greater effective participation of nationals. Congo provides that the timber resources are to be processed within the country and not exported as logs, and that processing should take place in the vicinity of the forest areas. Foreign participation in logging and wood processing should be reduced in accordance with the provisions of each utilization contract.

5.2 Long-Term Utilization Agreements as an Incentive for the Establishment or Expansion of Wood Processing Units

Practically all ATO member countries use the granting of long-term utilization contracts as an instrument to foster the development of forest industries. As a rule, contracts which cover larger forest areas may only be granted in connection with the expansion of already existing or the establishment of new timber processing units. The Forest Law of Gabon provides, for example, that new timber harvesting licences, covering forest areas of more than 15 000 ha, may only be granted to companies operating a forest industry. In Ivory Coast the redistribution of temporary logging permits in the established wood supply areas is to be made in favour of the existing or planned industries. Similar regulations exist in Cameroon, Congo, Equatorial Guinea and Zaire.

5.3 Minimum Percentages of Produced Log Volumes to be Converted Locally

The Forest Regulations of 1983 of Cameroon stipulate that at least 60% of all wood cut in concessions has to be transformed locally. In concessions with less than 20 000 ha of granted forests, the minimum volume for local processing has to be respected, but the concessionaire has an option of selling this volume to other industries or of operating his own processing unit. In concessions with more than 20 000 ha, the concessionaire has to establish a forest industry with a determined capacity. The required minimum of the annual log intake volume and the type of processing unit to be operated depend on the size of the granted concession.

The forest legislation of Congo provides that wood transformation contracts may only be granted if at least 40% of the logging volume is processed locally. This minimum rule applies, however, only in the northern part of the country. Concessions for forest areas in the south provide for 100% local transformation of Okoumé and at least 50% of other species. In the Central African Republic, minimum volumes for local wood processing have so far been determined for particular concession units only.

In Gabon, the minimum percentages of logs to be processed locally in new concessions is 75%

In Equatorial Guinea, the minimum of the installed processing capacity is calculated according to a formula based on the granted concession area. Zaire has taken a somewhat similar approach by varying the minimum volume for local processing with transport distance, location of the industry and previous processing performance. The permitted export volumes may vary between 15 and 45% of the total log production.

5.4 Restrictions of Log Exports and Log Supply Quota

Restrictions of log exports are of importance if large forest areas have already been granted to companies that are engaged in the log export trade. Their purpose is to improve the output of already existing wood processing units which operate below production capacity due to raw material supply problems, as well as to facilitate the establishment of additional industrial units. Log export restrictions may be:

- general and immediately effective such as the log export ban in Nigeria;
- general but phased in time such as the processing obligations of utilization contracts in Liberia;
- selective by species such as the export prohibition of determined timber species in Ghana;
and
- selective by species and related to a log supply quota as now practised in Ivory Coast.

Some utilization agreements of Liberia provide that companies have to process a minimum volume of their harvest locally starting at 20% during the first year and increasing by 20% every following year; after 5 years, the total log production from the concessions should be processed within the country.

In Ivory Coast a log supply quota was introduced in 1972, which obliged the concessionaires to deliver part of their log production to existing forest industries in the country. Log exports from species other than those defined as major export species were not considered in calculating the applicable quota. New regulations provide for some changes in the quota system. The applicable quota is now related to all industrial wood conversion. Permissible export quotas are tied up with the processing units and not any more with logging companies and log exporters.

The new regulations of Cameroon also provide for the possibility of establishing export quotas for logs and processed forest products, to be determined for each exporting company. National individuals and companies may only engage in log exports if they belong to the forestry profession, operate a logging enterprise, and are registered exporters. Non-nationals may only engage in the export trade if they fulfill the requirements for nationals and, in addition, operate a local wood-processing industry.

Ghana had introduced log export restrictions for the large wood processing units and in 1979 a log export ban on 14 principal timber species. In Nigeria, log export of major red-wood species had been prohibited during a first stage followed by similar restrictions for all other species.

5.5 Fiscal Incentives for Local Wood-Processing

Several countries have established fiscal incentives for wood-processing such as income tax exemption or reduction of import duties on logging and wood-processing equipment for companies that are prepared to establish new wood transformation industries. Incentives of this kind are usually dealt with by national investment legislation. In some cases, however, forest utilization contracts include specific provisions on this subject. As an example, the Liberian Timber Utilization Agreement of 1973 (Standard Form) may be mentioned.

A thorough evaluation of the economic justification of general investment incentives in the forest industry and a more selective application of such measures appear to be necessary.

5.6 Training of Forest Industry Personnel

The lack of a sufficient number of trained forest industry personnel is one of the most important obstacles in forest industry development. For this reason certain forest utilization contracts, and in particular those negotiated in Liberia, include a provision on training. The forest legislation of Congo also provides for specific clauses on training in industrial timber supply contracts.

On the whole, training requirements are not yet a standard feature of concession agreements in the ATO member countries. If particular clauses are in use, they tend to be formulated in a rather vague manner. But the responsibility for forest industry training cannot be transferred exclusively to the concessionaires. Effective training, in particular of forest workers and middle level technicians, will require combined training schemes in which the government and the industry participate. The basic training should be provided by specialized training centres, operated by national institutions, in which the concessionaires would participate by providing short-term courses and improved on-the-job training practices.

5.7 Measures in Favour of National Operators

The legislation of several ATO member countries provides for certain measures that are designed to support national operators in timber extraction and wood processing. Such measures refer, for instance, to the reserving of certain forest zones for exclusive timber harvesting by nationals (Gabon); the promotion of cooperatives among national operators (Gabon, Ivory Coast) and more favourable terms in concession granting under certain conditions (Ivory Coast); more favourable conditions for log exports in the case of nationals engaged in timber harvesting (Cameroon); more favourable conditions for national operators in dealing with required security deposits e.g. bank guarantees instead of payments to the Treasury (Cameroon); and technical assistance and credit facilities (Ivory Coast and Congo).

5.8 Regulations on Capital Participation

This is generally regulated by national investment codes and general business legislation. The Forest Law of Gabon includes, however, a particular provision which requires a special authorization by the Minister of Forests for any participation in the capital of a forest company as well as for the creation of a new company by any logging operator already established in the country. The purpose of the provision is to ensure that the maximum concession area of 200 000 ha cannot be by-passed by capital transactions.

The forest regulations in Cameroon provide for a similar authorization by the Minister in the case of any capital transaction of a forest company. It is also provided that a national company can only have a capital participation of any non-national of up to 30%, and that a company formed by non-nationals may only have a participation of up to 30% held by non-nationals which do not belong to the forestry profession.

5.9 Regulations on State Participation

Special provisions on state participation in the capital of timber harvesting and wood processing companies exist in particular in Ghana. The Forest Law of Cameroon provides that classified state forests shall be managed by the forest service with sales of standing volumes as standard practice; a state corporation or a company in which the state has at least 51% of the capital may, however, acquire timber utilization rights in such a forest.

5.10 Promotion of Lesser-Known or Unused Species of Technological Value

Some regulations or individual contract arrangements provide lists of lesser-known species whose removal is obligatory during timber harvesting operations. Such regulations are, however, difficult to implement if the economics of forest utilization are not favourable. This type of legal provision will not show good results unless it is supported by other fiscal and regulatory measures.

A very important measure of this kind is a system of forest revenue rates differentiated by groups of species. Its purpose is to reduce the considerably larger profit margins that exist usually for the current export species in comparison with those of the lesser-used species, the commercialization of which is to be promoted. Ivory Coast has continously and consistently followed such a policy during the last 10 years. The new regulations of Cameroon provide for the annual announcement of certain lesser-used tree species that are especially to be promoted. Rebates on the applicable forest revenue charges may be granted by the annual fiscal law for these species.

Another interesting measure for the promotion of lesser-known and so-far unused species is the determination of more favourable freight rates for certain groups of species. Such a policy has been included in recent regulations in the Ivory Coast.

5.11 Official Export Standards

Official grading and packaging according to standard sizes (e.g. in the case of sawnwood) improve the export quality and may be used in order to promote export of forest products and of lesser-known species. Several ATO member countries have made efforts in this respect. The legislation of most countries now includes provisions permitting the practice of official timber grading, and export quality control of logs and processed forest products. In Ivory Coast, sawnwood, shipped in standard sizes, is subject to an export duty which is 50% lower than the export duty on other sawnwood.

6. COLLECTION AND ASSESSMENT OF FOREST REVENUES

6.1 Importance of Forest Revenues

Public revenue from forest utilization is one of the most important components of the forest sector's value added. Forest revenue collected on primary species may represent between 20% and 50% of the log export value and over 50% of the total value added.

All ATO member countries and in particular Congo, Cameroon, Gabon, Ivory Coast, Liberia and Zaire have amended their legislation in order to insure more efficient forest revenue collection and assessment practices.

6.2 Forest Fees and Taxes as Elements of the Resource Rent

Tropical logs are by no means a uniform product. Log prices differ considerably by species and grades. Prices of high quality species and grades have generally increased faster than those of lower quality material.

The concept that the combined charges of all forest revenues should be related to the actual raw material value of standing timber, which represents, generally speaking, the resource rent due to the owner of the forests, has become more firmly established during recent years 1/. The aggregate of the applicable rates of taxes, levies and other charges on harvested timber should thus correspond to the timber value. It should, in particular, reflect the existing variations in prices and production costs of a given species and/or log quality.

6.3 Increasing Efforts to Amend the Applicable Laws and Regulations

Examining the amendments and revision of laws and regulations on forest revenue collection and assessment in some major export timber producing countries of ATO, the following conclusion may be drawn: The governments of the major timber producing countries have made considerable efforts in taking advantage of the rise of the raw material value and have tried to capture a greater proportion as governmental revenue. Additional measures, which at present are probably both of a regulatory of an implementing nature, will be required in order to follow this line more consistently within the various countries.

The principle of forest taxes and levies as part of the resource rent has, however, not yet been fully recognized in the legislation of the ATO member countries. A notable exception exists in the Forestry Ordinance and in the Forest Revenue Law of Congo, which stipulate that the aggregate of forest fees should correspond to the value of standing timber. The recent practice of modifying regularly the applicable tax rates in relation to changes of FOB prices and cost variations within the exported species and grades, in countries such as Ivory Coast, Cameroon and Gabon, reflects also to an increasing extent such an approach, even if the principle is not formally recognized in legislation.

6.4 Regular Revision of the Applicable Rates and of the Assessment Basis (Valeurs Mercuriales)

New provisions have been incorporated in forest legislation, which allow for a regular and more realistic assessment of forest fees and export taxes on logs and forest products. Ivory Coast, Cameroon and Gabon are using the possibilities of the annual fiscal laws and of the specific regulations to more frequently adjust the applicable tax rates of the various export species as well as the taxable base value of such species. Liberia has introduced new legal provisions that have simplified the number of different forest fees to 4 categories. Authority has been given to the Ministry of Finance and the Forest Development Authority to set the rates of the Industrialization Incentive Fee and to vary the rates of the other fees within certain limits. In Zaire, the rates of forest fees and taxes may, since 1979, be varied by governmental order.

6.5 Differentiation of Revenue Rates by Value of Species and Grades

Forest revenue collection has an important impact on the profitability of timber harvesting. Hence, revenue rates differentiated by species and log qualities can be used as an instrument to encourage better utilization of forest resources.

The basic structure of forest revenues levied in many tropical forests has, however, had a tendency to favour high grading. Species and grades, which were in highest demand

---

1/ For a detailed review of the concept of the forest resource rent and the problems related to the assessment and collection of forest fees, see: Gray, J.H. "Forest Revenue Systems in Developing Countries" FAO Forestry paper N° 43; 261 pp. Rome, 1983. This manual also presents a considerable amount of information on the existing revenue collection practices in a number of tropical countries

on world markets frequently had also the highest margin between FOB prices and the aggregate of collected revenues. Ivory Coast, recognizing this anomaly, has started to tax primary and middle-value species at much higher rates than low-value species. Since 1970, the prorata tax burden has more than tripled for high value species, increased moderately for middle-value species and remained fairly stable for the low-value species.

The tendency for an increasingly raw material value orientated revenue assessment can also be found in more recent forest fee regulations or fiscal laws of other ATO member countries such as Cameroon, Congo, Gabon and Liberia. In the latter country, the applicable export tax rates on logs have recently, however, been revised downward due to marketing difficulties.

6.6 Impact of Forest Fees and Taxes on the Promotion for Local Wood Processing

With the expansion of local industries, revenue assessment on processed forest products becomes of increasing importance. Revenue collection practices on processed timber must be compatible with policies to favour domestic wood processing and efficient use of timber resources.

The total tax burden per unit of the processed product should not be higher than the total charge against the comparable raw material exported in the form of logs. It may even be justified, during an initial period, to assess processed products at a somewhat lower rate than exported logs. On the other hand, an arbitrarily low level is not reasonable as this would foster inefficiency.

On the whole, one may state that the pattern of forest revenues that exists at present in the major wood exporting ATO member countries is compatible with these principles. A closer examination of the question of whether the combined burden of forest revenues on processed forest products is sufficiently differentiated in comparison with the charges on log exports, would, however, be appropriate in those member countries that are large producers and exporters of tropical timber.

6.7 Assessment Procedures on Raw Material Processed within the Country

Revenue charges on processed timber may be levied on the roundwood entering the mill or on the final product value, in particular, in form of export taxes. Whereas the latter is generally more easy to implement, the assessment on the raw material entering the production unit is more consistent with improved resources utilization. The levy of revenues on the raw material discourages wasteful processing standards. For all local conversion, assessment should consequently be based as a matter of principle on the value of the raw material (by species and log grades) and on the unit's average conversion rate. This is the procedure, that has, for instance, been provided for in the Congolese Forest Revenue Law.

6.8 Forest Fees and Taxes Related to the Granting of Utilization Contracts and Permits

The laws and regulations of all ATO member states provide for the payment of fees and taxes in connection with the granting, the renewal and, where permitted, the transfer of forest utilization contracts, or cutting permits. The proceeds from such payments generally represent only a small proportion of state revenue from timber harvesting.

6.9 Assessment of Area Tax

The assessment of an area tax on forest land which has been granted under various forms of utilization agreements, is standard practice in all ATO member countries. The principal purpose of this type of levy is to acknowledge the tenurial status of the forests and to prevent the operators from applying for too large concession areas.

6.10 Assessment of Taxes and Fees on Logs

Taxes and fees on logs (volume charges), both in form of logging fees, royalty and stumpage, generally assessed in the forests or at mill-entry, and in form of export taxes assessed in the ports, are the most important element of the revenue systems currently in force in the ATO member countries. These payments generally produce a large proportion of all state revenues from timber harvesting.

- Cameroon: A timber selling price and an export tax on logs are provided for. The applicable rates are fixed by the annual Finance Law. The rates of the export tax are fixed as a percentage of official check-prices, which are revised from time to time.

- Central African Republic: The rates of the logging tax are determined as a fixed amount per m^3 of wood and established by special laws.

- Congo: The tax on exported and locally processed logs is determined as a percentage of regularly established FOB prices. The applicable rates vary by species, grades, regional logging zones, and whether the logs are exported or processed within the country. Tax rates and assessment procedure are determined by a special forest revenue law.

- Equatorial Guinea: The rates of volume taxes are fixed by regulations.

- Gabon: The new forest law provides that the applicable tax rates and the method of assessment shall be fixed annually by the fiscal law, on the recommendation of the Minister of Forests. So far the rates of the log export tax have been fixed as a percentage of FOB, or of check-prices.

- Ghana: The royalty rates are determined as a fixed amount per harvested volume by special forest fees regulations, which may be revised from time to time. A log export tax is levied as a fixed amount per volume.

- Ivory Coast: The logging tax is determined by fixed rates per m^3 of wood. These have been varied fairly seldomly. The export tax on logs is determined as a percentage of check-prices for each species. The applicable rates and the assessment basis are changed by the annual fiscal law and special ordinances. Both are revised quite frequently.

- Liberia: The local stumpage fee is assessed as a fixed amount per m^3 of wood on all timber cut commercially on public and private land. The industrialization incentive fee is levied on all exported logs and assessed as a fixed amount per m^3 of wood. The rates of the industrialization fee may be revised by the Forest Development Authority in agreement with the Ministry of Finance.

- Nigeria: Logging fees are generally determined as a fixed amount charged per tree or m^3 of wood. Under certain conditions, the timber fees have recently been assessed on yield and the applicable rates fixed per hectare of harvested forest area.

- Tanzania: Royalties are generally determined as a fixed amount per removed volume.

- Zaire: A log export tax is levied as a fixed amount per m^3 of wood; no distinction has so far been made with regard to species and grades. The rates may be revised by governmental order.

6.11 Assessment of Taxes or Fees on Processed Forest Products

An export tax on processed forest products is levied in most ATO member countries. The tax is generally levied as a percentage of FOB values or officially determined check-price. In Liberia and Zaire, the export tax is levied as a fixed amount per volume of the respective product.

6.12 Participation of the Local Population in the Financial Benefits Accruing from Concession Granting

The general principle is that royalties and stumpage fees, which accrue from timber harvesting in constituted customary and local council forests are to be paid to the local entities. This is the case, for instance, in countries such as Ghana, Nigeria and Tanzania. In other countries, the forest areas are largely or exclusively under state tenure, or alternatively the timber harvesting rights of the unreserved forest land belong to the state. In these countries, the local population does not directly benefit from the forest fees and taxes that are levied in connection with the granting of concessions and logging permits. An exception from this rule exists, for example, in Cameroon and Ivory Coast. In Cameroon, the territorial tax, collected under forest licences arrangements, accrues to a separate account in favour of the Special Council Support Fund (FEICOM). An additional fee is levied as contribution to the execution of socio-economic infrastructures, which is utilized by the local councils in the area of timber harvesting.

7. FORESTRY FOR LOCAL COMMUNITY DEVELOPMENT (FLCD)

7.1 Forest Legislation as an Instrument to Support FLCD

Forestry for local community development has gained considerable momentum in national forest and rural development policies. Legislation is required in order to promote such policies. It should, in particular, offer a firm basis for assuring tangible benefits to local communities from forest areas and activities e.g. through methods of sharing revenues from state forests and protected areas, recognition of rights of local communities to forest usage (subject to necessary ecological controls), and according legal rights and powers to local communities to enable them to possess and manage local forests.

7.2 Present Situation of Forest Legislation

A review of the present forest legislation in the ATO member countries indicates that the content of the laws concentrates on state forestry, large scale timber harvesting, and on forest industry development. These aspects represent certainly very important segments of the sector's development. But modern forest legislation should equally reflect the new dimension of forestry, which is to make local communities benefit from their forest environment, and to establish new resources in order to meet basic rural needs. At present this is not yet the case and forest laws and regulations generally include little on the promotion of rural forestry. Some of their provisions even tend to exclude local communities from the use of trees and forests.

On the other hand, it should be mentioned that recent revisions of forest laws such as in Congo have specifically introduced the concept of FLCD and that customary and communal ownership of forest land have a long standing tradition in countries such as Ghana, Nigeria and Tanzania.

7.3 Important Aspects to be Considered in Future Revisions of Forest Legislation

The need to adapt legislation more closely to the specific requirements of rural forestry will grow in the future. Amendments may, in particular, be required with regard to:

- Provisions that establish firm and reliable rights of the local population over certain portions of forests and forest land. This refers especially to an effective protection of usage rights, as well as to mechanisms for the constitution and demarcation of communal and village forests.

- Provisions that facilitate the constitution of local communities or forestry cooperatives and that reinforce their decision making capacity with regard to the utilization and management of communal forest land.

- Provisions that facilitate a more comprehensive approach to technical assistance, vocational training, financial support and other production incentives. The effectiveness of these measures will depend on a close coordination among the various governmental agencies engaged in rural development.

7.4 Participation of the Population

Participation can be defined as the rural population taking part in and feeling responsible for forestry activities. The rural population can participate in forestry activities on its own land, on corporate, or on government lands; and in any phase of forestry development, from planting through to the consumption or sale of forest produce.

Participation requires:

- consultation with the people before decisions are taken;

- information and explanation to people of what will happen and why certain actions are taken;

- mobilizing support from below in order to convince higher authorities, to give priority to a certain area or project;

- involving people in work (nurseries, planting, etc.);

- allocating the whole or part of the proceeds to the people.

The forest legislation should provide for the necessary instruments that facilitate the participation of the rural population in all stages of planning, utilization and management of forests and forest land.

7.5 Land Tenure

If the prevailing tenure arrangements are not clear and cannot be easily understood, the local population may be discouraged from responding actively to rural forestry programmes. This is particularly the case if there are doubts whether a piece of land planted with trees by individuals may not be classified as state forest land. As long as the villagers run the risk of losing what they thought they had the right to expect, community forestry will not expand. The applicable legislation has to support the rights of the rural population.

Therefore, an important task for the government is to review the land-use and land tenure legislation on agricultural and forest lands if it wants to promote forestry for rural development. Appropriate forms of tenure are required in order to support the right kind of land-use, be it shelterbelts, trees on homesteads, different types of agroforestry, village wood-lots or communal forests. Appropriate land tenure is a way to guarantee to rural people who are involved in tree planting and in other forestry operations that they can receive the benefits of their labour. It is also a way to make people accept the responsibility for the use and management of forest land, and a basis for holding the land users accountable for such use.

7.6 Flexibility in Communal Forest Tenure

The legal problem is not necessarily one of changing the existing ownership pattern, but one of guaranteeing the rights to the proceeds of the land to the land users. Depending on customary rights and legal traditions in different countries, solutions to the problem of land tenure can differ. Also, in cases where the ownership of the land is not clear, and the land is not in use, the right to grow and harvest trees can still be granted to a certain group of people, for example a cooperative.

The point to be retained is that there are various juridical possibilities to introduce or promote communal forest tenure. The choice of a particular arrangement is very much a matter of the prevailing land tenure system and the applicable national legislation. It also appears that the actual utilization of communal forests by the local population depends much more on the factual modalities of the various tenure arrangements than on their respective formal and juridical structure.

7.7 Communal Forests on the Basis of Land Leases

Communal forest tenure may be constituted through land leases. Leases of forest land for a sufficiently long period are the logical solution in countries that have established predominant or exclusive state forest tenure. Leasing arrangements in favour of local communities are not very common in the tropical and subtropical forest zone. But they could offer, under certain conditions, a complementary and practical solution for the expansion of the communal forest areas.

7.8 Impact of Forest Legislation on Afforestation and Tree Planting

Afforestation of marginal agricultural land, or even planting of small wood lots or a few trees on adjacent border strips, may yield significant benefits to the farmer. This is one of the well established objectives of local FLCD programmes. Now it appears that the present tendency of the applicable legislation is to provide for a rather comprehensive definition of forest land. This means that planting of trees on private farm land may place such land immediately under the full regime of forest law. This again may imply a whole series of regulations, restrictions and authorizations which, from the farmer's point of view, can only be qualified as cumbersome and unnecessary.

Only to start with, he may be required to apply for a permit to cut down the trees, which he has planted a few years before, in order to get some fire wood for his family. Admittedly this may look somewhat theoretical and does presumably not occur too often in real life. Nevertheless, the point is that the planting of trees is subject to the

the provisions of the forest law and, this again usually means administration and rules for the local farmers.

7.9 Recognition of Specific Objectives of Tree Planting by Farmers and Villagers

The use of communal and private land for tree planting by local farmers and villagers is an integral part of a broader production system. This implies that the objectives of utilizing and promoting such land are much different from those of large-scale state forestry, or from industrial private plantations. Legislation on communal and private farm land in the more specific sense should take into account this situation and refrain as much as possible from regulating the establishment and utilization of trees and forests on such land. The basic assumption should be that only restrictions that are indispensable in order to safeguard specific public interests should be imposed on farmers and villagers who have engaged in forestry operations on their own land. Such restrictions may, for instance, be required for trees and forests with important protective functions.

7.10 Review of Forest Laws with Regard to Tree Planting and Village Wood Lots

The first step in analysing forest legislation in respect of planting trees or forests on farm land is therefore to ask whether the existing rules and regulations do not have deterrent effects on the initiative of rural people to engage in tree planting and complementary forestry activities. In order to avoid such an effect it is necessary that the applicable legislation distinguishes clearly and in a fairly liberal way between large scale forestry and the planting of trees and wood lots, that are part of the agricultural production system. The latter should be excluded from any restrictive provision that may be justified for forests and forest land in general. It may be useful to determine a minimum area or some other criteria for considering planted trees as forests in the legal sense. Moreover, afforestation on agricultural land should by no means change its tenurial status.

7.11 Promotion of Commercial Tree Planting through Special Convenants

The forest law of Tanzania has introduced the instrument of forestry dedication convenants in order to promote reforestation in rural areas. The forest administration may make such convenants with land owning individuals and institutions for the purpose of using certain parts of their land for commercial tree growing. Grants may be given to colleges and other public institutions for the establishment and management of forest plantations.

7.12 Financing Instruments for Rural Forestry

There are various mechanisms for channelling financial resources to rural forestry development programmes. Their suitability depends on a number of factors, e.g. the type of project to be financed, the financial possibilities, and the budgetary organization of the responsible agency. The following are among the most frequently used mechanisms:

- direct payments from a "special forestry fund" for planting and managing forestry resources in a difficult geographic zone;

- compensation through special financial arrangements from a "forestry fund", the national budget, or other sources, to cover the cost of planting or protecting forests in critical areas; i.e. erosion, watershed, or dam protection;
- participation, in which a public agency finances the cost of plantations, management, and extraction of forest products in forests located on communal or private lands, in return for a share in the revenue from the timber harvesting;

- rural development credits for agro-forest activities; credits are less suitable for larger forest plantations, since they require usually long periods of amortization;

- direct government investment, mainly for forests on communal lands, national parks, road infrastructures, training and extension.

7.13 Role of National Forest Services

Forest legislation needs to provide a legal basis for the provision of services and management by public forestry administrations to local communities in respect of communally owned forests, and agro-forestry practices by local farmers.

Executive management of state-owned lands is commonly entrusted to national forest departments or forest services. Their role should, however, expand to the support of communal, tribal or private land owners. Forest services should thus be put into a position in which they can respond in a more dynamic manner to the new demands of rural development. This may imply changes in the traditional forestry institutions. Decentralization of forest administrations, and the reinforcement of local field services, are of considerable importance in this context. Decentralization supports and facilitates multidisciplinary approaches and the active participation of local communities.

7.14 Emphasis on Motivating and Promotory Measures in Legislation

The concept of FLCD will only become a reality if local people feel that they will obtain tangible benefits from forestry activities. Legislation has thus to provide an adequate framework which determines clearly, what benefits may result to the population from rural forestry, who the beneficiaries,are what inputs the beneficiaries have to make, and how the benefits will be distributed.

On the whole, one may conclude that forestry for local community development needs principally legislative measures geared towards positive actions and assistance, rather that complicated regulatory provisions that tend to limit the initiative of rural people.

8. FOREST ADMINISTRATION

8.1 Impact of Changes in Legislation on Forest Administration

The modification of the existing as well as the introduction of new forest legisation requires a careful examination of the impact on forest administration and on the possibilities of enforcing the applicable laws and regulations. While improvements of the legal framework have been achieved in the ATO member countries, the most urgent problems arise at present from inadequate implementation. Coherent provisions and well-formulated legislation will fail to show practical results if the allocation of forest resources and their subsequent utilization cannot be planned and supervised in an appropriate manner. Legislative measures that encourage forestry for local community development cannot be put into practice if the necessary advice and assistance cannot be given to communal and private land owners.

The introduction of improved forest policies and their corresponding legislative provisions must be followed by a reinforcement and amelioration of the organizational structures of national forest services and other public agencies that are concerned with the sector's development.

8.2 Regulations on Forest Administration

The earlier generation of forest laws contained little or nothing on the role of the public administration concerned with forest resources management. This situation has changed and several of the newly adopted or revised laws now include specific provisions on the organization and responsibilities of public forest administrations. Among the recent texts, that define in some detail administrative and organizational issues, one can name those of countries like Liberia (1976), Ivory Coast (1966/1981), Cameroon (1981/1982), Ghana (1977/1982), Equatorial Guinea (1981), Congo (1974/1976) and Gabon (1982).

8.3 Principal Aspects of Legal Provisions

The new provisions related to the public forest administration generally cover the following aspects:

- The organizational status of the national forest administration as a full governmental service under the supervision of a ministry; as a semi-autonomous governmental commission; as a corporation.

- The range of responsibilities; the internal organizational structures; the interlinkage with other governmental and non-governmental agencies; and the power of recruiting agency personnel and managing budgetary funds.

- The financial arrangements for the functioning of the agencies, such as the transference of regular governmental budgetary funds; the levy of earmarked forest revenues for management and reforestation purposes; the establishment of special forest development funds; and the power to raise funds from extra-governmental sources.

8.4 Status of National Forest Administrations

Forest administrations at ministerial level (e.g. Congo, Gabon and Ivory Coast) or at a department level (e.g. Cameroon, Equatorial Guinea, Ghana, Nigeria, Tanzania, Zaire) exist in all ATO member countries with the exception of Liberia. In this country, the forest administration is organized as an autonomous governmental Forest Development Authority.

8.5 Responsibilities of National Forest Administrations

Obviously the specific responsibilities of a national forest administration vary in each country depending on the prevailing ecological, socio-economic and institutional conditions. In order to give an idea of the wide range of tasks that may have to be accomplished today by a public forest administration, the responsibilities of the Ministry for Forests and Water in Ivory Coast as established by Art. 1 of Decree Nr. 81-735 are summarized as follows:

- constitution, demarcation, conservation, regeneration, management and administration of the national forest estate;

- maintaining the integrity of the state forest domain and supervision of the utilization and administration of public and private forests;

- management of the state forests with a view to increasing their productivity and their protective functions with regard to environmental stability, and soil and water protection;

- execution of forest inventories in state forests as well as in other public and private forests;

- development of a national reforestation programme, and coordination and supervision of the corresponding operations;

- organization and supervision of timber harvesting in order to obtain improved productivity and improved utilization of raw material;

- supervision of forest products exports and their conformity with applicable export standards;

- research and development in the field of forest industries;

- supervision and collection of forest fees;

- preparation of statistics concerning the forest sector;

- constitution, demarcation, conservation, management and administration of national parks and other reserves;

- wildlife management and enforcement of legislation on hunting and protection of wild animals;

- protection of soils water and vegetation; and control of erosion and bush fires as required for a sustained utilization of natural resources;

- development of pisciculture and inland fishery resources;

- enforcement of applicable legislation and prosecution of offences;

- organization, administration, equipment and supervision of specialized education centres as related to forestry, wildlife management, pisciculture and inland fishery training;

- technical supervision of specialized governmental agencies operating under the authority of the Ministry of Forests and Water.

8.6 Changing Role of National Forest Services

National Forest Services are the principal governmental agencies for the management of the resources. They have an important policy setting function and are concerned with specific technical operations such as forest inventories and management plan preparation, silviculture and reforestation, protection and environmental forestry. Their role has changed significantly from a produce managing administration some 20 years ago, to that of an agency actively involved in integrated management of forest land for commercial timber production, local community development and environmental protection. Their functions in forest resources allocation may also have changed from agencies responsible for all aspects of forest utilization to that of more technically orientated forest services that have a leading, but not an exclusive, responsibility in granting timber harvesting rights.

8.7 Creation of Special Governmental Agencies

In several of the ATO member countries special governmental agencies that fulfill tasks complementary to those of the national forest service have been established. Under this category one may include, for instance, National Reforestation and Forest Development Agencies; Timber Harvesting and Wood Processing Corporations; and Timber Marketing Boards. Special laws, governmental decrees and regulations determine the constitution, responsibilities and operating procedures of these agencies.

8.8 National Reforestation and Forest Development Agencies

Whereas the policy setting functions as well as the application of regulatory legislation and enabling measures remain usually within the competence of the national forest administration, certain tasks of executive land management and, in particular reforestation, have been transferred to specialized governmental agencies. Such agencies operate under a semi-autonomous status providing greater organizational and budgetary flexibility as in the case of governmental forest departments. In a number of ATO member countries, reforestation and forest development agencies have been established, for example, in Congo, Cameroon, Central African Republic, Gabon and Ivory Coast.

8.9 State Forest Corporations

The functions of this type of governmental agencies are generally threefold:

- to act as governmental representative in joint ventures and other forms of business cooperation with private national and international investors;

- to provide managerial support, marketing expertise and training in timber harvesting and processing for smaller and medium-size operators and forestry cooperatives;

- and to engage as a state-owned enterprise in logging and wood processing.

State Forest Corporations exist, for instance, in Cameroon and Congo.

8.10 National Timber Marketing Agencies

These have been established principally with the following objectives:

- to control timber exports of the private timber harvesting companies;

- and/or to centralize log exports of certain species and/or forest products;

- to guarantee national standards in log grading and quality control of processed forest products, in order to improve the sector's competitiveness on export markets;

- to assist small- and medium-size enterprises in their activities;

- to promote timber and processed forest products exports in general.

The existing timber marketing agencies show a great variety in the relevancy and combination of these objectives. On the whole, one can say that the focus of their activity is presently centred on the guarantee of export standards and the marketing of species for which the agency has an export monopoly.

Timber marketing agencies exist in Congo, Gabon and Ghana.

8.11 Investment in Forest Management and in Reforestation

The forest must be viewed as a natural resource that requires financial investments like any other means of production - not as a mine to be merely exploited. Policy-makers are often ignorant of the nature and periodicity of forestry funding, which is linked to the biological production process of trees. This context needs to be clarified and forestry investments integrated with long-term development plans.

There is often a large discrepancy between the annually available funds, the departmental budgeting systems, and the requirements of silviculture and forest management. If corporate structures cannot be introduced to give greater financial flexibility, devices such as ear-marked levies and revolving funds can be instituted, with comparatively little change to institutions or legislation.

For these reasons, governmental policy in some countries includes "ear-marking" of a certain fraction of revenues from forests for reinvestment in forestry and/or for the development of services in the vicinity of the forest. This approach offers a balance between the Treasury principle of general revenues for national flexibility, and implementation of the important policy objective of sustained forest resources utilization.

8.12 Special Forest Levies and Forestry Development Funds

At present, the forest legislation of the following ATO member countries provides for special levies, or for forestry development funds, in order to finance reforestation and/or forest resources development in general:

- Cameroon: The regeneration fee and 55% of the timber selling price accrue to the state agency in charge of forest regeneration; 25% of the timber selling price accrues to the state agency in charge of forest inventories; in addition a special contribution to forestry development is levied which is to be used for equipment and control, forest management and for the promotion of wood.

- Central African Republic: 40% of certain forest fees accrues to the state agency in charge of forestry and fishery development; the proceeds of the reforestation and training tax accrue to the same agency.

- Congo: A special forest management tax is levied and paid into the Natural Resources Management Fund; a special reforestation tax is levied and paid into the Reforestation Fund. Proceeds from the management tax accrue in principle to the forest service, whereas the proceeds from the reforestation tax are to be used by the state agency in charge of reforestation.

- Gabon: A reforestation tax is levied, which accrues to the state agency in charge of reforestation.

- Ghana: A special silvicultural fee is levied and paid into the Forest Improvement Fund out of which the Chief Conservator may authorize payments for silvicultural works and payments to the traditional landowners in forest reserves.

- Ivory Coast: A reforestation tax is levied, which accrues to the state agency in charge of reforestation.

- Liberia: A reforestation fee is levied, which accrues to the Forestry Development Authority.

- Nigeria: There exists the possibility to assess a reforestation or afforestation levy.

9. GENERAL OBSERVATIONS AND CONCLUSIONS

9.1 Progress in Promulgating Sector Specific Legislation

During the last 20 years, several of the ATO member countries have made considerable efforts to develop a national sector specific legislation on the use, management and conservation of forests and forest land. Most of them have by now introduced a consolidated sector-specific legislation that consists of a basic forest law and of implementing regulations. Moreover, there has been a continous need, both in countries that had already promulgated forest laws at an earlier stage, as well as in countries with fairly recent forest laws, to revise, amend and expand their legislation.

9.2 Interface between Forest Law and other Economic Development and Natural Resources Management Legislation

The use of forests and forest land as well as the management of timber stands have become subject to a network of legal provisions, which are not any more exclusively regulated by specific forest legislation. This evolution leads to an increasing interdependence between forest laws on the one hand, and economic development laws as well as natural resources and environmental legislation on the other hand. The growing complexity of legislation will require in the future a thorough analysis of the compatibility of the various legislations.

This refers in particular to the interface between forest laws and industrial development legislation in countries such as Cameroon, Congo, Gabon, Ivory Coast and Zaire, which are strongly interested in building up a viable national timber processing industry. Since forestry for local community development is of concern to all ATO member countries, it appears also of great importance to review and coordinate more closely the provisions of the forest, agricultural and land-use legislation.

9.3 Adaptation to National Development Objectives

From a substantive point of view, one may say that the sector-specific legislation of many African countries, and in particular of several ATO member countries, has moved from the conventional standard format of forest laws that prevailed within the region at an earlier stage, to a more diversified system of national forestry codes and supplementary laws and regulations. This new legislation has become more adapted to the particular forest and forest industry conditions and more consistently embraces national development objectives of the various countries.

9.4 Management Orientation of Modern Forest Laws

The provisions of recent laws and regulations show a somewhat more pronounced orientation towards the technical aspects of forest resources management and a diminution of the merely repressive character of forest policing. Such legislation offers a more appropriate framework for adjusting the utilization of the forest resource to changing social and economic conditions. Considerable efforts will, however, still be necessary in order to continue this process.

9.5 Consolidation of the Forest Laws

Forest legislation must be comprehensive, well-structured and easy to understand. This implies, that the fundamental rules of forest utilization, management and conservation should ideally be assembled in one basic text, the Forest Ordinance, which can be referred to by the administration as well as by the citizens in all important cases. Considerable progress in this respect has been achieved in countries such as Congo, Cameroon, Gabon, Equatorial Guinea and Tanzania. In other ATO member countries, a similar effort may eventually be called for.

9.6 Regulatory Legislation and Promotive Measures

Forest legislation has traditionally been of a more repressive nature and is only expanding gradually to a more positive approach operating to induce and promote social behaviour in conformity with the existing or desired forest policies.

The move from regulatory legislation to promotive measures has important implications for economic and administrative efficiency as well as for social acceptance. Effective forest laws must thus be understood, be acceptable, and be beneficial to the majority of the people on whom they operate.

Within the ATO member countries, the promoting nature of forest legislation has principally the following three aspects:

- The provisions related to forest utilization, establishment of forest industries and forest concession management have to correspond to the economic reality of timber harvesting and processing. A particularly important role in this context is the assessment of the forest resource rent.

- Legislation is required to bring tangible benefits to local communities from forest areas, e.g. by methods of sharing revenues from state forests granted under timber harvesting rights; by recognition of usage rights subject to the necessary supervision for ecological reasons; and by according legal rights and powers to local communities to enable them to own and manage local forests.

- Legislation should establish a firm basis for the provision of services and management by public forestry administrations to communal and private forest owners.

9.7 Practicability of Forest Legislation

Modern and appropriate forest legislation demands a realistic balance between the content of legal norms and the possibilities of their implementation. Forest laws can only prove to be effective if their provisions are respected without requiring extraordinary and unjustifiable means of control and supervision. If one considers from this point of view the existing laws and regulations, one must admit that their provisions still appear to be formulated in a too complicated and perfectionistic manner. This situation is much in contradiction to the limited facilities in terms of personnel and equipment of the public forest administrations that are responsible for the application of these laws. In spite of certain advancements that may be sound in several of the ATO member countries, the practicability of forest legislation will require considerably more attention in future revisions and amendments.

9.8 Concentration of Legislation on Principal Policy Objectives and on Major Constraints

Forest legislation cannot regulate all the manifold problems that affect or may affect the development of the forest sector in any foreseeable detail. It should concentrate on the principal policy objectives and major constraints. Like in many other countries, the forest laws and regulations of the ATO member states include certain provisions, which from an overall aspect appear of less importance, but demand generally a considerable proportion of the technical and administrative capacity of forest administrations if they are to be implemented in the field. This refers in particular to a whole range of permits and licences which are required for the use of forests and forest produce by the local population, as well as to the collection procedures for certain forest fees.

It is important to review the applicable legislation in order to eliminate procedures that are likely to be difficult and time consuming in implementation but limited in their impact. This aspect should also be kept in mind when preparing new legislation. Without a systematic revision of the regulatory parts of the whole body of legislation, the forest administrations will not be in a position to assume the much needed role of promotionary agencies that are able to provide technical assistance and services to the rural population.

9.9 Flexibility of Forest Legislation

Clarity and practicability require that the provisions of the basic forest law are limited to determining general principles and subjects. More specific issues and such matters that need to be modified more frequently, should be determined by regulations, or where applicable, by administrative rules. Some of the recent forest laws show a greater degree of flexibility by more consistently making reference to specific regulations or by providing for general enabling clauses.

9.10 Presentation and Availability of Legislation

In general the formulation of legal provisions as well as the presentation of forest laws has improved during the last years but much remains still to be done. Particular attention should be paid to the logical sequence and length of provisions as well as to the working of individual clauses. Sub-titles for each section or paragraph facilitate the use of legislation.

A very practical problem results from the lack of availability of the existing laws and regulations to government officers and to the general public. This point appears of particular importance in view of the increasing number of legislative texts that have been adopted within the forest sector. The preparation of a consolidated collection of forest laws and regulations, and its regular updating, should consequently be of considerable concern to all ATO member countries.

9.11 Obstacles to the Implementation of Forest Legislation

Among the most obvious constraints that limit at present a coordinated development of forest resources and the implementation of improved national forest policies and corresponding legislation, the following should be mentioned:

- Lack of well-defined national land-use programmes, which could help to determine long-term forest land-use and management standards.

- Lack of management information in order to adjust the annual logging volume to the long-term production potential of certain forest areas, and to define minimum requirements with regard to logging and skidding practices, removal of commercial species, and construction of forest roads.

- Lack of a sufficient number of qualified field personnel who could effectively assist the rural population in order to benefit more directly from the sustained use of available forests, forest produce and wildlife, as well as support individual and village reforestation.

- Lack of coordination between agricultural and forestry programmes and activities that would allow a more integrated rural approach, in particular in accordance with the concepts of forestry for local community development.

- Lack of adequate information on timber processing and marketing. Without this information, it is impossible to decide what the best use of the raw material should be, what type of forest industries would be feasible, and what stipulation with regard to local wood-processing can be included in a particular agreement.

- Inadequate knowledge of the possible range of companies at the national level and/or from abroad, that would qualify for timber utilization contracts. It is thus difficult to select companies with experience in wood-processing and access to international markets.

- Inadequate information about the price/cost relationship in logging and wood-processing as required for a consistent assessment of forest revenues.

9.12 Strengthening National Forest Administrations as a Prerequisite for a More Consistent Application of Forest Laws

These shortcomings result to a large part from the weakness of forest administrations, suffering from shortage of professional and technical staff, scarcity of operational funds and inadequate organizational structures. The field organization of national forest departments in several countries is still little developed. Planning of forest utilization, as well as control and supervision of timber harvesting operations in the forest, are difficult. Institutional improvements of national agencies responsible for the forest sector's development, increased facilities for forest education, training and research, as well as additional funds will be required.

The strengthening of national forest administrations is therefore an indispensable prerequisite in order to obtain the full effectiveness of a modern forest legislation.

9.13 Need for Further Improvement of Forest Legislation

Law is one of the instruments for the implementation of policy. But sometimes legislation becomes outdated since it reflects policies of preceding generations. When changes in the sectorial objectives occur, e.g. when moving from a mere protection of forest resources to a multi-facet approach of production, social forestry and conservation, it is necessary to review forestry legislation and all other laws that have some impact on the utilization of forest resources.

The rapid evolution of forest laws and regulations, which can be noticed in the ATO member countries does, therefore, not imply that there is no need for further changes. Many forests laws still need improvement and consolidation. Others that appear to be quite modern at present will require further amendments in view of rapidly changing conditions. One can thus predict that the efforts of many governments in preparing and adapting legislation will continue and possibly gain even more momentum in the future.

PART II

COMMENTS ON THE APPLICABLE FOREST LAWS AND REGULATIONS AND ON THE SELECTED TEXTS

CAMEROON

A. Status of Forest Legislation

In 1973, the first unified forest law applicable for the whole territory of Cameroon was promulgated. The Forestry Law of 1973 was a well-balanced piece of legislation covering several points essential to modern forest utilization, including provisions related to national parks establishment, hunting and wildlife, and fisheries.

The Forestry Code of 1973 was replaced in 1981 by a new Forestry, Wildlife and Fishery Law. This law follows the principal lines already established by the preceding text, but appears to be more comprehensive in particular with regard to the management of various categories of forest land, timber allocation procedures, wildlife management, hunting control, and fisheries development.

Regulations replacing the various decrees under the previous forest law were made in 1982 and 1983. General forestry regulations implementing the Law of 1981 were adopted by Decree N° 169 (1983). The National Office for the Regeneration of Forests (ONAREF) was created by Decree N° 636 (1982). In 1981 another semi-autonomous agency, the National Centre of Forest Development (CENADEFOR) had already been established by Decree N° 223.

The applicable rates of forest fees and timber export duties are principally regulated by annual fiscal laws and the respective provisions have been modified on various occasions.

B. Presentation of the Selected Texts

Law N° 81-13 of 27 November 1981 to lay down forestry wildlife and fisheries regulations 1/ is a well-structured text that draws on previous experiences.

The general provisions of PART I provide a well-formulated statement on the objectives of the law and its subsequent regulations, which are the conservation, exploitation and development of the forest wildlife and fisheries resources of the country's forest estate and waterways. It establishes the principle of the joint application of the land tenure and state lands legislation as well as of the forest law in determining the system of forest ownership; and of the responsibility of national forestry, wildlife and fishery services for the management and protection of all categories of forest tenure. The first part presents a comprehensive definition of forests and forest lands; general rules on

---

1/ In most French-speaking countries, fishery development in rivers and fresh-water ponds has been traditionally associated with the forest sector and is under the responsibility of the national forest administration. Marine fishery development is generally subject to a separate specialized fisheries legislation.

the use of forests, wildlife and fishery resources on public and private land; provisions related to the applicable system for the collection of duties and taxes; and to the status of the national forests and wildlife administration.

PART II of the law is concerned with the various categories of the forest domain; forest utilization and management; and the promotion and marketing of timber and other forest produce.

The officially constituted state forests may be classified under a whole range of functional categories such as production, protection and recreation forests; integral nature reserves and national parks; wild animal and wild plant sanctuaries; and state game ranches. State forests shall cover at least 20% of the national territory. A management plan is to be prepared for each state forest and national park in accordance with the designation of the particular area as defined in the reservation decree. Local council forests are forests constituted by decree for the benefit of local councils, or originate from forest plantations, that have been made by these entities. Sec. 19 refers to private forests, which are forests planted by individual persons on land owned in compliance with the applicable land tenure legislation. This provision is new in the forest law and should facilitate individual tree planting and combined agro-forestry production. The use of local council and private forests may be decided upon by their owners within the limits of certain rules laid down by regulations.

All forests which have not yet been constituted as state or local council forests, or which are not under private tenure, are considered as communal forests. The forest produce in communal forests, with the exception of produce from trees planted by private individuals or local councils, belongs to the state. Timber harvesting rights may, however, be granted to the local population under conditions determined by regulations.

Forest exploitation in any forest zone is subject to the preparation of forest inventories. The total area granted through permits and licences to one single operating company and its affiliates may not exceed 200 000 ha. State forests sensu stricto shall be harvested as a matter of principle under state management and under the supervision of the administration in charge of forests, or by sales on the stump. A timber harvesting licence may, however, be granted to state corporations and companies in which the government has a majority participation. Timber harvesting on communal forest land is generally to be undertaken either by the sale of standing volume or by the grant of licences. Several provisions regulate the conditions under which timber harvesting licences may be issued and the obligations attached to their use.

Sections 31 and 33 determine the various payments and forest fees in connection with timber harvesting rights. These are the regeneration fees; the territorial tax; a contribution to forestry development; the selling price of the forest produce; and a contribution to the execution of socio-economic infrastructures. The rates of these taxes and fees are to be determined by the Annual Finance Law. The regeneration fee accrues to the state agency in charge of reforestation; the contribution to forestry development is to be used for forest equipment and control, forest management and for the promotion of forest products. 80% of the revenues from the selling price of forest produce are also earmarked for the funding of the state agencies in charge of forest inventories and forest regeneration. The funds from the contribution to the execution of socio-economic infrastructures accrue entirely to the local councils in the area of timber harvesting.

Log exports may only be carried out by nationals individually or grouped into companies, who have been granted timber harvesting rights. Other concessionaires have to operate a timber processing industry in accordance with the specific terms of the agreement in order to be entitled to certain log exports. Export quotas for the various types of forest produce may be fixed by the forest administration. Measures to promote the use of unknown or lesser-known timber species may be determined by regulations.

PART III consolidates the previous provisions on the exercise of hunting rights and on wildlife conservation and management. New in particular is the possibility to create game ranches operated by the state, and to declare buffer zones around protected areas. Additional sections, especially Sec. 69, provide for more effective measures of wildlife and environmental protection. Wildlife species are now classified into 3 classes, the first class being totally protected. The setting of fire is subject to regulations in order to avoid environmental destruction. The felling of trees in certain critical areas invaded by the desert, and the destruction of the vegetation along a water course and around a water source are prohibited.

PART IV on fisheries development 1/ is presumably that part of the new law that has been amended and increased most substantially. This part now includes clear and comprehensive definitions of the various fishery resources, fishing activities, and fish processing and commercialisation operations. It regulates the exercise of fishing rights for traditional, commercial and industrial uses; the management and protection of the fishery resources in general; the operating of sea farming and fish farming; the installation of fish processing establishments; and the sanitary inspection, packaging and transportation of fishery products.

The general regulations on forestry matters, adopted in 1983 by Decree N° 169, complement Part I and Part II of the new forestry law. The regulations are principally concerned with forest utilization. Chapters 1 and 2 of TITLE II determine, in particular, the objectives of forest regeneration; the exercise of certain customary rights in state forests; the procedure for constituting reserved forest areas and for subsequent changes in land-use; the preparation of forest inventories and management plans in state forests and the applicable rules for timber harvesting in such forests.

The following sections regulate in considerable detail the granting of timber harvesting licences; logging permits; and permits for other forest produce in communal forest areas. Of particular interest are the provisions related to the obligatory preparation of forest inventories by a specialized agency (Sec. 22) 2/ prior to the granting of licences; and to the constitution of a technical commission examining all applications for registration as forest operator and for the granting of harvesting rights for timber and other forest produce. The same refers to the newly introduced procedure for publishing the proposals of a new forest industry project, retained by the technical commission, in the particular area concerned; for inviting the population to make comments or appropriate objections; and for examining the project in a special meeting in which local and regional representatives of the public administrations and of the population as well as the prospective investor may participate.

TITLE III and TITLE IV regulate the particular requirements and conditions for authorized log exports; the submission of an annual report on exports of logs and processed products; and the publication of annual lists indicating lesser-used species to be promoted that may benefit from certain reductions of the forest tax subject to the

---

1/ Fisheries in the sense of the law includes the fishery resources of public waterways and coastlands (Sec. 2). A special fishery service is responsible for the management and protection of these resources (Sec. 6).

2/ Such as, in particular, the National Centre for Forest Development (CENADEFOR).

provisions of the Fiscal Law. Sec. 75 and Sec. 76, respectively, submit the participation in the transfer of capital within the logging industry to an official authorization, and establish certain rules for such financial transactions.

Decree N° 82/636 creates the National Office for the Regeneration of Forests ((NAREF), which assumes the role previously attributed to the National Forestry Fund. The decree determines the status of the new organization as an industrial and commercial agency with its own financial autonomy and defines its functions and responsibilities in the field of forest regeneration, reforestation, soil conservation and rehabilitation. Chapters II to IV regulate the composition and functions of its board; its internal organization; the applicable budgetary procedures; and the supervisory role of the financial commission.

Decree N° 81/223 determines the status of the National Centre for Forest Development (CENADEFOR) and defines its role and responsibilities in the fields of forest inventory and management; forest industry development and training; standardization and grading of logs and processed forest products; and promotion of wood consumption within the country. The subsequent provisions on the internal organization and budgetary affairs of the agency are similar to those of the previous text.

C. General Forest Legislation

- Décret N° 72/438 du 1er septembre 1972
  portant organisation du Ministère de l'Agriculture

  (Decree N° 72/438 of 1 September 1972
  providing for the organization of the Ministry of Agriculture)

- Ordonnance N° 73/18 du 22 mai 1973
  fixant le régime forestier national; Abrogé par Loi N° 81/13

  (Ordinance N° 73/18 of 22 May 1973
  setting up the National Forestry Legislation. Repealed by Law N° 81/13)

- Décret N° 73/257 du 22 mai 1973
  fixant la composition de la commission des mercuriales douanières

  (Decree N° 73/257 of 22 May 1973
  establishing the composition of the official customs price-list Commission)

- Arrêté N° 104/CAB/PR du 16 juillet 1973
  fixant les valeurs mercuriales de certains produits à l'exportation

  (Regulation N° 104/CAB/PR of 16 July 1973
  fixing the official prices of certain export goods)

- Décret N° 74/54 du 26 janvier 1974
  portant réorganisation du Ministère de l'Agriculture; Abrogé par Décret N° 83/169

  (Decree N° 74/54 of 26 January 1974
  providing for the reorganization of the Ministry of Agriculture. Repealed by Decree N° 83/169)

- Décret N° 74/035 du 15 novembre 1974
  portant application de l'Ordonnance 73/18 du 22 mai 1973 créant un Fonds National Forestier et Piscicole; Abrogé par Décret N° 82/636

  (Decree No 74/035 of 15 November 1974
  implementing Ordinance 73/18 of 22 May 1973 providing for the setting up of a National Forestry and Fisheries Fund. Repealed by Decree N° 82/636)

- Loi N° 75/3 du 28 juin 1975 (Loi de Finances)
  modifiant les droits et taxes sur les permis d'exploitation forestière

  (Law N° 75/3 of 28 June 1975 (Finance Law)
  amending duties and taxes on logging permits)

- Arrêté N° 74/CAB/PM du 14 novembre 1975
  fixant les valeurs mercuriales de certains produits à l'exportation

  (Regulation N° 74/CAB/PM of 14 November 1975
  fixing the official prices of certain export goods)

- Décret N° 76/256 du 1er juillet 1976
  portant réorganisation du Ministère de l'Agriculture

  (Decree N° 76/256 of 1 July 1976
  providing for the reorganization of the Ministry of Agriculture)

- Arrêté du 17 août 1977
  fixant les valeurs mercuriales de certains produits à l'exportation

  (Regulation of 17 August 1977
  fixing the official prices of certain export goods)

- Décret N° 81/223 du 9 juin 1981
  portant création du Centre National de Développement des Forêts

  (Decree N° 81/223 of 9 June 1981
  setting up the National Centre for Forestry Development)

- Loi N° 81/13 du 27 novembre 1981
  portant Régime des Forêts, de la Faune et de la Pêche

  (Law N° 81/13 of 27 November 1981
  laying down the forestry, wildlife and fisheries regulations)

- Décret N° 82/636 du 8 décembre 1982
  portant création de l'Office National de Régénération des Forêts

  (Decree N° 82/636 of 8 December 1982
  setting up the National Office for the Regeneration of Forests)

- Décret N° 83/169 du 12 avril 1983
  fixant le régime des forêts

  (Decree N° 83/169 of 12 April 1983
  laying down the forestry regulations)

CENTRAL AFRICAN REPUBLIC

A. Status of Forest Legislation

The Central African Republic's Forest Law dates from 1962; some small amendments of the law were made in 1962 and 1971. A draft law presenting a consolidated and slightly modified version of the Forestry Code has been prepared in 1981.

Timber harvesting agreements, that include a considerable proportion of standard clauses, supplement the provisions of the forest law concerned with forest utilization.

The payment of forest fees and taxes and their applicable rates are determined by the Forest Revenue Law of 1962 and several ordinances adopted during the period 1974-1981. Ordinance N° 69/37 and its implementing Decree N° 69/293 establish the National Forestry Office and regulate its organization and functions.

Wildlife and hunting, nature protection and fishery development are subject to specialized legislation. The basic texts were promulgated in 1960 and 1961. Numerous regulations and administrative rules have been adopted in the meantime, in particular with regard to hunting control and the constitution of national parks, game reserves and other protected areas. A consolidation of the numerous texts appears to be necessary. A draft law on wildlife conservation and the exercise of hunting activities, and draft regulations have been prepared in 1978 and 1979.

B. Presentation of the Selected Texts

The Forestry Code of 1962 is quite representative of the first generation of forest laws that have been promulgated in the French-speaking countries of West and Central Africa after independence. The law is principally concerned with the constitution of the various categories of the forest estate and the determination of timber harvesting procedures. Whereas its general structure appears to be well-organized, certain sections are extensively long and tend to include provisions, that would generally be settled by regulations. Moreover little is said in the law on forest management and reforestation in order to sustain the long-term production capacity of the forests.

The FIRST TITLE of the Forestry Code presents a rather conventional definition of forests, which refers only to its production function. The SECOND TITLE deals in considerable detail with the constitution of permanent state forests and protected forests of communal and collective tenure. The forest reservation procedures include provisions for the acknowledgment and, if necessary, limitation of local usage rights. Timber harvesting in the reserved and protected forests may be carried out in the form of state operated logging or through auctioning of standing timber.

Forests which have not been constituted as reserves or protected forests belong to the customary domain in which the collective owners or individual may freely exercise their customary rights. Timber harvesting in these forests is, however, subject to state control; in certain cases the forest administration is acting as an agent between the customary owners and the logging companies. Industrial timber harvesting is principally carried out by granting temporary exploitation permits; the area of a permit may range between 500 ha and 20 000 ha. In the case of larger companies with timber processing installations, special agreements with a duration of 15 years have been made that provide for the granting of several temporary exploitation permits.

The various forest taxes and fees that are to be levied in connection with timber harvesting are determined in the THIRD TITLE of the Forestry Code. A certain proportion (40%) of the revenues from timber auctioning and temporary logging permits accrue to the National Forestry Office for financing its operations.

Law N° 61/282 establishes the applicable rates of forest fees and taxes, which are to be levied under the forest law. These are, in particular, the area tax, a tax for special permits, and the forest production tax applicable on log exports, sawnwood and processed forest products. The rates of the forest production tax were revised in 1974 by Law N° 74/014. In 1979, Ordinance N° 79/025 established a "Reforestation and Training Tax" which has to be paid by all companies irrespective of the terms of their particular agreement.

Ordinance N° 69/49 establishes the National Forestry Office and defines its principal functions. In the field of forestry the responsibilities of the Office include forest management and watershed conservation; plantation establishment; promotion of wood utilization; and training. The functions of the Office also refer to inland fishery development.

C. General Forest Legislation

- Loi N° 61/273 du 5 février 1962
  portant création du Code forestier centrafricain

  (Law N° 61/273 of 5 February 1962
  setting up the Central African Forestry Code)

- Loi N° 61/282 du 5 février 1962
  fixant le taux des redevances forestières

  (Law N° 61/282 of 5 February 1962
  fixing the rates of forest fees)

- Loi N° 62/341 du 7 décembre 1962
  portant modification des Art. 43 et 65 de la Loi N° 61/273 du Code forestier centrafricain

  (Law N° 62/341 of 7 December 1962
  amending Sections 43 and 65 of Law N° 61/273 (Central African Forestry Code))

- Arrêté N° 2306/65 du 12 novembre 1965
  portant création du Centre forestier pilote (en annexe les Statuts du Centre Forestier Pilote)

  (Regulation N° 2306/65 of 12 November 1965
  providing for the creation of the Pilot Forestry Centre (with Appendix containing Rules and Regulations of the Pilot Forestry Centre))

- Ordonnance N° 67/64 du 25 novembre 1967
  portant attribution d'un permis temporaire d'exploitation forestière au Centre pilote forestier

  (Ordinance N° 67/64 of 25 November 1967
  providing for the allocation of a temporary logging permit to the Pilot Forestry Centre)

- Convention entre le Gouvernement centrafricain représenté par le ministre du Développement et NN. (attribution des permis temporaires d'exploitation); Modèle 1968

  (Agreement between the Government of the Central African Republic, represented by the Minister of Development, and NN concerning the allocation of temporary logging permits. 1968 Form)

- Cahier des charges (Modèle Standard)
  concernant les permis temporaires d'exploitation de bois d'oeuvre

  (Specifications (Standard Form)
  regarding temporary timber harvesting permits)

- Ordonnance N° 69/010 du 26 mars 1969 (Art. 4)
  fixant le montant de la taxe sur pied dans les permis speciaux

  (Ordinance 69/010 of 26 March 1969 (Section 4)
  fixing the rate of the standing timber tax in special permits)

- Ordonnance N° 69/37 du 4 juillet 1969
  portant création d'une forêt communale classée dans le périmètre urbain de Bangui

  (Ordinance N° 69/37 of 4 July 1969
  providing for the creation of a protected Communal Forest in the urban area of Bangui)

- Ordonnance N° 69/049 du 23 septembre 1969
  portant création de l'Office national des forêts

  (Ordinance N° 69/049 of 23 September 1969
  providing for the creation of the National Forestry Office)

- Ordonnance N° 69/293 du 23 septembre 1969
  portant approbation des Statuts de l'Office national des forêts

  (Ordinance N° 69/293 of 23 September 1969
  approving the statutes of the National Forestry Office)

- Ordonnance N° 71/44 du 27 mars 1971
  modifiant l'Ordonnance N° 69/049 du 23 septembre 1969 créant l'Office national des forêts

  (Ordinance N° 71/44 of 27 March 1971
  amending Ordinance N° 69/049 of 23 September 1969 creating the National Forestry Office)

- Ordonnance N° 71/045 du 27 mars 1971
  portant abrogation et remplacement de l'Art. 66 du Code forestier centrafricain

  (Ordinance N° 71/045 of 27 March 1971
  repealing and replacing Section 66 of the Central African Forestry Code)

- Arrêté N° 009 du 19 février 1973
  portant fixation des valeurs mercuriales applicables à l'exportation

  (Regulation N° 009 of 19 February 1973
  fixing the official price-list applicable to export goods)

- Ordonnance N° 74/014 du 24 janvier 1974
  portant modification de la Taxe de Production applicable aux bois centrafricains exportés

  (Ordinance N° 74/014 of 24 January 1974
  amending the logging tax applicable to exported Central African timber)

- Ordonnance N° 79/025 du 8 mai 1979
  portant création d'une taxe dite "Taxe de Reboisement et de Formation" de 1 000 Francs le m^3 Export Sapelli et Sipo

  (Ordinance N° 79/025 of 8 May 1979
  levying a Reforestation and Training Tax of 1 000 Francs per m^3 of Sapelli and Sipo exports)

- Ordonnance N° 80/80 du 26 septembre 1980
  relevant la taxe de Reboisement et de Formation à 2 000 Francs le m^3 de Sapelli et Sipo

  (Ordinance N° 80/80 of 26 September 1980
  raising the Reforestation and Training Tax to 2 000 Francs per m^3 of Sapelli and Sipo

- Ordonnance N° 80/109 du 23 novembre 1981
  prévoyant le recouvrement de la taxe forestière par l'Office national des forêts

  (Ordinance N° 80/109 of 23 November 1981
  providing for the collection of forest fees by the National Forestry Office)

- Loi N° 81/... (<u>Projet</u>)
  modifiant les dispositions de l'Ordonnance N° 80
  modifiant partiellement les dispositions de la Loi N° 61/273 portant création du Code forestier centrafricain

  (Law N° 81/...(Draft)
  amending the provisions of Ordinance N° 80 partially amending the provisions of Law 61/273 setting up the Central African Forestry Code)

<u>Legislation on Nature Conservation, Hunting Control and Fishery Development</u> (Basic texts)

- Loi N° 60/140 du 19 août 1960 sur la Protection de la Nature

  (Law N° 60/140 of 19 August 1960 on Nature Protection)

- Loi N° 60/141 du 9 septembre 1960 réglementant l'Exercice de la Chasse

  (Law N° 60/141 of 9 September 1960 on Hunting Control)

- Loi N° 61/283 du 1er août 1961 réglementant l'Exercice de la Pêche

  (Law N° 61/283 of 1 August 1961 on Fishing Control)

CONGO

A. Status of Forest Legislation

In 1974 a major change of the country's forest legislation occurred. A new forestry code was adopted by Law N° 004/74 replacing the whole of the previous texts which had become much dispersed and obsolete. A partial revision of the new forestry code was made in 1982 by Law N° 32. The revision refers in particular to an amendment of the first chapter concerned with land tenure and forest reservations, as well as to some adjustments in the chapters on forest offences and prosecution. The central part of the law providing general rules on forest resources utilization and timber allocation procedures has remained unchanged.

The forest regulations adopted in 1974 by Decree N° 188, which implement the provisions of the forestry code, are well-structured, comprehensive and formulated in considerable detail. A redraft of this decree, with a few amendments but no major changes as far as its substance is concerned, has been prepared recently.

In accordance with the forestry code and its regulations, industrial timber processing contracts may be signed between the government and private or state operated companies. The standard format of these agreements includes clauses on the participating investors; the allocated quantities of raw material and its commercialization; the envisaged processing installations and production schedule; training obligations; and such management prescriptions as necessary in the particular forest zone, in which a raw material supply contract is to be granted.

Forest revenue assessment and collection are subject to the provisions of Law N° 005/74 on forest fees and taxes, which has so far remained unchanged.

Additional legislation adopted in 1974 and 1975 provides for the creation of 3 semi-autonomous government agencies, which are the Congolese Forest Office in Charge of Reforestation; the National Agency for Timber Harvesting; and the Congolese Wood Office in charge of commercialization.

Wildlife management and hunting control are not covered by the forestry code. A draft law on wildlife management and protection has been prepared in 1981.

B. Presentation of the Selected Texts

Law N° 004/74 as amended in 1982 provides an example of as modern and comprehensive national forestry code as may be required for the conditions of tropical forestry in Central Africa. Of particular interest appears the law's tendency to coordinate the granting of timber harvesting rights with the necessities of management, infrastructural planning, and forest industry development. The long-term utilization of the available forest resources, that avoids the constant migration of the exploitation activities from one forest zone to the other, is a fundamental objective of the law. The granting of timber harvesting rights is not necessarily linked to the allocation of determined forest areas, in which the concessionaire has exclusive exploitation rights. The new concept is, instead, to issue a forest contract that allows for the harvesting of a specified annual volume of raw material that is guaranteed for a determined period of years.

The FIRST TITLE of the Forestry Code is concerned with the constitution of the forest domain and the exercise of customary rights. The new Art. 3 provides the possibility for the declaration of production forests, protection forests, national parks, various forms of nature reserves, and of forests for local community development. The latter category of forests had not existed in the previous text and is of particular importance; especially since the law stipulates in its first article that all forests and reforestation areas are under state forest tenure. The provisions related to the usage rights of the local population confirm the principle that these rights may be practised freely in all forest areas including the ones that have been granted by timber utilization contracts. Usage rights that are incompatible with the objectives of certain classified areas (e.g. in the case of protection forests, national parks and nature reserves) may, however, be restricted or abolished against compensation. The constitution of the various categories of classified forests and the practice of usage rights are to be regulated by reservation decrees.

The SECOND TITLE establishes the principles of forest resources utilization, conservation and management, and of an economically sound policy for granting timber harvesting rights. It provides for local processing of the raw material; national control on forest exploitation; and the payment of forest fees and taxes on all forest land whether granted to private operators or to state agencies engaged in logging and processing. This part also contains provisions on the preparation of forest inventories and the division of forest areas into management units on all land on which timber harvesting rights are to be granted.

Reforestation operations are to be carried out by a specialized state agency; the expenditure for reforestation and forest management are financed by a forestry development fund.

The provisions on the economic utilization of the forest resources regulate the various categories of forest utilization contracts that may be granted; the specific conditions that govern timber harvesting and wood processing; and the procedures to be followed in applying and preparing such contracts up to their final signature. Four categories of contracts and permits are to be distinguished: contracts for industrial timber processing; contracts for timber harvesting; tree permits subject to auctioning; and special permits for local uses of wood and other forest produce.

The THIRD TITLE of the Forestry Code is concerned with forest offences and prosecution. The FINAL TITLE contained transitory provisions, that were necessary in 1974 when the law was promulgated; these provisions have been abolished by the Amendment Law N° 32/82.

The structure and list of clauses of the regulations that implement the Forestry Code are presented in the redrafted version of 1982. They are comprehensive and offer the advantage of assembling all relevant provisions in one single text. The regulations determine in particular:

- the professions relevant to timber harvesting and wood processing, and the applicable procedure for registration of the various groups of operators;

- the division of forests into sectors, zones and forest management units;

- the preparation of inventories and management plans, and the applicable methods for determining the maximum annual allowable cut in each unit;

- the demarcation of management and, where applicable, timber harvesting units, and the determination of minimum cutting limits for groups of species;

- the procedure for granting timber harvesting rights and the issue of annual logging permits;

- the regulation of the access to exploitation areas;

- the circulation of forest produce;

- statistical information to be supplied by the operating companies, and the control of export prices;

- special guidelines for forest management and wood processing as applicable in the various regions of the country.

Forest Revenue Law N° 005/74 completely reorganizes the system for the assessment and collection of forest fees and taxes. It has also accomplished considerable simplification in the structure of the applicable revenue charges by reducing the previously existing 12 categories of fees and taxes to only 4 kinds of levies. This law provides in particular:

- that, with the exception of customary uses, all timber harvesting operations by private companies as well as by the state agencies and their affiliates are subject to the payment of forest fees and taxes;

- that the aggregate of the various revenues should correspond to the actual value of standing timber of each timber species and grade;

- that the payments of forest fees and taxes have to be made by the exporters in the case of log exports, and by the local industry in the case of processed forest products.

Art. 9 of the law includes a table presenting the applicable rates of the log export tax as a percentage of regularly revised check prices (FOB values) per m^3 of exported logs. The rates are differentiated by species, grades, and for 8 production zones of the country. In the case of Okoumé and Limba, 3 log qualities, and in the case of the other main export species, 2 grades have been retained for assessment purposes. The 8 production zones are determined principally according to transport distance and costs, and are described in Art. 11. A similar table for roundwood processed locally, is presented in Art. 10, which determines the applicable rates of the processing fee. Again the rates are varied according to species and grades, but not according to regional variations in transport costs. The regularly revised check prices of log exports are also used as a reference basis in this case.

The Forest Revenue Law provides for the levy of a forest management and a reforestation tax on all harvested timber; the combined assessment rates vary between 2.5% and 3.5% of the reference log export values. The applicable rate is to be fixed in the agreement of the individual utilization contract. The LAST TITLE of the law establishes the Forest Resources Management Fund and the Reforestation Fund as stipulated by Art. 30 of the Forest Code.

C. General Forest Legislation

- Loi N° 004/74 du 4 janvier 1974
  portant Code forestier; modifié par Loi N° 32/82

  (Law N° 004/74 of 4 January 1974
  setting up the Forestry Code. Amended by Law N° 32/82

- Loi N° 005/74 du 4 janvier 1974
  fixant les redevances dues au titre de l'exploitation des ressources forestières

  (Law N° 005/74 of 4 January 1974
  establishing forest resources harvesting fees)

- Loi N° 006/74 du 4 janvier 1974
  portant création de l'Office congolais des forêts (OCF)

  (Law N° 006/74 of 4 January 1974
  setting up the Congolese Forest Office)

- Loi N° 007/74 du 4 janvier 1974
  portant création de la Société nationale de l'exploitation des bois (SNEB)

  (Law N° 007/74 of 4 January 1974
  setting up the National Agency for Timber Harvesting)

- Décret N° 74/142 du 1er avril 1974
  portant organisation de l'Office congolais des forêts (OCF)

  (Decree N° 74/142 of 1 April 1974
  providing for the organization of the Congolese Forest Office)

- Décret N° 74/143 du 1er avril 1974
  portant organisation de la Société nationale de l'exploitation des bois (SNEB)

  (Decree N° 74/143 of 1 April 1974
  providing for the organization of the National Agency for Timber Harvesting).

- Décret N° 74/188 du 5 mai 1974
  portant application du Code forestier

  (Decree N° 74/188 of 5 May 1974
  providing for the implementation of the Forestry Code)

- Arrêté N° 315
  définissant les unités forestières d'aménagement dans le secteur forestier Nord

  (Regulation N° 315
  defining the forest management units in the North forest sector)

- Arrêté N° 316
  définissant les unités forestières d'aménagement dans le secteur forestier Sud

  (Regulation N° 316
  defining the forest management units in the South forest sector)

- Arrêté N° 00741/MEF/DEFRN du 24 mai 1974
  portant appel d'offres pour la mise en exploitation de 6 unités d'exploitation forestières de l'Unité d'aménagement N° 3, Zone 1 dans la région de Ouesso

  (Regulation N° 00741/MEF/DEFRN of 24 May 1974
  concerning an invitation for bids for the opening to logging of 6 forest units in Management Unit N° 3, Area 1 of the Ouesso region)

- Arrêté N° 00742/MEF/DEFRN du 24 mai 1974
  portant appel d'offres pour la mise en exploitation de 6 unités forestières d'aménagement dans le secteur Sud

  (Regulation N° 00742/MEF/DEFRN of 24 May 1974
  concerning an invitation for bids for the opening to logging of 6 forest units in the South sector)

- Décret N° 74/280 du 19 juillet 1974
  portant approbation des Statuts du Centre forestier de formation professionnelle et de démonstration (CFFPD) de Mossendjo

  (Decree N° 74/280 of 19 July 1974
  approving the statutes of the Forestry Professional Training and Demonstration Centre of Mossendjo)

- Décret N° 75/182 du 14 avril 1975
  déterminant les attributions des départements ministériels

  (Decree N° 75/182 of 14 April 1975
  laying down the responsibilities of Ministry departments)

- Décret N° 75/191 du 18 avril 1975
  portant création et organisation de l'Office congolais des bois (OCB)

  (Decree N° 75/191 of 18 April 1975
  providing for the creation and organization of the Congolese Wood Office)

- Loi N° 32/82 du 7 juillet 1982
  portant modification du Code forestier

  (Law N° 32/82 of 7 July 1982
  amending the Forestry Code)

- Décret N° ..... (Projet)
  portant application du Code forestier

  (Decree N° ..... (Draft)
  providing for the implementation of the Forestry Code)

- Contrat de transformation industrielle des bois entre le Gouvernement de la République populaire du Congo et NN.; Modèle Standard 1982

  (Industrial timber processing contract between the Government of the People's Republic of the Congo and NN. 1982 Standard Form)

EQUATORIAL GUINEA

A. Status of Forest Legislation

Prior to the promulgation of the first national forest law in 1981, the forest legislation of the country was much dispersed and had become to a large extent obsolete. The old legislation had consisted of a series of decrees and administrative rules that had been adopted before independence. Law N° 14/1981 introduces a consolidated basic forest law, which represents a considerable progress for forestry development in the country.

Standard Forms for concession permits are currently in use.

B. Presentation of the Selected Texts

The FIRST TITLE of the new consolidated Forest Law N° 14/1981 establishes principles for a rational and long-term utilization of forest resources. It defines forests, wildlife, and the various categories of forest produce, and confirms the importance of the forestry cover for soil, water and wildlife conservation. Art. 4 classifies the activities of the forest industry and provides for a control of certain types of wood processing units.

Forest tenure includes state, communal and private forests. The state forests comprise, in particular, the zones of primary tropical forests in which timber harvesting may be organized by the forest service or by granting forest concessions with a maximum duration of 20 years. Special permits for subsistence agriculture and domestic timber uses may be issued free of charge. Timber harvesting in communal forests may be undertaken by the community itself, by the state or by granting concessions. Revenues that accrue from communal forest land are reserved for the local entities; 20% of all revenues are withheld in order to finance improvements in these forests. Private forests result from reforestation on private land and are subject to state supervision. The last chapter of this Title contains the basic provisions for wildlife conservation, management and hunting control. It establishes the regulations for the declaration of national parks, and various forms of public and private hunting districts.

The SECOND TITLE of the law determines in detail the granting of forest concessions. Concessions may comprise public forest land and private forests, the latter being included on the basis of special agreements. The general method for allocating timber harvesting rights is auctioning. Chapter II provides the administrative framework for application, selection of offers and granting of concession titles. It determines the obligations of the concessionaire, the payments of revenues and fees, the restrictions applicable to the exercise of the granted rights, and the reasons for terminating the concession. The technical requirements include regulations on minimum diameter cutting limits, preparation of forest inventories and logging plans and provisions on local wood processing. The concession duration varies between 4 years for areas of less than 25 000 ha; 5 years with the possibility of renewal for areas from 25 000 to 50 000 ha; 10 years for areas up 80 000 ha; and larger periods not exceeding 20 years for concessions above this limit. Chapter III establishes the possibility to negotiate certain concession arrangements instead of using the auctioning method for allocation. This provision appears to be of particular importance for the granting of large concession areas for periods that exceed the established 5 year limit.

The THIRD TITLE is concerned with the overnmental functions and responsibilities related to forest resources management and with the organization of a national forest service

The Standard Form, as presently in use for Granting of Forest Concessions, is a brief document that describes the boundaries of the granted area, establishes a prohibition of log exports and the obligation of constructing and operating a sawmill, and provides for the respect of local rights and the enforcement of the applicable legislation. In the case of long-term concessions that cover large forest areas, the elaboration of a more detailed and comprehensive document will probably be necessary in the future.

C. General Forest Legislation

- Decreto-Ley Forestal N° 14/1981 de fecha 29 Septiembre

  (Forest Decree-Law N° 14/1981 of 29 September)

- Decreto N° ..... (Modelo 1981) de fecha .....
  port que se otorga une Concesion Forestal de ..... Has al subdito .....

  (Decree N° ..... (1981 Standard Form) of .....
  granting a forest concession of ..... ha to Citizen .....)

GABON

A. Status of Forest Legislation

In Gabon, part of the laws and regulations adopted prior to 1960, remained in force after independence, but had been amended subsequently on various occasions.

In 1968 the modification of certain parts of Decree N° 46/116 had brought an important advancement in the evolution of the country's forest legislation. Thus the modified Art. 28 provided the possibility to grant industrial timber harvesting permits by special contractual arrangements to wood processing units. The maximum duration of such permits and contracts was 30 years. Timber harvesting in the granted areas was subject to special conditions fixed in an annex to the agreement.

The contractual arrangements determined, in particular, the area and limits of the granted forest zone; the applicable arbitration procedure in the case of conflict between the two parties; and the duration of the concession (generally 20 years). Other clauses referred to infrastructural improvements; the establishment of processing installations; the execution of inventories; and the submission of annual logging programmes. The interesting point with this type of agreement was that it represented one of the first examples in the French-speaking countries of West and Central Africa for timber allocation procedures that were not unilateraly based on legislative and administrative rules, but included contractual elements.

Several decrees, adopted during the period 1968-1980, provided regulations on forest industry development and log exports. The forest fees and taxes were determined by a series of ordinances and regulations. On the whole, one can say that the Gabonese forest legislation had become more and more complicated and difficult to apply. A complete rearrangement had become indispensable.

The reorganization of the basic forest legislation was accomplished by a new Forest law promulgated in 1982. As its Title (Loi d'Orientation) indicates, this law has been conceived as a broad legislative framework that provides an orientation on fundamental issues of forestry, wildlife and fisheries development. Its adoption means a major progress in consolidating the Gabonese forest legislation at the present stage. Moreover the new forest law provides a flexible basis for implementing and more detailed provisions. Several articles contain a reference to the elaboration of specific regulations, and Art. 117 provides a general enabling clause allowing for the making of regulations on all measures that are necessary for the law's application. The preparation of such texts will contribute to a further modernization and consolidation of the forest legislation as a whole.

B. Presentation of the Selected Texts

Law N° 1/82 offers another example for the new texts of modern forest resources legislation, which have gradually replaced more conventional laws that had existed in the French-speaking countries of West and Central Africa two decades ago.

The more management orientated character of the 1982 forestry law is demonstrated by its introductory provisions. TITLE I presents a comprehensive list of forest development objectives that includes the rational use and the renewal of the forest resources; improved utilization of the available raw material; establishment of wood processing industries within the country; and an increased national participation in the sector's activities. The broad sectorial orientation of the law is also documented by the fact that it refers to forestry, wildlife as well as to fisheries development of the river and coastal domain.

TITLE I also contains provisions referring to forest ownership as determined by the land tenure and the forest legislation; to the exercise of customary rights; and to the organization of the national forest service and the responsibilities of its agents.

TITLE II is concerned with the classification, constitution and utilization of the forestry domain. It distinguishes between the reserved state forests, which form part of the public domain, and the protected state forests, which are considered as part of the state's private domain. The reserved forests may be classified according to various categories of functional uses such as permanent production forests and reforestation zones; national parks; protection forests, recreation forests; and special wildlife management zones. Art. 13 stipulates that the reserved state forests should cover at least 40% of the national territory.

TITLE II includes important provisions on the granting of timber harvesting rights and, in particular, a clause which stipulates that no new permits with an area exceeding 15 000 ha may be issued if the timber harvesting operation is not attached to a local wood processing unit. At least 75% of the logs produced under such permits have to be processed locally. Logs produced for export from industrial permits are subject to a special tax, the rate of which shall be determined by the annual fiscal law. The principle already established by previous legislation, that the country is subdivided in logging zones, and that in the first zone (principally the coastal area) logging is reserved to nationals, has been retained in the new law. The first zone now includes by definition a five kilometre strip on both sides of railways, rivers and major traffic roads. Timber harvesting in forests in the vicinity of villages is reserved to the villagers.

New and important provisions are presented in the Art. 24 to 31. The maximum concession held by one single person or company and its affiliates may not exceed 200 000 ha. No new concession may be granted in cases in which the total concession area is actually beyond this limit. Logs and processed forest products are subject to official grading as determined by regulations. The granting, renewal and transfer of concessions rights, as well as the production and export of logs and processed products, involve the payment of fees and taxes, the rates of which are to be determined by the Annual Fiscal Law on proposal of the Minister in charge of forestry affairs.

TITLE III (which amounts to approximatively half of the text of the law) is exclusively concerned with wildlife and related matters. It presents in a clear and well-organized manner a large number of provisions that determine the constitution of wildlife management zones and of protected wildlife areas; the creation of national parks; fully protected nature and wildlife reserves; and of similar areas. It regulates further the administration and utilization of wildlife management zones; the exercise of hunting activities; the issue of hunting permits; and the possession, transport and export of game and trophies.

The use and development of the fisheries resources (coastal and inland fishery) are determined by the provisions in TITLE IV.

TITLE V on economic and financial issues confirms the principle that the Fiscal Law shall determine annually, on proposal of the Minister in charge of forestry, the rates and the assessment basis for all payments, fees and taxes levied or charged in relation to the utilization of timber, forests, wildlife and fisheries resources. This provision represents considerable progress in relation to the previous situation and will facilitate a regular adjustment of all public financial charges applicable in the forest sector, in accordance with changes in production costs and product prices.

Title V provides finally, subject to further regulations, for the creation of a state reforestation agency and for the promotion of cooperative arrangements of nationals in order to facilitate their greater participation in forest resources utilization.

Law N° 10/75 established the National Forestry School (ENEF) under the administrative authority of the Ministry of Forests and under the academic authority of the Ministry of National Education. ENEF is, not only for Gabon but also at the regional level, of considerable importance in the field of higher and technical forestry education, and should serve as a reminder of the fact that the formation of human resources is one of the key issues for the advancement of the forest sector in the member countries of ATO, and that it is essential to create an appropriate legislative and institutional framework for forest education.

The introductory part of the law determines the joined authority of the competent ministries; defines the options of forest education (medium and university level) to be provided for by ENEF; establishes the possibility for foreigners to participate in the offered formation; and regulates the financial resources of the School. The following chapters determine the responsibilities of the Director; the composition and functioning of the Advisory Council; and the categories of Diploma that are delivered. The composition of the Advisory Council appears of particular interest since it allows for a large representation of the various governmental services and other agencies, including the private sector, which are concerned with forestry develpment.

C. General Forest Legislation

New texts

- Loi N° 1/82 du 22 juillet 1982
  dite Loi d'Orientation en matière des eaux et forêts

  (Law N° 1/82 of 22 July 1982
  Known as Forest Orientation Law)

Previous Texts

- Loi N° 31/66 du 19 décembre 1966
  modifiant les taux de la taxe de superficie et de transfert

  (Law N° 31/66 of 19 December 1966
  amending the rates of area and transfer taxes)

- Loi N° 40/66 du 31 décembre 1966
  déterminant les droits et taxes applicables à l'exportation des placages et contre-plaqués

  (Law N° 40/66 of 31 December 1966
  fixing duties and taxes applicable to veneer and plywood exports)

- Loi N° 42/66 du 31 décembre 1966
  modifiant les taxes et droits de sortie (Okoumé LM 5,5%)

  (Law N° 42/66 of 31 December 1966
  amending export taxes and duties (Okoumé LM 5.5%)

- Ordonnance N° 21/68 du 9 avril 1968
  portant modification de la réglementation forestière

  (Ordinance N° 21/68 of 9 April 1968
  amending the Forestry Regulations)

- Arrêté N° 00367/SF/MEF/PR du 15 avril 1968
  définissant la procédure d'attribution des permis industriels en application des Art. 37 et 38 bis (nouveaux) du Décret validé du 20 mai 1946

  (Regulation N° 00367/SF/MEF/PR of 15 April 1968
  defining industrial permits allocation procedures as laid down in Srt. 37 and 38 bis (new) of the ratified Decree of 20 May 1946)

- Décret N° 68/282/PR du 6 juin 1968
  portant adjudication des droits de coupe pour 1968 en 2e zone

  (Decree N° 68/282/PR of 6 June 1968
  providing for the award of felling rights for 1968 in the 2nd zone)

- Arrêté N° 57/MEF/SF du 30 octobre 1968
  déterminant le diamètre d'exploitabilité des essences Ilomba et Agba

  (Regulation N° 57/MEF/SF of 30 October 1968
  defining the minimum cutting limits of Ilomba and Agba species)

- Arrêté N° 65/MEF du 27 novembre 1968
  déterminant les modalités d'adjudication des droits de coupe

  (Regulation N° 65/MEF of 27 November 1968
  defining felling rights allocation procedures)

- Ordonnance N° 64/68 du 30 décembre 1968
  déterminant les taux de droit et taxes de sortie applicables aux bois bruts

  (Ordinance N° 64/68 of 30 December 1968
  fixing taxes and duties applicable to rough timber exports)

- Arrêté N° 0089 du 25 janvier 1969
  portant Cahier général des charges pour l'exploitation forestière des bois d'oeuvre

  (Regulation N° 0089 of 25 January 1969
  laying down General Specifications for industrial timber harvesting)

- Décret N° 69/180 du 4 mars 1969
  réglementant la délivrance des permis spéciaux (application de l'Art. 33 nouveau du Décret du 20 mai 1946)

  (Decree N° 69/180 of 4 March 1969
  regulating the delivery of special permits (in pursuance of new Section 33 of the Decree of 20 May 1946)

- Ordonnance N° 50/71 du 15 septembre 1971
  exemptant de droits d'enregistrement et de timbre les contrats d'octroi de PTE dans la zone d'attraction du chemin de fer

  (Ordinance N° 50/71 of 15 September 1971
  exempting from registration and stamp fees contracts granting temporary logging permits in the railway area (where logging is reserved to nationals))

- Ordonnance N° 37 bis du 1er novembre 1971
  fixant les redevances applicables dans la zone d'attraction du chemin de fer

  (Ordinance N° 37 bis of 1 November 1971
  fixing the logging fees applicable in the railway area

- Ordonnance N° 57/71 du 1er novembre 1971
  créant un Office national des bois du Gabon

  (Ordinance N° 57/71 of 1 November 1971
  providing for the creation of the National Timber Office of Gabon)

- Modèle de contrat entre l'Etat gabonais et NN.

  (Standard contract form between the Gabonese Government and NN.)

- Loi N° 10/75 du 18 décembre 1975
  portant création de l'Ecole nationale des eaux et forêts

  (Law N° 10/75 of 18 December 1975
  setting up the National Forestry School)

Legislation on National Parks, Wildlife and Hunting (Previous Texts)

Législation sur les Parcs nationaux, la Faune et la Chasse (Textes antérieurs).

- Loi N° 46/1960 (en partie)
  réglementant la chasse et l'usage des armes de chasse

  (Law N° 46/1960
  partly regulating hunting and the use of hunting weapons)

- Décret N° 00084/1961
  portant application de la Loi N° 46/1960 en matière de chasse

  (Decree N° 00084/1961
  providing for the implementation of Law N° 46/1960 on hunting)

- Loi N° 01/1964
  réglementant la chasse

  (Law N° 01/1964
  on hunting control)

- Ordonnance N° 03/1967
  réglementant les taux des permis de chasse

  (Ordinance N° 03/1967
  regulating the fees for hunting permits)

- Décret N° 00172 du 23 avril 1971
  Réglementant le statut des aires d'exploitation rationnelle

  (Decree N° 00172 of 23 April 1971
  regulating the status of rational management areas)

GHANA

A. Status of Forest Legislation

The forest legislation of Ghana is presently contained in several ordinances, acts and decrees. The Forestry Ordinance (1927), supplemented by the Forest Reserves Regulations (1927), is its principal element. The Forestry Ordinance has been amended on various occasions and sections 22 to 33, on forest offences and prosecution, have been substituted in 1974 by the provisions of the Forest Protection Decree.

The Trees and Timber Ordinance of 1950, and the Protected Timber Lands Act of 1959 had been the main complementary elements to the Forestry Ordinance. Both texts were replaced in 1974 by the provisions of the Trees and Timber Decree (Decree N° 273). Various regulations referring to protected areas outside forest reserves, to the control of cutting, log measurement and log exports, that had been made during the period 1958 to 1960 under the Trees and Timber Ordinance, and under the Protected Timber Lands Act, have, however, remained in force. The Forest Improvement Fund Act of 1960 is concerned with silvicultural operations in forest reserves.

The Concessions Act of 1962 regulates, in its Sec. 16, the granting of timber harvestings rights. The Timber Leases and Licences Regulations of 1962 and the Forest Fees Regulations of 1976 have been made under this Act. Standard Forms for the granting of timber felling agreements and for timber leases exist and determine the rights and responsibilities of the logging companies.

The larger timber processing industries are subject to the provisions of the Timber Operations (Government Participation) Decree of 1972 and the forest industry, as a whole, is covered by certain provisions of the Investment Policy Decree of 1975. The Timber Industry and Ghana Timber Marketing Board Amendment Decree of 1977 contains important provisions on log exports and on the commercialization of log and forest products.

In 1980, the Ghana Forestry Commission Act had reorganized various governmental agencies that are in charge of certain aspects of forestry and forest industries development. This law has recently been repealed by the Provisional National Defence Council Law N° 42 of 1982; the new text also provides for the functioning of a forestry commission but more of an advisory nature.

Wildlife and hunting are regulated by a specialized legislation; the basic texts date from the years 1961/62.

B. Presentation of Selected Texts

The provisions of the Forests Ordinance (1927 and amendments) represent a rather typical example of the older generation of forest laws. The Ordinance is principally concerned with the constitution and management of forest reserves and with certain measures of forest protection. Of particular interest appears SECTION 18, which stipulates that the constitution of a forest reserve shall not alter the ownership of the land. It offers an option for the management of such reserves, either by the owners under the direction of the forest service, or by the government for the benefit of the owners. Within the given limits, the Ordinance has had a considerable impact on the establishment of the country's permanent forest estate on land under customary tenure.

The Forest Improvement Fund Act (1960) supports this policy, and allows channelling the payments from timber harvesting and other fees, which are due to the customary land owners, as well as grants and other payments from the government, into the established forest improvement fund. The director of the forest service is entitled to make payments for costs incurred in connection with timber harvesting and silvicultural improvements undertaken in the forest reserves, and to authorize the necessary expenditure for royalties due to the landowners. As stated in SECTION 8, the director of the forest service is acting in this respect on behalf of the customary authorities.

The context of the Concessions Act (1962) is a much broader one; it had been adopted in order to provide an opportunity for the government to vary the terms of existing agreements (e.g. the terms of mineral concessions) by renegotiation after independence. SECTION 16 deals specifically with forest reserves and timber concessions and establishes principles determining the relation between the state and the customary land owners. The first one is that all forest reserves which had already existed at the promulgation of the Act, as well as all reserves to be constituted at a later stage, are vested in the President acting in trust for the customary land owners. This trusteeship does not affect the existing customary and other rights, subject to the prevailing laws and regulations. The second principle is that land outside forest reserves, on which timber harvesting rights (concessions) have been granted, as well as the rights to timber and trees on any other land, are also vested in the President acting in trust for the customary owners. Revenues that accrue from forestry lands and rights on timber and trees are to be collected by the government and disbursed in accordance with the provisions of the Administration of Lands Act. SECTION 14 allows the respective fees to be varied by regulation.

The third point, established by SUB-SECTION (8) of SECTION 16, is that the Forest Ordinance shall apply to any land outside forest reserves on which timber harvesting rights are granted. On the whole, one may conclude that the provisions of the Concession Act on forestry matters have considerably reinforced the government's position with regard to forest land management and allocation of timber harvesting rights, but have maintained the fundamental position of the customary entities as land owning units.

The Trees and Timber Decree (1974) complements the provisions of the Forests Ordinance. PART I determines the registration of locality marks and of property marks identifying the production regions and the log producers; the marking of logs and stumps in the forests; and the marking of logs for export. PART II provides for the declaration of protected areas on forest land that is not included in forest reserves; the objective being to avoid unnecessary waste of trees and timber on such land. Permits for farming in protected areas may be issued. Regulations may be made determining the obligations of concessionaires in protected areas; the appointment of forest guards; and the fees to be levied on the concessionaires in connection with such appointments. PART III provides for the possibility to make general regulations related to silvicultural measures such as the prescription of minimum girth limits, and to the control of transit or export of timber.

The Forest Fees Regulations (1976) consolidate the various categories of payments related to timber harvesting under concessions and permits. The royalty rates, determined in SCHEDULE 1, substitute the rates previously fixed in individual leases and agreements; they may be revised from time to time. The regulations provide for a silvicultural fee, to be paid annually and assessed on the basis of the actual logging area both within and outside forest reserves. The payments, which accrue from the fee, are credited to the Forest Improvement Fund. The Trees and Timber Control of Measurement Regulations (1960 and 1976) prescribe official log scaling and the delivery of measurement certificates.

The Timber Industry and Ghana Timber Marketing Board Amendment Decree (1977) strengthens the powers of the marketing Board and establishes that the Board shall be the only agent for exports of timber and timber products over land; that the Board may fix the export prices of timber and timber products; and that the Board may determine the percentage to be exported, or to be marketed within the country. The decree provides for the possibility to make regulations on the marketing of timber and timber products, the registration of exporters, and on related matters. The Economic Plants Protection Decree (1979) offers an example of the need for protecting agricultural crops (Cocoa) from timber harvesting operations, and provides for the restriction of such rights and the payment of compensation, if the specified agricultural crops suffer from logging.

In 1980, the Ghana Forestry Commission Act had been adopted, in order to create the Ghana Forestry Commission as an autonomous governmental agency with far reaching responsibilities. The Ghana Timber Marketing Board; the Forest Products Research Institute; the Forest Department; and the Department of Game and Wildlife had been placed as Divisions under the authority of the new organization. As a result of SECTIONS 34, 64 and 65 of the Proclamation Law (1982), the Act of 1980 is revoked and the Ghana Forestry Commission dissolved. The Forestry Department, and the Department of Game and Wildlife are reestablished as governmental departments, and the Ghana Timber Marketing Board as a statutory body. The Proclamation Law retains, however, the concept of a common platform for the agencies concerned with various aspects of the sector's development by creating a Forestry Commission, that has largely monitoring, advisory and consultative functions. The composition of the new Forestry Commission shall be determined by the Provisional Natural Defence Council; the Forest Products Research Institute remains attached to the new Commission.

C. General Forest Legislation/Législation Forestière Générale

- The Forest Ordinance (Cap. 157) of 30 March 1927
  Section 22 to 33 have been repealed by Forest Protection Decree NRCD 243
  - The Forest Reserves Regulations 1927 L.N. 31 of 1927
- Trees and Timber Decree 1974 NRCD 273
  - The Trees and Timber (Control of Cutting) Regulations 1958 L.N. 368/58
  - The Trees and Timber (Measurement) Regulations 1958 L.N. 388/58
  - The Trees and Timber (Control of Measurement) Regulations 1960 L.I. 23/60
  - The Trees and Timber (Control of Measurement) Amendment Regulations 1976 of 31 August 1976
  - The Trees and Timber (Control of Export of Logs) Regulations 1961 L.I. 130/61
  - The Timber Lands (Protected Areas) Regulations 1959 L.N. 311/59
- The Forest Improvement Fund Act
  1960 N° 12 of 1960

- The Concessions Act, 1962 (Act. 124); Section 16 (11) repealed by NRCD 243

  . The Timber Leases and Licences Regulations 1962

  . The Concessions Regulations 1962

  . Forest Fees Regulations 1976 (L.I. 1089)

- The Forest Protection Decree 1974 NRCD 243

- Timber Operations (Government Participation) Decree 1972 NRCD 139/72

- Investment Policy Decree 1975

- Timber Industry and Ghana Timber Marketing Board (Amendment) Decree 1977

- Sections 4(1), 4(2) and 4(3) of Economic Plants Protection Decree 1979 AFRCD 47

- Provisional National Defence Council (PNDC) Law 42; 1982 Sections 34(1), 64, 65 and 67

- Standard Form for Timber Felling Agreements, Ministry of Lands

- Standard Form for Timber Leases, Ministry of Lands

Wildlife and Hunting Legislation/Législation Faune et Chasse

- The Wild Animals Preservation Act, 1961 N° 43 of 1961

  . Wild Animals Preservation (Game Reserves) Regulations L.I. 171/62

  . Wild Animals Preservation (Close Season and Restriction of Hunting) Regulations, 1962 L.I. 212/62

  . Wild Animals Preservation (Hunting Licences) Regulations L.I. 213/62

IVORY COAST

A. Status of Forest Legislation

The Forestry Code of Ivory Coast was adopted in 1965 and supplemented by a set of forest regulations issued during the period 1966/67. Several decrees regulate the classification of forest land and forest protection; the prosecution of forest offences; timber harvesting practices; and the status of logging companies, wood exporters and forest industries. So far the 1965 Forestry Code, together with its implementing regulations, have proved to be sufficiently flexible to allow for an adaption of the country's forest legislation to the rapidly changing conditions within the forest sector. Some smaller amendments of the Forestry Code are presently under consideration.

An additional series of forest regulations were adopted during the period 1972 to 1974. These are concerned with the establishment of regional wood supply areas; annual log supply obligations to local timber conversion units; a special annual authorization to start or continue logging operations; the subletting of forest concession areas; and cooperative arrangements among the holders of permits in particular areas. The log supply quota system, originally introduced by Decree N° 72-543 and the complementary rules of Order N° 2044/1972, was modified in 1975 and 1982. Another interesting regulation, introduced for the first time in 1978, and modified subsequently on several occasions, is the official determination of certain standard freight rates for the export of unused or little-used but technologically valuable species, in order to facilitate their promotion and commercialization on international markets.

Decree N° 78-231 presents important provisions on the management and utilization of the forest domain. The organization of the forest administration has been subject to various regulations. The most recent texts are Decree N° 81-735 determining the responsibilities and organization of the Ministry of Forestry, and regulations concerning the establishment of the field services of the Ministry. The basic texts that have created and organized the National Reforestation Agency (DOFEFOR), date from 1966 and 1970; several amendments have been made in the meantime.

The assessment and collection of forest revenue and timber export taxes are determined by special laws and regulations as well as by provision of the annual fiscal laws; these provisions have been subject to numerous changes and amendments.

National parks, wildlife management and hunting control are subject to a specialized legislation; the basic texts were adopted in 1965/1966.

B. Presentation of the Selected Texts

The Forestry Code, promulgated by Law N° 65-425, follows to a considerable extent the format of the basic forest legislation that existed in several French-speaking countries of West and Central Africa after independence. The Forestry Code is fairly short and well-structured, and its provisions appear to be defined in a precise manner and well formulated. The INTRODUCTORY PART of the Law defines forests, protection areas and reforestation zones, as well as the various categories of tenure that may exist within the forest domain. The SECOND TITLE is concerned with the constitution of reserved and protected forests, the exercise of customary rights and the granting of timber harvesting rights in forests under state ownership. The content of customary rights is described and their exercise limited to protected forests. Art. 11 provides, in order to compensate for this limitation, that the constitution of state forest reserves shall leave sufficiently large areas of protected forests allowing for the practice of customary rights as required by the population. Timber harvesting may be undertaken directly by the forest service, by special permits, and by temporary logging permits. The current practice has been the granting of 5-year renewable logging permits.

The THIRD TITLE deals with forests owned by private and collective owners. It provides for the possibility of setting aside certain parts of state land for reforestation by private and local groups, subject to administrative rules and regulations. One should, however, note that the law has no special provisions on the creation of permanent forest land under communal or customary tenure, and on the utilization and management of such forests. TITLE IV deals with forest fires and the remainder of the law with forest offences and prosecution.

Decree N° 66-422 creates the Agency for the Developpment of Forest Plantations (SODEFOR) as a semi-autonomous governmental agency. The principal activities of SODEFOR are the execution of large scale forest plantations; the preparation of forest inventories and management plans; and the elaboration of studies and programmes in particular in the field of timber processing and marketing promotion of lesser used species. The responsibility for forest demarcation, which had been assumed as a task by SODEFOR since 1970, was retransferred to the Ministry of Forestry in 1981. The Decree regulates the responsibilities of the agency; its corporate capital and the funds available for its activities; the composition and functioning of its supervisory board; the applicable financial controlling procedures; and approves the regulations for its internal functioning.

Decree N° 72-114 on the creation of raw material supply zones for certain wood processing industries is the first of several texts that aim at a reorganization of timber harvesting areas, allocated by temporary logging permits. The Decree defines 26 supply zones, that cover practically the whole of the country's utilizable forests, and facilitates the regrouping of the existing permits within these zones. Art. 3 and 4 provide that the industries established in each zone shall have a priority for raw material supply. New permits may only be granted to already established or newly planned industries, or to associations of logging companies linked to such industries through raw material supply agreements.

Decrees N° 72-125 and N° 72-606 adopt measures in order to facilitate a closer integration between timber harvesting and wood processing, and to stop certain anomalies in the use of logging permits. The FIRST DECREE authorizes the sub-contracting of permits under certain conditions subject to official approval. The subcontractor becomes fully responsible for the payment of forest fees, and has to submit a logging plan. The administration may determine special clauses in order to avoid unnecessary waste of raw material. The SECOND DECREE provides a basis for the association of individuals or logging companies subject to official approval. The permits of the members of the group may be organized in order to allow a more efficient and economical operation. The duration of the permits is 5 years, starting from the date when the group of operators is established. Additional areas may be granted, if the group participates in timber processing activities.

An important adjustment of the forest reservation concept, as determined in the Forestry Ordinance, to the country's profoundly changed land-use pattern of today, is made by Decree N° 78-231. Its FIRST TITLE distinguishes between two principal categories of the forest domain: the permanent forest domain of the state (or classified state forest reserves), and the rural forest domain of the state. The permanent forest domain serves for long-term wood production and the maintenance of an ecological equilibrium.

The rural forestry domain presents a land reserve for the expansion of agriculture. This category of forest land shall be protected and available for timber harvesting, until it is needed for agricultural development.

The SECOND TITLE is concerned with the permanent forest domain of the state. It provides that 3 million hectares of undergraded forests shall be classified within the high forest zone of the country; another 1.7 million ha of natural forests are to be classified as permanent in the Savanna region. Special measures shall be taken for the demarcation and protection of the permanent forest domain, for a rational and long-term utilization of the available timber stands, and for reforestation.

The THIRD TITLE refers to the rural forestry domain of the state and provides that such land may only be released if the start of agricultural development operations has been scheduled by a decision made jointly by the Ministries of Forestry, Agriculture and Finance. All commercially usable timber has to be removed by salvage logging before the land can be released. The ANNEX to the decree presents a detailed list of forest reserves, classified prior to 1978, which are to be incorporated into the permanent forest domain of the State. A second list identifies those of the previous forest reserves, which are declassified and transferred to the state's rural forestry domain.

Decree N° 81-735 determines the responsibilities and the organization of the Ministry of Forestry and offers an example for the range of provisions that may be necessary in the field of forest administration. Art. 1 defines the responsibilities of the Minister with regard to forestry, fisheries, wildlife, and soil and water conservation. The description of the responsibilities is made in considerable detail and gives an excellent idea of the numerous and complicated tasks that may have to be accomplished by a modern public forest administration. The following articles establish the divisions at ministerial level, and the field organization which is subdivided into regions, inspections and districts. Art. 7 announces the intention of establishing a special fund in order to finance operations that are required for the implementation of the national forest policy.

Decree N° 82-70 presents new important provisions on raw material supply obligations to the forest industry and on registration of exporters of logs and processed forest products. It replaces, in particular, the provisions regulating the log supply quota, introduced by Decree N° 72-543, and the regulations on timber exporters, adopted by Decree N° 78-234.

The new decree confirms the principle that log exports are subject to a quotas system. The quota determines the quantity of permitted log exports as a percentage of the quantity which is processed within the country. Whereas the previous regulations had established such quota only for certain species and for the primary processing industry, the new system takes into account all species and all stages of processing.

Each wood processing company thus generates a certain export quota, which is authorized annually by the Ministry. A confirmed log export quota may freely be transferred between various timber processing companies; between a processing company and a registered log exporter; and among various log exporters.

Log exports may be undertaken by officially registered companies and cooperatives within the limits of the authorized quota. Mere trading companies and wood processing enterprises may qualify for registration as a log exporters, whereas the previous regulations had limited registration to timber processing companies only. The formal registration of exporters is made by ministerial decision of the Minister of Commerce on proposal of the Minister of Forestry.

The percentages of permitted log export volumes in relation to the volumes processed locally are differentiated by species and determined by ministerial decision. The timber processing companies have to inform the Ministry every 3 months on raw material intake; output and kind of produced products; and the log quota transferred to other companies. The exporters have to provide every 3 months information on the quota transferred from or to other companies and a justification of the exported log volumes.

Art. 6 establishes an interministerial advisory commission with representatives from the ministries in charge of Forestry, Industry, Finance, Commerce, Agriculture and Naval Affairs. The commission is responsible for controlling the log export quota system and advising on matters related to the promotion of industrial timber processing.

The decree also modifies the applicable customs classification for log exports by species.

Ordinance N° 82-71 and Law N° 81-127 represent recent examples for the determination of the applicable rates of export taxes on logs and processed forest products. Ordinance N° 82-71 indicates the applicable tax rate for each species as a percentage of the export value. The reference export values (check prices) are determined from time to time by regulations. The level of the export tax; the considerable differentiation between the tax rates applicable to high value, as compared to lower value species; and the fact that approximatively 50 species are presently identified in the export declarations, should be noted. Law N° 81-127 determines the export tax rates on sawnwood, veneer and plywood, which are lower, even if the cumulative effects of the recovery rate and value added are taken into account. The Law also modifies the rates of the logging tax; the rates are differentiated by 3 groups of species, and whether the logs are exported or processed locally.

C. General Forest Legislation

- Loi N° 65-425 du 20 décembre 1965
  portant Code forestier

  (Law N° 65-425 of 20 December 1965
  establishing the Forestry Code)

- Décret N° 66-50 du 8 mars 1966
  réglementant la profession d'exploitant forestier

  (Decree N° 66-50 of 8 March 1966
  regulating the profession of logging operator)

- Décret N° 66-51 du 8 mars 1966
  portant attribution des permis temporaires d'exploitation

  (Decree N° 66-51 of 8 March 1966
  providing for the allocation of temporary logging permits)

- Décret N° 66-52 du 8 mars 1966
  fixant les modalités de mises à feu autorisées

  (Decree N° 66-52 of 8 March 1966
  fixing the procedures for authorized burnings)

- Décret N° 66-122 du 31 mars 1966
  déterminant les essences forestières dites protégées

  (Decree N° 66-122 of 31 March 1966
  determining the so-called protected forest species)

- Décret N° 66-410 du 13 septembre 1966
  portant attribution des permis temporaires d'exploitation

  (Decree N° 66-410 of 13 September 1966
  providing for the allocation of temporary logging permits)

- Décret N° 66-420 du 15 septembre 1966
  portant réglementation des industries du bois

  (Decree N° 66-420 of 15 September 1966
  regulating timber industries)

- Décret N° 66-421 du 15 septembre 1966
  réglementant l'exploitation des bois d'oeuvre et d'ébénisterie, de service de feu et à charbon complété par Décret N° 73-346 du 11 juillet 1973

  (Decree N° 66-421 of 15 September 1966
  regulating the harvesting of industrial, furniture, fuel and charcoal wood. Completed by Decree N° 73-346 of 11 July 1973)

- Décret N° 66-427 du 15 septembre 1966
  portant répartition du produit net des amendes, confiscations, restitutions, dommages-intérêts, contraintes et transactions en matière de police forestière

  (Decree N° 66-427 of 15 September 1966
  providing for the apportionment of the net proceeds deriving from fines, confiscations, refunds, damages, destraints and transactions in matters of forest police)

- Décret N° 66-428 du 15 septembre 1966
  fixant les procédures de classement et de déclassement des forêts domaniales

  (Decree N° 66-428 of 15 September 1966
  establishing the classifying and declassifying procedures of state-owned forests)

- Arrêté N° 1165 du 26 septembre 1966
  précisant les dispositions du Décret N° 66-410

  (Regulation N° 1165 of 26 September 1966
  specifying the provisions of Decree N° 66-410)

- Arrêté N° 1166 du 26 septembre 1966
  précisant les dispositions du Décret N° 66-51

  (Regulation N° 1166 of 26 September 1966
  specifying the provisions of Decree N° 66-51)

- Arrêté N° 1399 du 4 novembre 1966
  fixant les modalités d'application du Décret N° 66-421 du 15 septembre 1966 réglementant l'exploitation des bois d'oeuvre et d'ébénisterie, de service de feu et à charbon. Complété (Art. 22) par Arrêté N° 743 du 30 novembre 1976

  (Regulation N° 1399 of 4 November 1966
  establishing the implementation procedures of Decree N° 66-421 of 15 September 1966 regulating the harvesting of industrial, furniture, fuel and charcoal wood. Completed (Section 22) by Regulation N° 743 of 30 November 1976)

- Décret N° 66- 536 du 17 novembre 1966
  fixant les modalités de représentation de l'Administration devant les tribunaux répressifs et la procédure des transactions en matière forestière

  (Decree N° 66-536 of 17 November 1966
  establishing the procedures for the representation of the Administration before repressive courts and those for forestry transactions)

- Arrêté N° 1577 du 5 décembre 1966
  fixant les modalités d'application du Décret N° 66-420 du 15 septembre 1966, portant réglementation des industries du bois

  (Regulation N° 1577 of 5 December 1966
  establishing the implementation procedures of Decree N° 66-420 of 15 September 1966 regulating timber processing industries)

- Décret N° 67-78 du 16 février 1967
  portant attribution des permis temporaires d'exploitation

  (Decree N° 67-78 of 16 February 1967
  regarding the allocation of temporary logging permits)

- Arrêté N° 243 du 1er mars 1967
  rectifiant l'Arrêté N° 1577 du 5 décembre 1966 fixant les modalités d'application du Décret N° 66-420 du 15 septembre 1966 portant réglementation des industries du bois

  (Regulation N° 243 of 1 March 1967
  amending Regulation N° 1577 of 5 December 1966 establishing the implementation procedures of Decree N° 66-420 of 15 September 1966 regulating timber processing industries)

- Décret N° 67-521 du 28 novembre 1967
  portant attribution des permis temporaires d'exploitation

  (Decree N° 67-521 of 28 November 1967
  concerning the allocation of temporary logging permits)

- Décret N° 67-522 du 28 novembre 1967
  portant suspension d'attributions de permis d'exploitation forestière

  (Decree N° 67-522 of 28 November 1967
  suspending the allocations of logging permits)

- Décret N° 67-576 du 15 décembre 1967
  réglementant la profession d'exportateur de bois ou de produits ligneux
  Abrogé par Décret N° 78-234

  (Decree N° 67-576 of 15 December 1967
  regulating the profession of timber or wood products exporter
  Repealed by Decree N° 78-234)

- Arrêté interministériel N° 50-85 AEF du 24 janvier 1968
  portant application des dispositions du Décret N° 67-576 du 15 décembre 1967, réglementant la profession d'exportateur de bois
  Abrogé par Arrêté N° 001 MINEFOR/COM du 2 février 1979

  (Interministerial Regulation N° 50-85 AEF of 24 January 1968
  implementing the provisions of Decree N° 67-576 of 15 December 1967 regulating the profession of timber exporter.
  Repealed by Regulation N° 001 MINEFOR/COM of 2 February 1979)

- Décret N° 69-310 du 4 septembre 1969
  portant attribution de permis temporaires d'exploitation forestière

  (Decree N° 69-310 of 4 September 1969
  regulating the allocation of temporary logging permits)

- Décret N° 72-114 du 9 février 1972
  portant création de 26 périmètres d'approvisionnement en matière ligneuse des industries du bois

  (Decree N° 72-114 of 9 February 1972
  providing for the setting up of 26 raw material supply zones for wood processing industries)

- Décret N° 72-116 du 9 février 1972
  soumettant à l'autorisation préalable du Secrétaire d'Etat chargé de la reforestation la mise en exploitation effective des permis forestiers concédés à titre temporaire

  (Decree N° 72-116 of 9 February 1972
  subjecting logging operations by holders of temporary logging permits to previous authorization from the Secretary of State in charge of Reforestation)

- Décret N° 72-125 du 9 février 1972
  portant création d'un contrat de fermage pour certains permis temporaires d'exploitation

  (Decree N° 72-125 of 9 February 1972
  establishing a lease contract for certain temporary logging permits)

- Décret N° 72-126 du 9 février 1972
  portant autorisation d'exploitation sur les permis temporaires d'exploitation forestière

  (Decree N° 72-126 of 9 February 1972
  granting logging authorization for temporary logging permits)

- Arrêté N° 21 SER du 8 juin 1972
  attribution des permis temporaires d'exploitation forestière
  Abrogé par Arrêté N° 34 MINEFOR du 8 juin 1982

  (Regulation N° 21 SER of 8 June 1972
  regarding the allocation of temporary logging permits
  Repealed by Regulation N° 34 MINEFOR of 8 June 1982)

- Décret N° 72-543 du 28 août 1972
  portant obligation aux exportateurs de bois agréés d'assurer l'approvisionnement des usines locales
  Abrogé par Décret N° 82-70

  (Decree N° 72-543 of 28 August 1972
  stating the obligation for registered log exporters to ensure supply of raw material to local industries
  Repealed by Decrée N° 82-70)

- Arrêté N° 2044 du 8 septembre 1972
  portant obligation aux exportateurs de bois agréés d'assurer l'approvisionnement des usines locales
  Abrogé par Arrêté N° 927 du 19 décembre 1975

  (Regulation N° 2044 of 8 September 1972
  stating the obligation for registered log exporters to ensure supply of raw material to local industries
  Repealed by Regulation No. 927 of 19 December 1975

- Décret N° 72-606 du 18 septembre 1972
  portant création de sociétés civiles de groupement d'exploitants forestiers

  (Decree N° 72-606 of 18 September 1972
  providing for the creation of civil companies grouping logging operators)

- Décret N° 72-695 du 31 octobre 1972
  fixant la procédure d'attribution des agréments d'exploitants forestiers et des permis temporaires d'exploitation

  (Decree N° 72-695 of 31 October 1972
  regulating the registration procedures for logging operators and the allocation of temporary logging permits)

- Arrêté N° 0028 du 16 janvier 1973
  pour l'application du Décret N° 72-543 du 28 août 1972 portant obligation aux exportateurs de bois agréés d'assurer l'approvisionnement des usines locales
  Abrogé par Arrêté N° 927 du 19 décembre 1975

  (Regulation N° 0028 of 16 January 1973
  providing for the application of Decree N° 72-543 of 28 August 1972 stating the obligation for registered log exporters to supply raw material to local industries
  Repealed by Regulation N° 927 of 19 December 1975

- Décret N° 73-341 du 5 juillet 1973
  portant fixation des règles applicables en matière de signalisation et d'enlèvement des billes tombées sur le domaine public ou déposées sur les aires des ponts-bascules

  (Decree N° 73-341 of 5 July 1973
  regulating the reporting and removal of logs fallen on public ground or stacked on weigh-bridge areas)

- Décret N° 73-346 du 11 juillet 1973
  modifiant et complétant certaines dispositions du cahier des charges annexé au Décret N° 66-421 du 15 septembre 1966 réglementant l'exploitation des bois d'oeuvre et d'ébénisterie, de service de feu et à charbon

  (Decree N° 73-346 of 11 July 1973
  amending and completing certain provisions of the specifications attached to Decree N° 66-421 of 15 September 1966 regulating the harvesting of industrial, furniture, fuel and charcoal wood)

- Arrêté N° 1192 du 25 septembre 1973
  pour l'application du Décret N° 72-543 du 23 août 1972 portant obligation aux exportateurs de bois agréés d'assurer l'approvisionnement des usines locales
  Abrogé par Arrêté N° 927 du 19 décembre 1975

  (Regulation N° 1192 of 25 September 1973
  implementing Decree N° 72-543 of 23 August 1972 providing for the obligation for registered log exporters to ensure raw material supplies to local industries
  Repealed by Regulation N° 927 of 19 December 1975)

- Décret N° 73-490 du 11 octobre 1973
  portant obligation aux usines de première transformation d'assurer l'approvisionnement du marché local

  (Decree N° 73-490 of 11 October 1973
  stating the obligation for primary processing industries to ensure the local market supply)

- Arrêté N° 096 du 12 octobre 1973
  pris pour l'application du Décret N° 72-543 du 28 août 1972 portant obligation aux exportateurs agréés d'assurer l'approvisionnement des usines locales

  (Regulation N° 096 of 12 October 1973
  for the implementation of Decree N° 72-543 of 28 August 1972 stating the obligation for registered log exporters to ensure raw material supplies to local industries)

- Arrêté N° 132 du 11 décembre 1973
  fixant certaines modalités d'application du Décret N° 66-420 du 15 septembre 1966 portant réglementation des industries du bois

  (Regulation N° 132 of 11 December 1973
  fixing certain implementation procedures of Decree N° 66-420 of 15 September 1966 regulating timber processing industries)

- Décret N° 75-385 du 6 juin 1975
  portant constitution du périmètre papetier
  Modifié par Décret N° 77-15 du 7 janvier 1977

  (Decree N° 75-385 of 6 June 1975
  providing for the setting up of a pulp and paper production area
  Amended by Decree N° 77-15 of 7 January 1977)

- Arrêté Interministériel N° 927 MINEFOR/MEF/MC du 19 décembre 1975
  pris pour l'application du Décret N° 72-543 du 23 août 1972, portant obligation aux exportateurs de bois agréés d'assurer l'approvisionnement des usines locales

  (Interministerial Regulation N° 927 MINEFOR/MEF/MC of 19 December 1975
  providing for the implementation of Decree N° 72-543 of 23 August 1972 concerning the obligation for registered log exporters to supply raw material to local industries)

- Décret N° 76-281 du 20 avril 1976
  déterminant les conditions d'entrée en Côte-d'Ivoire des marchandises étrangères de toutes origines et de toute provenance, ainsi que les conditions d'exportation et de réexpédition des marchandises à destination de l'étranger

  Annexe B Chapitre 44 Bois Bruts: Liste des Essences
  Modifié par Décret N° 82-70

  (Decree N° 76-281 of 20 April 1976
  regulating the import into the Ivory Coast of foreign goods of any origin or provenance as well as the procedures for the export and re-export of goods abroad

  Appendix B Chapter 44 (Rough Timber): List of species
  Amended by Decree N° 82-70)

- Arrêté N° 743 MINEFOR/DAB du 30 novembre 1976
  portant modification de l'Arrêté N° 1399 du 4 novembre 1966 fixant les modalités d'application du Décret N° 66-421 du 15 septembre 1966

  (Regulation N° 743 MINEFOR/DAB of 30 November 1976
  amending Regulation N° 1399 of 4 November 1966 establishing the implementation procedure of Decree N° 66-421 of 15 September 1966)

- Décret N° 77-15 du 7 janvier 1977
  portant extension de la superficie du périmètre papetier de San-Pedro et définissant ses nouvelles limites

  (Decree N° 77-15 of 7 January 1977
  providing for the extension of the San-Pedro pulp and paper production area and defining its new boundaries)

- Décret N° 78-231 du 15 mars 1978
  fixant les modalités de gestion du domaine forestier de l'Etat

  (Decree N° 78-231 of 15 March 1978
  establishing the management procedures of state-owned forests)

- Décret N° 78-234 du 20 mars 1978
  réglementant la profession d'exportateur de bois et produits ligneux
  Abrogé par Décret N° 82-70

  (Decree N° 78-234 of 20 March 1978
  regulating the profession of timber and wood products exporters
  Repealed by Decree N° 82-70)

- Arrêté N° 05-78 MINIMAR/MC/MEF du 18 septembre 1978
  établissant la liste des essences technologiquement valables qui ne sont pas ou très peu exploitées, et portant homologation des taux de fret du bois
  Abrogé par Arrêté N° 06-79 du 22 mars 1979

  (Regulation N° 05-78 MINIMAR/MC/MEF of 18 September 1978
  establishing the list of little-used or unused technologically valuable species and fixing their standard freight rates
  Repealed by Regulation N° 06-79 of 22 March 1979)

- Arrêté N° 001/MINEFOR/COM du 2 février 1979
  portant application des dispositions du Décret N° 78-234 du 20 mars 1978, réglementant la profession d'exportateur de bois ou de produits ligneux

  (Regulation N° 001/MINEFOR/COM of 2 February 1979
  providing for the implementation of the provisions of Decree N° 72-234 of 20 March 1978 regulating the profession of timber and wood products exporters)

- Arrêté Interministériel N° 06-79 MINIMAR/MC/MEF du 22 mars 1979
  établissant la liste des essences technologiquement valables qui ne sont pas ou très peu exploitées, et portant homologation des taux de fret du bois
  Abrogé par Arrêté N° 11-79 MINIMAR/MC/MEF

  (Interministerial Regulation N° 06-79 MINIMAR/MC/MEF of 22 March 1979
  establishing the list of little-used or unused technologically valuable species and fixing their standard freight rates
  Repealed by Regulation N° 11-79 MINIMAR/MC/MEF)

- Arrêté N° 11-79 MINIMAR/MC/MEF du 28 septembre 1979
  établissant la liste des essences technologiquement valables qui ne sont pas ou très peu exploitées, et portant homologation des taux de fret de bois
  Abrogé par Arrêté N° 08-81 MINIMAR/MC/MEF

  (Regulation N° 11-79 MINIMAR/MC/MEF of 28 September 1979
  establishing the list of little-used or unused technologically valuable species and fixing their standard freight rates
  Repealed by Regulation N° 08-81 MINIMAR/MC/MEF)

- Arrêté N° 17 MC/DCE du 17 février 1981
  portant agrément ou renouvellement d'agrément des exportateurs de bois en grumes pour l'année 1981

  (Regulation N° 17 MC/DCE of 17 February 1981
  concerning log exporters registration or renewal of registration for 1981)

- Arrêté Interministériel N° 08-81 MINIMAR/MC/MEF du 1er octobre 1981 établissant la liste des essences technologiquement valables qui ne sont pas ou très peu exploitées et portant homologation des taux de fret du bois

  (Interministerial Regulation N° 08-81 MINIMAR/MC/MEF of 1 October 1981 establishing the list of little-used or unused technologically valuable species and fixing their standard freight rates)

- Décision N° 22 MINEFOR/DCF du 9 décembre 1981 fixant les conditions de circulation du bois en grumes

  (Decision N° 22 MINEFOR/DCF of 9 December 1981 regulating the transit of logs)

- Décret N° 82-70 du 13 janvier 1982 fixant les conditions d'approvisionnement en bois des industries locales et d'exportation de bois et de produits ligneux, et abrogeant les Décrets N° 72-543 du 28 août 1972, portant obligation aux exportateurs de bois agréés d'assurer l'approvisionnement des usines; et N° 78-234 du 20 mars 1978 réglementant la profession d'exportateur en bois et en produits ligneux

  (Decree N° 82-70 of 13 January 1982 regulating the supply of raw material to local industries and the export of timber and wood products, and repealing Decrees N° 72-543 of 28 August 1972 stating the obligation for registered log exporters to ensure supplies to local industries, and N° 78-234 of 20 March 1978 regulating the profession of timber and wood products exporters)

- Décision N° 32 MINEFOR/DCFC du 8 juin 1982 portant institution d'un bordereau de route homologué valable pour tous les exploitants forestiers de Côte-d'Ivoire

  (Decision N° 32 MINEFOR/DCFC of 8 June 1982 setting up a standard road way-bill valid for all Ivory Coast logging operators)

- Arrêté N° 34 MINEFOR/DCFC du 27 juillet 1982 portant modification de l'Arrêté N° 21 SER du 8 juin 1972, portant création d'une commission consultative d'attribution des permis temporaires d'exploitation forestière

  (Regulation N° 34 MINEFOR/DCFC of 27 July 1982 amending Regulation N° 21 SER of 8 June 1972, setting up an advisory commission responsible for the allocation of temporary logging permits)

- Décision N° 1506 MINEFOR/DPF du 7 septembre 1982 portant interdiction d'exploitation forestière en zone de savanne de Côte-d'Ivoire

  (Decision N° 1506 MINEFOR/DPF of 7 September 1982 prohibiting timber harvesting in the savanna area of the Ivory Coast)

- Décision N° 1284 MINEFOR/DCFC du 28 octobre 1982 portant institution d'un livre-journal dans les parking à bois en grumes-export

  (Decision N° 1284 MINEFOR/DCFC of 28 October 1982 providing for the institution of a daybook in export logs parking areas)

Forest Revenues

- Loi N° 59-250 du 31 décembre 1959
  instituant le droit unique de sortie sur le bois

  (Law N° 59-250 of 31 December 1959
  establishing a Single Log Export Duty)

- Loi N° 62-61 du 10 février 1962
  créant la contribution nationale sur le bois

  (Law N° 62-61 of 10 February 1962
  establishing a National Timber Contribution)

- Ordonnance N° 62-216 du 26 juin 1962
  remplaçant la majoration du droit unique de sortie sur le bois au titre de la Contribution nationale par une majoration des taux des taxes et redevances forestières

  (Ordinance N° 62-216 of 26 June 1962
  replacing the increase on the Single Log Export Duty payable as national contribution by a rate increase of forest taxes and fees)

- Loi N° 64-127 du 11 mars 1964
  portant changement du taux de la contribution nationale

  (Law N° 64-127 of 11 March 1964
  amending the rate of the national contribution)

- Ordonnance N° 66-626 du 31 décembre 1966
  portant fixation du montant des redevances forestières en matière d'exploitation des bois d'oeuvre et d'ébénisterie, et instituant une taxe de reboisement

  (Ordinance N° 66-626 of 31 December 1966
  fixing the level of forest fees for industrial, furniture, fuel and charcoal wood harvesting and establishing a reforestation tax)

- Loi N° 67-588 du 31 décembre 1967
  portant Loi des Finances pour l'exercice 1968 (modification d'assiette de recouvrement et de contentieux concernant la taxe d'abattage)

  (Law N° 67-588 of 31 December 1967
  establishing a Finance law for financial year 1968 (modifying the assessment of collection and litigation concerning the logging tax))

- Décret N° 69-522 du 30 septembre 1969
  fixant les valeurs mercuriales devant servir de base pour le calcul des droits ad valorem des bois à l'exportation

  (Decree N° 69-522 of 30 September 1969
  establishing the official price-list to be used as reference for levying ad valorem duties on export logs)

- Ordonnance N° 69-523 du 30 décembre 1969
  portant modification du tarif des douanes à l'entrée et à la sortie (majoration de la Contribution nationale sur les grumes à 4%, taux de la taxe de reboisement et de délimitation du domaine forestier à 3%)

  (Ordinance N° 69-523 of 30 December 1969
  amending import and export customs duties (increasing the national contribution on logs by 4% and fixing the rate of the forest domain reforestation and demarcation tax at 3%))

- Loi des Finances 1971 du 28 décembre 1970
  augmentant le prélèvement au profit du Conseil ivoirien des chargeurs à 0,30%

  (1971 Finance Law of 28 December 1970
  establishing a 0,30% levy increase in favour of the "Conseil ivoirien des chargeurs" (Ivory Coast Loaders Board))

- Ordonnance N° 71-72 du 16 février 1971
  augmentant la Contribution nationale et le droit de sortie pour le Niangon

  (Ordinance N° 71-72 of 16 February 1971
  increasing the national contribution and export tax for the Niangon)

- Décret N° 71-73 du 16 février 1971
  portant modification des valeurs mercuriales à l'exportation pour 6 essences

  (Decree N° 71-73 of 16 February 1971
  amending the official export prices of 6 species)

- Loi des Finances 1972 N° 71-683 du 28 décembre 1971
  portant augmentation des taux des taxes de reboisement, changement de nom de la taxe de reboisement

  (1972 Finance Law N° 71-683 of 28 December 1971
  increasing the rate of the reforestation tax and changing its name)

- Décret N° 72-222 du 22 mars 1972
  fixant les valeurs mercuriales devant servir de base pour le calcul des droits et taxes ad valorem de certains produits et marchandises à l'importation et à l'exportation

  (Decree N° 72-222 of 22 March 1972
  establishing the official price-list to be used as reference for levying ad valorem duties and taxes on certain import and export goods and products)

- Loi des Finances 1973
  portant augmentation du droit unique de sortie sur le bois en grumes de 11 à 15% et de 13 à 18%

  (1973 Finance Law
  increasing the Single Log Export Duty from 11 to 15% and from 13 to 18%)

- Décret N° 73-174 du 27 avril 1973
  portant modification des valeurs mercuriales des bois en grumes à l'exportation

  (Decree N° 73-174 of 27 April 1973
  amending the official price-list for export logs)

- Décret N° 73-201 du 21 mai 1973
  portant modification des valeurs mercuriales des bois en grumes à l'exportation

  (Decree N° 73-201 of 21 May 1973
  amending the official price-list for export logs)

- Ordonnance N° 73-315 du 3 juillet 1973
  modifiant les droits de sortie sur les bois (remaniement général du tarif des droits de sortie sur le bois en grumes et les produits de bois, fusion de l'ancien droit de sortie, de la Contribution nationale et de la taxe de reboisement en un seul droit unique de sortie: regroupement des essences en 3 catégories)

  (Ordinance N° 73-315 of 3 July 1973
  amending log export duties (general revision of export duties on logs and wood products, blending of the former export duty, the national contribution and the reforestation tax into a single export duty; grouping of the species into 3 categories))

- Loi des Finance 1974
  (nouvelle taxe de 0,30% applicable à la valeur en douane de marchandises soumises à l'importation et à l'exportation)

  (1974 Finance Law
  (new 0,30% tax applicable to the customs prices of imported and exported goods))

- Décret N° 74-116 du 14 mars 1974
  portant modification des valeurs mercuriales des bois en grumes à l'exportation

  (Decree N° 74-116 du 14 March 1974
  amending the official price-list for export logs)

- Arrêté Interministériel N° 0614 MEF/CAB du 17 avril 1973
  fixant les modalités de perception de l'indemnité forfaitaire due par les exploitants forestiers au titre des travaux d'intérêt général

  (Interministerial Regulation N° 0614 MEF/CAB of 17 April 1973
  laying down the collection procedures of the agreed indemnity exigible from logging operators for works of public interest)

- Décret N° 76-186 du 11 mars 1976
  fixant les valeurs mercuriales servant de base à la liquidation des droits et taxes à l'importation et à l'exportation

  (Decree N° 76-186 of 11 March 1976
  establishing the official price-list used as reference for the settlement of import and export duties and taxes)

- Ordonnance N° 76-187 du 11 mars 1976
  portant modification du tarif des droits d'exportation concernant certaines essences forestières

  (Ordinance N° 76-187 of 11 March 1976
  modifying the tariff of export duties on certain forest species)

- Décret N° 76-893 du 31 décembre 1976
  fixant les valeurs mercuriales servant de base à la liquidation des droits et taxes à l'importation et à l'exportation

  (Decree N° 76-893 of 31 December 1976
  fixing the official price-lists used as a reference for the settlement of import and export duties and taxes)

- Ordonnance N° 79-10 du 5 janvier 1979
  portant modification du tarif des droits d'entrée et de sortie

  (Ordinance N° 79-10 of 5 January 1979
  amending the tariff of export and import duties)

- Loi N° 79-401 du 21 mai 1979
  portant modification du tarif des droits d'entrée et de sortie

  (Law N° 79-401 of 21 May 1979
  amending the tariff of export and import duties)

- Décret N° 80-117 du 25 janvier 1980
  fixant les valeurs mercuriales des essences exportées en grumes

  (Decree N° 80-117 of 25 January 1980
  fixing the official price-list for species exported in logs)

- Loi N° 81-127 du 31 décembre 1981
  Budget général de fonctionnement pour l'exercice 1982 (Art. 9 droit unique de sortie sur les bois/Taxe d'abattage)

  (Law N° 81-127 of 31 December 1981
  General work budget for financial year 1982 (Art. 9: Single Log Export Duty/Logging Tax))

- Ordonnance N° 82-71 du 13 janvier 1982
  portant modification du tarif des droits de sortie des bois en grumes

  (Ordinance N° 82-71 of 13 January 1982
  amending the tariff of the log export duties)

Other Forest Legislation (Organization of the Forest Administration)

- Décret N° 66-422 du 15 septembre 1966
  portant création d'une société d'Etat, dénommée "Société pour le développement des plantations forestières" (SODEFOR)

  (Decree N° 66-422 of 15 September 1966
  setting up a government agency called "Agency for the development of forest plantations" (SODEFOR)

- 1966
  Statuts de la Société pour le développement des plantations forestières (SODEFOR)

  (1966
  Statutes of the Agency for the development of forest plantations (SODEFOR))

- Arrêté N° 250 AGRI du 21 février 1970
  portant organisation interne de SODEFOR

  (Regulation N° 250 AGRI of 21 February 1970
  establishing the internal organization of SODEFOR)

- Décret N° 71-476 du 23 septembre 1971
  portant attribution du Ministère de l'agriculture

  (Decree N° 71-476 of 23 September 1971
  establishing the responsibilities of the Ministry of Agriculture)

- Décret N° 71-477 du 23 septembre 1971
  portant organisation du Ministère de l'agriculture

  (Decree N° 71-477 of 23 September 1971
  establishing the organization of the Ministry of Agriculture)

- Décret N° 71-478 du 23 septembre 1971
  fixant les attributions du Secrétaire d'Etat chargé des parcs nationaux et portant organisation du Secrétariat d'Etat

  (Decree N° 71-478 of 23 September 1971
  establishing the responsibilities of the Secretary of State in charge of national parks and providing for the organization of the Secretariat of State)

- Décret N° 71-479 du 23 septembre 1971
  fixant les attributions du Secrétaire d'Etat chargé de la reforestation et portant organisation du Secrétariat d'Etat

  (Decree N° 71-479 of 23 September 1971
  fixing the responsibilities of the Secretary of State in charge of reforestation and providing for the organization of the Secretariat of State)

- Décret N° 71-621 du 23 novembre 1971
  complétant le Décret N° 71-479 du 23 septembre 1971 susvisé

  (Decree N° 71-621 of 23 November 1971
  completing Decree N° 71-479 of 23 September 1971 referred to above)

- Décret N° 73-77 du 13 février 1973
  fixant les attributions du Secrétaire d'Etat chargé de la reforestation et portant organisation du Secrétariat d'Etat

  (Decree N° 73-77 of 13 February 1973
  fixing the responsibilities of the Secretary of State in charge of reforestation and providing for the organization of the Secretariat of State)

- Arrêté N° 022 du 13 mars 1973
  portant organisation du Secrétariat d'Etat chargé de la reforestation

  (Regulation N° 022 of 13 March 1973
  providing for the organization of the Secretariat of State responsible for reforestation)

- Arrêté N° 123 SER du 20 novembre 1973
  portant organisation interne de SODEFOR

  (Regulation N° 123 SER of 20 November 1973
  providing for the internal organization of SODEFOR)

- Décret N° 74-749 du 3 décembre 1974
  fixant les attributions du Ministère des eaux et forêts et portant organisation du Ministère
  Abrogé par Décret N° 78-689

  (Decree N° 74-749 of 3 December 1974
  establishing the responsibilities of the Ministry of Forestry and providing for its organization
  Repealed by Decree N° 78-689)

- Arrêté N° 55 MINEFOR/CAB/D du 23 janvier 1976
  portant organisation territoriale du Ministère des eaux et forêts

  (Regulation N° 55 MINEFOR/CAB/D of 23 January 1976
  providing for the territorial organization of the Ministry of Forestry)

- Arrêté N° 72 MINEFOR du 9 février 1976
  portant modification de l'organisation interne de SODEFOR

  (Regulation N° 72 MINEFOR of 9 February 1976,
  amending the internal organization of SODEFOR)

- Arrêté N° 112 MINEFOR/CAB/D du 19 février 1976
  portant organisation administrative des régions, inspections, cantonnements et postes forestiers

  (Regulation N° 112 MINEFOR/CAB/D of 19 February 1976
  providing for the administrative organization of regions, divisions, districts and forest stations)

- Décret N° 77-150 du 9 mars 1977
  portant attribution et organisation du Ministère de la protection de la nature et de l'environnement
  Abrogé par Décret N° 78-689

  (Decree N° 77-150 of 9 March 1977
  establishing the responsibilities and organization of the Ministry of Nature Protection and Environment
  Repealed by Decree N° 78-689)

- Décret N° 78-689 du 18 août 1978
  fixant les attributions du Ministère des eaux et forêts et portant organisation du Ministère
  Abrogé par Décret N° 81-735

  (Decree N° 78-689 of 18 August 1978
  establishing the responsibilities of the Ministry of Forestry and providing for its organization
  Repealed by Decree N° 81-735)

- Arrêté N° 136 MINEFOR/CAB du 27 mars 1981
  portant modification de l'organisation interne de SODEFOR et transfert des activités d'une direction

  (Regulation N° 136 MINEFOR/CAB of 27 March 1981
  amending the internal organization of SODEFOR and transferring the activities of one of its divisions)

- Décret N° 81-735 du 2 septembre 1981
  fixant les attributions du Ministre des eaux et forêts et portant organisation du Ministère

  (Decree N° 81-735 of 2 September 1981
  establishing the responsibilities of the Minister of Forestry and providing for the organization of the Ministry)

- Décret N° 81-736 du 2 septembre 1981
  fixant les attributions du Ministre de l'environnement et portant organisation de son Ministère

  (Decree N° 81-736 of 2 September 1981
  establishing the responsibilities of the Minister of Environment and providing for the organization of his Ministry)

Legislation on National Parks, Wildlife and Hunting

- Loi N° 65-255 du 4 août 1965
  relative à la protection de la faune et à l'exercice de la chasse

  (Law N° 65-255 of 4 August 1965
  concerning wildlife protection and hunting control)

- Loi N° 65-425 du 20 décembre 1965
  portant statut et réglementation de la procédure de classement et de déclassement des réserves naturelles intégrales ou partielles et des parcs nationaux

  (Law N° 65-425 of 20 December 1965
  establishing the status and regulating the classification and declassification procedures of integral or partial nature reserves and national parks)

- Décret N° 66-423 du 15 septembre 1966
  fixant le régime des permis de chasse et les modalités de leurs attributions en République de Côte-d'Ivoire

  (Decree N° 66-423 of 15 September 1966
  regulating hunting permits and their procedures of allocation in the Ivory Coast)

- Décret N° 66-424 du 15 septembre 1966
  relatif à la licence de guide de chasse

  (Decree N° 66-424 of 15 September 1966
  concerning hunting guides licences)

- Décret N° 66-425 du 15 septembre 1966
  réglementant le trafic, la circulation, l'importation, l'exportation des trophées d'animaux protégés et spectaculaires et leurs dépouilles

  (Decree N° 66-425 of 15 September 1966
  regulating the trading, circulation, import and export of trophies and hides of protected and spectacular animals)

- Décret N° 66-433 du 15 septembre 1966
  portant statut et réglementation de la procédure de classement et de déclassement des réserves naturelles intégrales ou partielles et des parcs nationaux

  (Decree N° 66-433 of 15 September 1966
  establishing the status and regulating the classification and declassification procedures of integral or partial nature reserves and national parks)

- Arrêté N° 1712 du 29 septembre 1966
  fixant les conditions d'élimination ou d'éloignement des animaux nuisibles

  (Regulation N° 1712 of 29 September 1966
  establishing the elimination or removal procedures of harmful animals)

- Arrêté N° 68 du 23 janvier 1967
  fixant les tarifs des taxes et redevances en matière de chasse et de capture des animaux sauvages

  (Regulation N° 68 of 23 January 1967
  laying down the rates of taxes and fees exigible for hunting and capturing wild animals)

- Arrêté N° 621 du 29 mai 1967
  réglementant la destination des produits de chasse

  (Regulation N° 621 of 29 May 1967
  concerning the destination of hunting products)

- Arrêté N° 1068 du 29 septembre 1967
  réglementant la chasse des crocodiles et varans dans un but commercial

  (Regulation N° 1068 of 29 September 1967
  concerning the hunt of crocodiles and monitor lizards for commercial purposes)

- Arrêté N° 1069 du 29 septembre 1967
  réglementant la détention d'animaux vivants par des particuliers

  (Regulation N° 1069 of 29 September 1967
  concerning the detention of live animals by individuals)

- Décret N° 72-544 du 28 août 1972
  portant création du Parc national de Toi
  Abrogé par Décret N° 77-348

  (Decree N° 72-544 of 28 August 1972
  providing for the creation of the National Park of Toi
  Repealed by Decree N° 77-348)

- Décret N° 72-545 du 28 août 1972
  portant création de la réserve partielle de faune du N'Zo

  (Decree N° 72-545 of 28 August 1972
  providing for the creation of the partial Game Reserve of N'Zo)

- Décret N° 76-215 du 19 février 1976
  portant création du Parc national du Mont-Sangbé

  (Decree N° 76-215 of 19 February 1976
  providing for the creation of the National Park of Mont-Sangbé

- Décret N° 77-348 du 3 juin 1977
  portant redéfinition des limites du Parc national de Toi et création d'une zone périphérique de protection

  (Decree N° 77-348 of 3 June 1977
  redefining the boundaries of the National Park of Toi and establishing a surrounding buffer zone)

- Décret N° 81-218 du 2 avril 1981
  portant création du Parc national d'Azagny avec une zone périphérique de protection

  (Decree N° 81-218 of 2 April 1981
  providing for the creation of the National Park of Azagny with a surrounding buffer zone)

LIBERIA

A. Status of Forest Legislation

The Forestry Act (1953) together with the Supplementary Act for the Conservation of Forests (1957) provide the framework for the use of forests and wildlife resources, and for the creation of national parks. The reorganization of the forest administration was accomplished under the Forest Development Authority Act (1976). A series of regulations on logging and timber transport; timber export sales contracts; control of non-concession logging operations; and forest land management of private owners, have been made under this Act.

In 1973, a modified standard format for a Timber Concession Agreement, prepared by the government's Concession Secretariat, was introduced to facilitate concession management. Standard forms for forest exploration and timber harvesting permits are also in use.

The principal provisions on the assessment and collection of forest fees and taxes are provided in Chapter 20 of the Revenue and Finance Law (1977). The applicable rates of the fees and taxes have been modified by regulations in 1979, 1981 and 1982.

B. Presentation of the Selected Texts

The Forest Act of 1953 is a fairly short law on forest conservation. It is primarly concerned with the establishment of a national forest administration; the determination of policies and objectives of conservation; and the constitution of government and native authority forest reserves, communal forests, and national parks. The Supplementary Act of 1957 focuses on the utilization aspects and the various methods of granting rights for the harvesting of timber and other forest produce. It sets out in detail the procedures for granting permits and forest concessions; the rights and obligations of the operators; and the control of transported logs with property marks. The Act of 1957 also contains some basic provisions on hunting and wildlife conservation, including the possibility to create protected wildlife areas.

The Timber Concession Agreement (Standard Format), introduced in 1973 and slightly modified in the meantime, is a comprehensive and detailed document for the granting of long-term forest utilization contracts. It contains a number of clauses, which elsewhere, in particular in the French-speaking countries of West and Central Africa, would be determined by the general regulations on timber harvesting. It also covers certain financial and organizational issues which in other countries are generally regulated by investment and corporate business laws. The list of clauses covers the principal matters covered by the agreement. These are, in particular, the general terms of the concession; the rights and obligations of the concessionaire in timber harvesting; the obligations related to forest protection, reforestation and reporting; the fiscal obligations and those on corporate business organization; the employment conditions; and the applicable laws, penalties and arbitration procedure.

The Forest Management Plan (Standard Format), as approved by the forest administration, is part of the concession agreement and refers in particular to logging methods; timber harvesting by an annual coupe system; reforestation by the concessionaire, or by the forest service; construction of logging roads; and log scaling.

The Act creating the Forestry Development Authority of 1976 is again a comparatively short text. This law makes an important contribution to the evolution of the country's general body of forest legislation and has so far been of a considerable impact. The FIRST 2 SECTIONS repeal certain parts of the National Resources Law and establish the Forest Development Authority (FDA) as a corporate body under the Public Authorities Law.

SECTION 3 and 4 define the responsibilities and powers of the new agency. These are far reaching and comprise practically all aspects of modern forest and wildlife resources management. Among the specifically mentioned responsibilities are the coordination of forestry development with other land-uses; forest research; training and technical assistance to those engaged in forestry activities; and the promotion of a general knowledge on natural resources conservation throughout the country. The powers given to the agency are threefold: management of forests and forest land; regulatory functions; and business undertakings. FDA is, for instance, entitled to create and manage forest reserves and national parks; to adopt rules as required for the implementation of its policies and objectives; to issue, amend and rescind forestry regulations; to enforce laws and regulations on the conservation and development of forest resources; and to operate as a business corporation by engaging in commercial undertakings as a principal or in conjunction with other partners.

The remaining sections of the Law (SEC. 6 - 16) determine the composition and functions of the board of directors; the appointment and responsibilities of the agency's management; auditing and reporting procedures; the special powers of forest officers employed by FDA; and the creation of Advisory Conservation Committees in each country that ensure the liaison between the local population and FDA.

Regulation N° 5 on Assistance to Owners of Private Forest Lands, made under the FDA Act, offers an interesting example of certain aspects of private forests, and the role of a public forest administration in supporting their utilization and management. The obligation to provide technical assistance is, as explained in the introductory part of the regulation, related to the fact that commercial timber harvesting on private land is, since 1977, subject to the payment of forest fees and taxes.

The adopted principle is that the owners of private forest land may use the resource within the limits of the general legislative provisions, but have to inform FDA on their intended operations. The agency is then in a position to make suitable proposals and offer appropriate assistance. The owners again, as clearly specified in SEC. 3(2), may freely decide whether they want to implement FDA's recommendation. PART III of the regulation provides the possibility to place certain areas under continuous forest production with FDA offering assistance for the preparation of a management plan; to organize salvage logging operations on other land in order to recuperate all usable raw material with FDA offering to prepare a plan of operation; and to contract a concessionaire in the vicinity of the owner's forests or any operator for logging and extraction with FDA acting as an advising agent.

FDA Regulation N° 6 is concerned with timber harvesting outside forest concessions. It establishes a framework for timber harvesting on smaller areas for which the rules and regulations on large forest concessions do not fit. The regulation provides for the issue of Forest Survey Permits; for the granting of Forest Exploitation Permits, provided that other timber allocation procedures are not judged more appropriate by FDA; and for the determination of specific clauses, in particular for the possibility to determine a supply obligation to local processing units.

Chapter 20 of the 1977 Revenue and Financial Law reorganizes and consolidates the system of forest fees. It establishes a fairly simple and flexible framework for forest revenue assessment, which is compatible with the policy objective of increased local timber processing. It also allows for regular ajustments of the applicable rates to changing conditions in log prices and production costs. All fees are to be assessed by FDA.

The local stumpage fee is to be paid on all timber cut for commercial uses in reserved forests or on privately owned land. The industrialization incentive fee is specifically designed for the encouragement of wood processing and is levied on all exported logs (similar to a log export tax). The rates and the basis of assessment of both fees shall be determined and regulated from time to time by FDA together with the Ministry of Finance.

The forest products fee is levied on exported processed forest products (in particular on sawnwood); the applicable rate may vary within certain limits (between 5% and 15% of the FOB market value) as determined by FDA. The reforestation fee is charged as a fixed amount per m^3 of wood harvested for commercial use, except if the logging operator carries out a reforestation programme by himself. The chargeable amount may be varied by FDA in accordance with the overnment within certain limits (between $ 1.50 and $ 4.50 per cubic metre).

FDA Regulations N° 7 and N° 10 offer examples of the rates of fees as established in 1979, and 1982. The considerable variation of the rates of the industrialization incentive fee for species with different commercial values, as shown in Regulation N° 7, appears to be of interest. The corresponding rates, determined in 1982, are lower due to difficulties in export markets, but the rate of the most valuable species (Sipo) has been maintained.

C. General Forest Legislation/Législation Forestière Générale

- An Act for the Conservation of the Forests of the Republic of Liberia approved April 17, 1953
- Supplementary Act for the Conservation of the Forests of the Republic of Liberia approved February 28, 1957
- An Act creating the Forestry Development Authority (FDA) approved November 1, 1976
- Act adopting a Revenue and Finance Law (Chapter 20: Stumpage and Forest Products Fee) approved May 24, 1977
- FDA Regulation N° 1 on Reduction of Waste of Forest Resources April 15, 1978
- FDA Regulation N° 2 on Registration of Timber Export Sales Contracts August 31, 1978
- FDA Regulation N° 3 on the Issue of Way bills; November 24, 1978
- FDA Regulation N° 4 on Control of Non-Concession Forest Operations April 23, 1979
- FDA Regulation N° 5 on Assistance to Owners of Private Forest Lands April 23, 1979

- FDA Regulation N° 6 on Exploitation Permits for Non-Concession Public Forest Land
  September 1, 1979

- FDA Regulation N° 7 on Revised Forest Fees and Taxes
  December 7, 1979

- FDA Regulation N° 8 on Revised Industrialization Incentive Fees
  July 31, 1981

- FDA Regulation N° 9 Enabling a Special Trade Depression Allowance on Certain Forest Fees
  March 23, 1982

- FDA Regulation N° 10 Enabling a Further Reduction of Certain Forest Fees
  November 8, 1982

- Permit for Forest Survey (Standard Form)
  Forest Development Authority

- Forest Salvage Permit (Standard Form)
  Forest Development Authority

- Permit for Non-Concession Operators (Standard Form)
  Forest Development Authority

- Timber Concession Agreement (Standard Form)

- Forest Management Plan (Standard Form) as Applicable to Forest Products Utilization Contracts

NIGERIA

A. Status of Forest Legislation

The body of the country's forest legislation was originally formed by the Forest Ordinance of 1937; the Forestry Regulations, made under Section 46 of the Ordinance; and certain rules and by-laws, adopted by local authorities, such as the Forestry Rules, made for the Southern and for the Northern Provinces. The Forestry Ordinance and its subsidiary legislation had been subject to numerous amendments.

After independence, the Ordinance was modified, principally with regard to certain formal aspects adjusting the text to the newly created governmental structures. The substantive elements of the forest legislation have, however, remained to a large extent unchanged. New forestry laws, firstly applicable in the major regions of the country, and subsequently in the principal States of the Nigerian Federation, have thus been introduced by amendments to the Forestry Ordinance. The Forestry Amendment Edicts of 1969 and 1973, promulgated for the Western State of Nigeria, offer, for instance, typical examples for the adaptation of State forest legislation.

Similar developments have occurred with regard to the forestry regulations and certain local authority rules and by-laws. The substantive parts of the regulations, similar to the provisions of the Forestry Ordinance itself, have generally remained in force. The local authority rules have been modified to a somewhat larger extent; in some cases they have been abolished.

On the whole, it appears that the Nigerian forest legislation is, at present, still fairly uniform and homogeneous and relies largely, as far as its substantive content is concerned, on the structure and provisions of the Forestry Ordinance and its implementing regulations. This in spite of the fact that numerous amendments have been made in the meantime and that the number of States, as well as the governmental organization in the States, has experienced a dynamic evolution.

The Forestry Ordinance and the State forest laws generally do not include wildlife conservation, hunting control and national parks management. These subjects are covered by specialized legislation.

B. Presentation of the Selected Texts

The Northern Nigerian Forestry Ordinance, in its revised 1960 edition, appears to be representative of the forest law of the whole of the former Northern Nigeria. A significant aspect of the Forestry Ordinance is the dual system of land tenure and forest reservation in favour of the government on one hand, and of local authorities and communities on the other hand. This, together with the fact that forestry is a State and not a Federal matter, allows for a considerable participation of regional and local forces in the decision making, as well as in the implementing process on the use of forest resources.

The FIRST PART of the Ordinance presents a detailed list of definitions as used in the text of the law. It also provides for the possibility that native authority and local government councils may appoint forestry personnel for the purposes of the law, subject to approval by the Minister. PART II of the Ordinance is concerned with the constitution of government forest reserves and protected forests. It determines in detail the procedures for carrying out an inquiry on a proposed reservation; the constitution of a forest reserve; the determination of usage rights that may be practised within its boundaries; the power to modify such rights and to exclude certain areas over which claims are admitted; and the power to de-reserve previously declared forest reserves.

The following parts (PART III, IV and V) contain provisions on the constitution of Native Authority and Local Government Council forest reserves and protected forests. The provisions, related to the preparatory inquiry and to the reservation, are largely similar to those in Part II. The principle is that the initiative for such reservation and the constituting order itself shall emanate from the local authority, subject to approval by the Minister. The local authorities have the power to revise or modify the reservation order and to de-reserve certain areas, again subject to ministerial approval. There exists also the possibility to convert, under certain conditions, government forest reserves to native authority and Local Government Council reserves.

Protection, control and management of Native Authority and Local Government Council forests are, as set out in PART VI, to be undertaken by the responsible local entity. Its activities are, however, subject to supervision and control by the State forest service. In the case of serious shortcomings and deficiencies, a reserved forest may be placed temporarily under the guidance and direction of the Chief Conservator, who then becomes responsible for its proper protection, control and management. In this case, the necessary measures are to be taken by the Chief Conservator on behalf and for the benefit of the local entity.

PART VII provides for the possibility to declare the reservation of communal forest areas. Such declaration may be made by the Native Authority or Local Government Council on request of the interested community. The same local entity has also the power to vary or cancel a previous declaration. The use and management of a communal forest area is the responsibility of the community, acting on advice of the local authority. The latter may make rules on the utilization of communal forests, in particular with regard to the taking of forest produce; the limitation of sale or export of certain categories of forest produce; the establishment of nurseries; reforestation; and the protection of forest areas and forest products.

The general provisions (PART VIII) of the Ordinance refer to the entry upon forest land for the marking of boundaries; the prevention of damages to trees; the prevention of offences; the use of forest produce from local reserves and protected forests for public purposes; the payment of forest fees to the Treasury and to Local Government Council; and land acquisition for public purpose within the Native Council or Local Government Council forests. This part also has a detailed list on forestry matters that may be subject to governmental regulations. An additional section (SECTION 46A) has been inserted, which allows for regulations on restricting the export of forest produce. Native Authorities and Local Government Councils are empowered to make rules on certain forestry matters relevant to the area within their jurisdiction, subject to the approval of the governmental authority.

The Forestry Amendment Edict of 1969 (Western State) modifies and adapts the Forestry Ordinance as State forest legislation. This Edict introduces two important new elements.

SECTION 4 provides for the creation and functioning of a Forestry Advisory Commission, formed by members with particular knowledge on forestry matters; representatives of the timber trade and the industry; communal forest owners; and the Chief Conservator of Forests. The principal functions of the advisory commission are the formulation of short- and medium-term policies on all aspects of forestry and forest industry development, and the making of proposals for securing funds necessary for the implementation of such policies. A new Schedule, attached to the forest law, determines in detail the constitution and proceedings of the advisory commission. It provides, in particular, for the tenure of office of members (3 years); eligibility for re-appointment; temporary membership and co-option of persons; the procedures for voting; and the power to make standing orders for regulating internal proceedings.

SECTION 18 establishes a Forestry Trust Fund for the promotion of regeneration and afforestation, not only in forest reserves, but generally within the territory of the State. This section provides also for the possibility to transfer, under certain conditions, forestry staff of local council and joint boards to the public service of the State.

The structure and content of the Forestry Regulations are presented with the consequential amendments as adopted for the former Western Region. One may conclude from the resumed provisions that the regulations are little concerned with forestry land-use and management, these aspects being covered in considerable detail in the law itself. Their purpose is, in fact, to regulate timber cutting and transport, and the various methods of allocating harvesting rights for timber and other forest produce. The regulations thus deal with general prohibitions and exemptions in using forest produce: the issue of permits and licences on forestry land in general, in protected forests and within forest reserves; the payment of compensation; the deposit of a security bond; and the procedure for determining fees and royalties.

An example of the current practice of revenue assessment is presented with the Fees and Royalty Regulation of the Ondo State, adopted in 1981. The SCHEDULES A to C determine the applicable rates either per harvested tree (stumpage rate) in the case of felling permits outside forest reserves; or based on yield and assessed on the total logging output per hectare; or related to out-turn volume and assessed per m^3 of harvested timber. Rebates for salvage logging in certain areas are provided for.

The assessment of a fixed sum per hectare regardless of the volume actually harvested has been introduced in order to induce a maximum recovery of raw material. This method of revenue assessment is, however, only compatible with improved forest utilization if practised under certain conditions: the principal ones being that a maximum removal of logs is desirable (for instance in the case of subsequent reforestation); that the forest service has full inventory information on the harvestable volumes in the given area; and that the rate of the fee or royalty per hectare is realistically estimated in relation to prevailing production costs and selling prices.

C. General Forest Legislation/Législation Forestière Générale

- The Forestry Ordinance 1937 (Chapter 75) with consequential amendments and modifications

- The Forestry Regulations (Enacted under Sec. 46 of the Forestry Ordinance), 1943 with consequential amendments

- The Forestry (Southern Provinces Native Authorities) Rules, 1943 with consequential amendments

- The Forestry (Northern Provinces Native Authorities) Rules, 1951 with amendments

- The Timber Revenue Collection (Native Authorities) Rules

- The Forestry (Northern Region Native Authorities) Rules 1955 with amendments
- The Eastern Region Forest Law, 1955
- The Forestry Regulations Eastern Region 1956
- The Forestry Ordinance with Amendments; Northern Region 1960
- The Forestry Law (Eastern State), Cap. 38
- The Forestry (amendment) Edict; Western State 1969
- The Forestry (amendment) Edict; Western State 1973
- Forestry Regulations (Ondo State) on Fees and Royalties OD.S.L.N. 5 of October 28 1980

Other Legislation/Autre Législation

- The Wild Animals Preservation Law (Cap. 132)
- Land and Native Ordinance (Cap. 105)
- Land Use Decree 1978

TANZANIA

A. Status of Forest Legislation

The Forest Ordinance of Tanzania dates from 1921 with subsequent revisions and amendments. In 1982 a new Forestry Ordinance was prepared, which consolidates the previous legislation and introduces important additional concepts on forest land management and communal forestry. The new officially approved text has been presented to the legislature for enactment.

The harvesting of timber and other forest produce has generally been made under permits, licences and timber sales agreements, for which standard clauses have been written.

Wildlife conservation and management, hunting practices and control, and national parks management are subject to specialized legislation.

B. Presentation of Selected Texts

The structure and list of provisions of the new Forestry Ordinance, as presented to the legislature, demonstrate the comprehensive character of the law and its orientation towards an active management of forest resources. PART I sets out an interpretation of the principal terms used in the law, and establishes the responsibility of the national forest administration for all aspects of forestry as covered by the subsequent provisions. It also provides the possibility of appointing honorary forest officers for the purpose of the Ordinance. As indicated in the interpretation their status is similar to the status of officers within the forest service.

PART II contains principles on the creation of tate forest reserves, provisional state forest reserves, and local authority forest reserves. One of the notable points is the fact that all forests to be reserved, whether under state or local authority tenure, are to be dealt with in the same manner. This approach appears to be somewhat different from those taken in the legislation of some other countries, which focus generally on state forest reservation.

SECTION 5 and 6 provide for considerable flexibility in forest tenure and reservation. On the one side, the government is empowered to exchange part of a state forest reserve with any other land, the exchanged parts receiving the legal status of the adjacent land; to transfer local authority forest under certain conditions (e.g. in the case of dissolution of the authority) to state tenure; and to unify, divide or modify the area of state forest reserves. On the other hand a local authority may, in consultation with the Director of Forestry, declare any area in state controlled forest reserves or any area within its jurisdiction a local authority forest reserve.

The following sections of Part II deal in great detail with the applicable procedures for constituting reserved forests, and in particular with customary and other rights that exist or are claimed to exist in the areas publicly announced for reservation. After the formal declaration of the reservation order, the boundaries of the reserved forests have to be demarcated on the ground. Maps of state forests, provisional state forests, and local authority forests must be certified and registered by the Director of Surveys.

The state forest reserves shall be deemed to be permanent and their utilization shall proceed in accordance with appropriate forest management standards. The declaration of reserved local authority forests may only be revoked or modified if the written approval of the Director of Forestry has been given.

The management of local authority forests and of state forest reserves is regulated in PART III of the Ordinance. SECTION 10 firmly establishes the principle of full responsibility of the local authorities for the forest reserved on their behalf. A local authority is in charge of the maintenance, control and management of all local forests under its jurisdiction. The management costs have to be met by the authority and the fees, collected on forest produce, form part of its revenues. The utilization of local authority forest reserves is, however, subject to supervision and restrictions.

The authority shall use the advice of the Director of Forestry. The latter is entitled to make written statements and proposals, and to appear personally or by his representative before the local authority, in order to explain such statements and proposals. If a local authority mis-uses its forests and if it is in the public interest to stop such mis-use, the Director of Forestry may be nominated by the Minister to be in charge of the management for the particular local forests. In this case, the Director of Forestry shall exercise the powers established under the Forest Ordinance with regard to these forests, and shall manage them on behalf and for the benefit of the local authority. The authority shall receive the net income from forest utilization, and shall bear the financial loss that may occur.

Utilization and development of state forest reserves is subject to the provisions of working or management plans; these plans are to be prepared in accordance with the objectives detailed in the reservation declaration. The Director of Forestry may lease, in accordance with the provisions of the Land Ordinance, land in state forests that is required for the construction and installations of wood processing plants. He may also lease land in tate forests for recreational and other uses in the public interest.

The purpose of the provisions in PART IV on forestry dedication convenants is to foster reforestation in rural areas. The Director of Forestry may make convenants with land-owning individuals and institutions, with the objective to use certain portions of land for commercial tree growing, in accordance with the rules of sound forestry practices. The convenants are enforceable against the convenator and his successors under certain conditions. Grants may be given to colleges and other public institutions for the establishment and management of forest plantations.

PART V and VI deal with the protection of forests and forest produce and the granting of licences. The various activities, prohibited in state forest reserves without a licence or other authority, are determined. National trees, which benefit from a special protection on all lands, may be declared by order of the Minister. SECTION 18B is of particular interest. It provides for an assessment of natural forest resources by the Director of Forestry, especially in water catchment areas on private land, in order to identify the necessary conservation measures.

Licences for all purposes of the law, and in particular for the harvesting of forest produce, may be granted by the Director of Forestry on state forest land, and by the local authorities for the local forests under their jurisdiction. Licences may from time to time be cancelled or suspended by the appropriate authority.

The remainder of the Forestry Ordinance deals with the powers of forest officers; forest offences; compounding of certain offences, prosecution; and other related matters. ART 30 of the final part provides powers to make rules either of general application, or in respect of a particular state forest, or in respect to any forest produce. The list of enumerated purposes for which such rules may be made, is very detailed and refers to practically all aspects of forestry and forest industry development. By delegating the regulation of a considerable proportion of forestry matters to its subsidiary legislation, the Forestry Ordinance thus retains the character of a basic law and refrains from too many provisions within the law itself. The focusing of the Ordinance rather on principles of forest resources utilization than on details of their application, should allow for a considerable amount of flexibility in the further evolution of the country's forest law.

C. General Forest Legislation/Législation Forestière Générale

- Forests Ordinance (Cap. 389) 1921 revised and amended
- Timber Sale Agreement (Standard Format)
- Exclusive Licence to Take Trees and Timber or Other Forest Produce (Standard Format)
- Forest Ordinance (Cap. 389) 1982

Other Legislation/Autre Législation

- Fauna Conservation Ordinance, Cap. 302; Revised and amended (1958)
- Fauna Conservation (Amendment) Ordinance N° 9, 1960
- Fauna Conservation (Amendment) Ordinance N° 8, 1961
- National Parks Ordinance N° 12, 1959
- National Parks Ordinance (Amendment) Act N° 37, 1962
- National Parks Ordinance (Amendment) Act N° 44, 1964
- Declaration of National Parks Proclamation (C.N. N° 237 and N° 502) 1960
- College of African Wildlife Management Act N° 8, 1964
- Land Ordinance
- Mining Ordinance

ZAIRE

A. Status of Forest Legislation

In Zaire the Forestry Decree of 1949 is still in force and forms the basic text on forestry matters. Certain of its parts, such as Art. 28 and 29, have been modified, or replaced by amendments. Other Ordinances and regulations, adopted during the period 1947 to 1955, complement the provisions of the 1949 Forestry Decree. This refers, for instance, to Ordinance N° 187 of 1947, on the granting of rights on the state's private domain; to Ordinance N° 52/110 of 1951 regulating timber cutting; and to Ordinance N° 52/175 of 1953 on the prevention of forest fires.

In 1975, an Interdepartmental Order, regulating log exports, was issued. Ordinance N° 79/244 of 1979 contains new provisions on taxes and fees to be collected on timber and other forest produce. The presently applicable rates have been fixed by a Departmental Order issued in 1983.

The responsibilities and the organizational structure of the Department of Environment, Nature Protection and Tourism, which is presently in charge of forestry matters, are determined by Ordinance N° 231 of 1975 and subsequent amendments. This refers, in particular, to Ordinance N° 002 of 1977, which transfers several directions and services (including certain services concerned with forestry affairs) to that Department.

Several important measures that concern reforestation programmes, applications for logging permits, and the granting of raw material supply contracts to timber processing industries, have been introduced during the last years by Administrative Circulars. Some corresponding regulations have been prepared but not yet formally adopted.

Wildlife management, hunting control and national parks developement are regulated by separate texts. The Zaire Institute for Nature Conservation operates under Law N° 023 of 1975 with subsequent amendments.

On the whole, it appears that several important measures have been taken during recent years, in order to adjust Zaire's forest legislation to the requirements of large scale forest utilization. This refers, in particular, to certain provisions on local wood processing, forest revenue assessment, and reforestation. The main body of forest legislation is, however, scattered in numerous texts and its principal elements date from the period 1948 to 1952. A complete review of this legislation and its consolidation in a new basic forestry law, the provisions of which correspond to the present reality of the forest sector, would certainly be appropriate and facilitate the management and conservation of the country's sizable forest resources.

B. Presentation of Selected Texts

The Forestry Decree of 1949 is another typical example of the first generation forestry laws, that have been adopted in several parts of West and Central Africa. Its structure and provisions follow the general format of other texts of this period. The Forestry Decree presents in its FIRST ARTICLE a definition of forests and of the forest regime. It regulates the constitution of reserved and protected forest; the practice of forest usage rights, both for subsistence and commercial uses; the granting of logging permits; the assessment of forest fees; the payment of a reforestation fee; certain restrictions on forest clearings on private land; and forest offences and prosectution.

A somewhat exceptional and fairly modern provision for a legislative text of that period is presented in Art. 15. This article stipulates that forest inventories are to be prepared with the objective of determining the annual allowable cut for timber harvesting in a particular forest. If certain management information is not available, a provisional allowable cut may be fixed.

Interdepartmental Order N° 01059 of 1975 contains important provisions on log export restrictions. It replaces the previously existing regulations on export prohibitions, and establishes a log export quota within the range of 15 to 45% of the total production volume. The administration is authorized to determine annually the permitted export quota for each company on the basis of the following criteria: transport distance to the port from which the logs are exported; location of logging and processing units in the same region; and volume of logs processed in the previous year. The volume of exported logs of Limba, Wenge and Afrormosia may not exceed for each species 10% of the total authorized log export volume. The Order provides also for certain quality standards to be respected for exported logs.

The applicable forest fees and taxes are determined in Art. 2A and TITLE II of Ordonnance N° 244 promulgated in 1979. The Ordonnance establishes in particular an area tax in the case of raw material supply contracts, a log production tax, the rates of which are varied according to 3 groups of species and to transport distance; and an export tax on logs and processed forest products. Art. 11 provides for the possibility to modify the rates of all taxes and fees by Departmental Order.

New rates of forest taxes and fees have been determined by Departmental Order N° 1 issued in 1983. The Order increases the area tax for raw material supply contracts; the base values of the log production tax, and certain fees on other forest produce. The rates of the export tax on logs and processed forest products are maintained at the some level as fixed by the Ordonnance N° 244 in 1979.

Administrative Circular N° 1640, issued in 1980, summarizes the Department's position with regard to the recently adopted policy of granting raw material supply contracts to already existing or planned wood processing industries. The purpose of such contracts is to ensure a continuous supply of timber for a sufficiently long period of time, in order to allow an adequate return on the invested capital.

The object of the contract is a determined annual volume of logs of certain species that may be harvested within a particular production forest. The maximum permittable duration of any contract is 20 years. Under certain conditions a letter of intention, which confirms the possibility of eventually granting such a contract, may be signed by the government during the preparatory phase of an investment project. The Circular determines in detail the documents and information to be submitted in connection with the application for a raw material supply contract.

Administrative Circular N° 1986, distributed in 1980, defines the Department's position with regard to the implementation of the adopted policy on reforestation. Two possibilities are foreseen in order to carry out reforestation. Either the company carries out reforestation by its own within the zone in which the raw material supply contract is granted, or the company maintains a forest nursery and provides infrastructural support to the reforestation units of the forest administration that operate in the same region.

Each company has to prepare a 3 year or 5 year reforestation programme in collaboration with the regional directions to which the reforestation units belong. The granting and renewal of logging permits are only possible if the reforestation programme is implemented. The reforestation obligation does not exempt the logging operators from the payment of the forestry tax.

C. General Forest Legislation

- Ordonnance N° 187 du 16 juin 1947
(Règles à suivre pour l'obtention des droits d'emphytéose et de superficies sur le domaine privé de l'Etat)

(Ordinance N° 187 of 16 June 1947
stating the procedures for the award of hereditary lease and area rights on the state's private domain)

- Ordonnance N° 41-131 du 14 avril 1948
(Conditions d'exportation des bois)
modifiée par les Ordonnances N° 41-170 du 25 mai 1957 et N° 41-2 du 3 janvier 1958

(Ordinance N° 41-131 of 14 April 1948
(Timber export procedures)
amended by Ordinances N° 41-170 of 25 May 1957 and N° 41-2 of 3 January 1958)

- Décret du 11 avril 1949 sur le régime forestier

(Decree of 11 April 1949 on Forestry Legislation)

- Ordonnance N° 52-208 du 15 juin 1950
(Règles à suivre et redevances à payer pour coupes de bois par les concessionnaires de mines et les titulaires de permis de traitement)

(Ordinance N° 52-208 of 15 June 1950
establishing logging regulations and fees concerning mining concessionnaires and holders of processing permits)

- Ordonnance N° 52-205 du 15 juin 1950 (régime forestier)

(Ordinance N° 52-205 of 15 June 1950 (Forestry Legislation))

- Ordonnance N° 52-371 du 28 octobre 1950
(Règles d'exploitation des bois dans les forêts soumises au régime forestier)
modifiée par l'Ordonnance N° 52-507 du 19 juin 1952

(Ordinance N° 52-371 of 28 October 1950
(Logging regulations in forests subject to Forestry Legislation)
amended by Ordinance N° 52-507 of 19 June 1952)

- Ordonnance N° 52-119 du 2 mai 1951
(Règles à suivre dans les coupes de bois autorisées par le Décret du 11 avril 1949 sur le régime forestier

(Ordinance N° 52-119 of 2 May 1951
(Regulations on cuts authorized by Decree of 11 April 1949 on Forestry Legislation))

- Ordonnance N° 52-175 du 20 mai 1953
(Incendies en forêt)

(Ordinance N° 52-175 of 20 May 1953
(Prevention of forest fires))

- Ordonnance N° 52-289 du 29 août 1955
  (exportation des bois)

  (Ordinance N° 52-289 of 29 August 1955
  (Log exports))

- Ordonnance N° 75-231 du 22 juillet 1975
  fixant les attributions du Département de l'environnement, conservation de la nature et tourisme

  (Ordinance N° 75-231 of 22 July 1975
  stating the responsibilities of the Department of Environment, Nature Protection and Tourism)

- Arrêté Interdépartemental N° 01059 du 22 octobre 1975
  portant réglementation sur l'exportation de grumes

  (Interdepartmental Regulation N° 01059 of 22 October 1975
  regulating timber exports)

- Ordonnance N° 77-022 du 22 février 1977
  portant transfert des directions et services au Département de l'environnement, conservation de la nature et tourisme (Transfert des responsabilités pour le développement forestier)

  (Ordinance N° 77-022 of 22 February 1977
  providing for the transfer of the management and services to the Department of Environment, Nature Protection and Tourism (transfer of responsibilities for forestry development))

- Ordonnance N° 79-244 du 16 octobre 1979
  fixant le taux et règle d'assiette et de recouvrement des taxes et redevances en matière administrative, judiciaire et domaniale perçues à l'initiative du Département de l'environnement, conservation de la nature et tourisme

  (Ordinance N° 79-244 of 16 October 1979 fixing the rates and rules of assessment and collection of taxes and fees connected with administrative, legal and state property matters levied through the Department of Environment, Nature Protection and Tourism)

- Arrêté N° 0001 CCE/ADRE/83 du 26 janvier 1983
  portant modification de certains taux des taxes et redevances perçues à l'initiative du Département de l'environnement, conservation de la nature et tourisme

  (Regulation N° 0001 CCE/ADRE/83 of 26 January 1983
  amending the rates of certain taxes and fees levied through the Department of Environment, Nature Protection and Tourism)

Administrative Circulars

- Note Circulaire N° 054 DECNT/BCE/78 du 18 janvier 1978
  (Problèmes des feux de brousse et de l'incendie des forêts)

  (Administrative Circular N° 054 DECNT/BCE/78 of 18 January 1978
  (Problems connected with bush and forest fires))

- Note Circulaire N° 0637 SE/DECNT/78 du 13 juillet 1978
  (Programme de reboisement pour 5 ans à fournir par les exploitants)

  (Administrative Circular N° 0637 SE/DECNT/78 of 13 July 1978
  (5-year reforestation programme to be implemented by logging operators))

- Note Circulaire N° 0489 SE/DECNT/79 du 28 février 1979
  (Modalités des programmes de reboisement à être effectués par les exploitants)

  (Administrative Circular N° 0489 SE/DECNT/79 of 28 February 1979
  (Procedures relating to reforestation programmes to be implemented by logging operators))

- Note explicative des formulaires de demandes de permis de coupe de bois destinée aux agents chargés de l'attribution de permis de coupe (1980)

  (Explanatory Note for logging permits application forms intended for agents in charge of logging permits delivery (1980))

- Note Circulaire CCE/DECNT/80
  Procédure pour présentation des dossiers de demandes de permis de coupe de bois (1980)

  (Administrative Circular CCE/DECNT/80
  laying down procedures for the presentation of logging permits application files (1980))

- Note Circulaire N° 1640 SG/DECNT/80 du 5 juin 1980
  (Conditions pour accorder une garantie d'approvisionnement en matière première)

  (Administrative Circular N° 1640 SG/DECNT/80 of 5 June 1980
  (Establishing the conditions required for the allocation of raw material supplies))

- Note Circulaire N° 1986 DECNT/CCE/80
  (Modalités d'exécution pour les programmes de reboisement à la charge des exploitants forestiers)

  (Administrative Circular N° 1986 DECNT/CCE/80
  (Procedures for the implementation of reforestation programmes to be carried out by logging operators))

- Note Circulaire N° 1253 CCE/DECNT/80
  concernant les contrats de vente du bois à l'étranger

  (Administrative Circular N° 1253 CCE/DECNT/80
  concerning Timber Export Sales Contracts)

- Note Circulaire N° 1124 DECNT/CCE/81
  concernant la prospection forestière

  (Administrative Circular N° 1124 DECNT/CCE/81
  concerning forest prospecting)

Other Forest Legislation

- Ordonnance-Loi N° 72-012 du 21 février 1972
  portant modification des status et de la dénomination de l'Institut national pour la conservation de la nature
  Abrogé par Loi N° 75-023

  (Ordinance-Law N° 72-012 of 21 February 1972
  amending the status and denomination of the National Institute for Nature Conservation
  Repealed by Law N° 75-023)

- Loi N° 75-023 du 22 juillet 1975
  portant statut de l'Institut zaïrois pour la conservation de la nature

  (Law N° 75-023 of 22 July 1975
  establishing the status of the Zaïre Institute for Nature Conservation)

- Loi N° 82-002 du 28 mai 1982
  portant réglementation de la chasse

  (Law N° 82-002 of 28 May 1982
  concerning Hunting Control)

PART III

RECOMMENDATIONS FOR FURTHER ACTIVITIES OF THE AFRICAN TIMBER ORGANIZATION IN THE FIELD OF FOREST LEGISLATION AND ADMINISTRATION

The Role of the African Timber Organization

The African Timber Organization may assume an important role in promoting the advancement of the forest sector's institutional framework within its member countries. Its contribution could, generally speaking, be:

- to draw the attention of member countries on the role of forest legislation, administration and of human resources formation for implementing national policies on improved forest utilization;

- to facilitate the exchange of experiences in these fields and to disseminate the already available information; and

- to initiate and engage analytical and comparative work on such institutional problems, that are of particular relevance and common interest to the member countries.

Distribution of the Forest Law Collection to Member Countries

It is recommended that this Forest Law Collection, which contains a considerable number of forest laws and regulations, should be distributed in its finalized version to the national administrations, forestry schools and research institutes of all member countries.

Regular Updating of the Forest Law Collection and Preparation of Supplementary Volumes

It is suggested that the African Timber Organization in collaboration with its member countries should continue this work by preparing regularly (for instance every 2 years) supplementary volumes to the Forest Law Collection. The continuation of this work would offer an opportunity to keep the member countries informed on further evolutions, by presenting newly adopted laws and regulations, and important amendments to the already existing texts.

In addition to forestry and forest industry development, the next volume should also cover that part of the wildlife, national parks and fishery development legislation, which has not been included in the now available collection.

Organization of Regular Meetings on Forest Legislation and Administration

The review of the evolution of forest law in the ATO member countries demonstrates, that there exists considerable expertise on forest legislation within the various countries It appears to be of considerable interest to facilitate the transfer of experience and an increased collaboration among the countries, as well as to encourage the exchange of ideas and concepts between lawyers, foresters and other specialists of the natural resources sector. It is this cross-fertilization that has a practical and rapid pay off in tuning together the legislative texts related to forestry and natural resources development. The transfer of experience and regional concertation is particularly relevant as far as legislative provisions on timber utilization systems and forest revenue assessment are concerned.

It is recommended that regular sessions of national experts from the member countries under the auspices of the African Timber Organization should present an opportunity to exchange the views on forest legislation both in its formal and substantive aspects. Such sessions should be open not only to specialists on forest legislation, but also to senior officials in national forest administrations concerned with the formulation of forest policy and planning.

Basic Documentation on Specific Subjects of Common Interest in the Field of Forest Institutions

In addition to the dissemination of already available information among member countries, complementary activities are recommended to ATO. This refers in particular to the preparation of comparative studies on forest institutions. Among the many aspects, that might be of common interest within the region, three problems have been singled out that appear, at present, of an immediate concern to the ATO member countries.

1) Continuation of the Analytical Work on the Evolution on Forest Legislation

There is a striking difference between the rapidly evolving legal practice and the limited amount of analytical work on forest legislation, that has been undertaken during the last years. Among the issues that merit a more systematic investigation the following aspects should be considered:

- the evolution of forest legislation under the conditions of comparable ecological conditions and comparable legal systems;
- the adequacy of legislation norms and the possibilities of their implementation in relation to the actual situation of forestry and forest industry development;
- the interdependence of forest laws and other natural resources legislation.

It is recommended that this work should be continued by the African Timber Organization, in particular, by committing funds for engaging specialists from member countries (e.g. staff from universities), who can prepare specific comparative studies.

2) Preparation of an ATO Document on the Economic Impact of Forest Revenue Systems

The forest law of most countries now provides considerable flexibility to adjust the applicable rates of forest fees and taxes to changing market conditions. The practical problem is, however, to evaluate the economic impact of changes in the assessment basis on forest industry development at the country level and within the region. It is therefore suggested that a document should be prepared by ATO which examines the effect of the combined forest fees and taxes on the exports of logs and processed forest products. This document should also identify the necessary additional information that is required for improving and monitoring forest revenue systems at country level.

The preparation of a study of this kind is time consuming since it implies an in-depth knowledge of the forest conditions of the various member countries. Its launching needs advanced preparation.

3) Preparation of an ATO Document on the Evaluation of Manpower Requirements for Forestry and Forest Industry Development in the Member Countries

The lack of a sufficient quantity of trained personnel is one of the principal bottlenecks in implementing improved policies and legislation on forestry and forest industry development. It is suggested that an exchange of experience in this respect might help to foster the process of human resources formation within the member countries. It is recommended that ATO should elaborate a basic document on manpower requirements in forestry and forest industries, and on the needs for improved forest education and training. A considerable amount of material on these subjects has been made available during the last year by the Forestry Department of FAO, part of which could be used for the preparation of the document.

BIBLIOGRAPHY/BIBLIOGRAPHIE

Gordon, W.A. Obstacles à la foresterie tropicale. Le mode d'occupation des terres. Unasylva, Vol. 15 (1); 1961

FAO Politique en matière de réserves forestières et droit d'usage en Afrique FAO/Afr. Timber Tr. Conférence intergouvernementale sur les tendances et perspectives de la production, de la consommation et du commerce du bois en Afrique; 1965

Gordon W.A. Tenurial consideration in tropical forestry. The evolution of Forest-hold Tenures. Madrid, 6 CFM; 1966

Owusu, I.G.K. Forest laws of Ghana and court procedure. Accra; 1966

Mifsud, F.M. Droit foncier coutumier en Afrique. Rome, Collection FAO: Série législative N° 7, 106 pp.; 1967

François, T. La législation forestière au Gabon. Rome, FAO/PNUD, Rapport technique; 1968

Oka Koffi, T. Manuel de droit forestier à l'usage des ingénieurs et contrôleurs des Eaux et Forêts. Abidjan. Tome I & II; 1969

Chollet, A.P. et Taylor C.J. Politique, législation et administration forestières. Rome, FAO/PNUD, FO:SF/CMR/6, Rapport technique 4; 1972

King; K.F.S. Législation et administration forestières. Buenos Aires, 7° Congrès forestier mondial; 1972

Uhart, E. Législation forestière. Libreville, FAO/PNUD, FO/DP/GAB/68/50 Rapport technique 20; 1973

Schmithüsen, F. Concession d'exploitation, fiscalité sur le bois et législation forestière. République Populaire du Congo. Rome, FAO/PNUD, FO:PRC/71/515 Document de travail

Adeyoju, S.K. Land Use and Tenure in the Tropics. Unasylva Vol. 28, N° 112/113, 1976

Schmithüsen, F. Contrats d'exploitation forestière sur domaine public. Rome, FAO, seconde édition 1977

du Saussay, Ch. Le droit de la protection de la faune en Empire Centrafricain. Document de Travail N° 12, 118 pp CAR/72/010. FAO, Rome; 1978

Schmithüsen, F. La législation forestière dans les principaux pays producteurs de bois de l'Afrique Occidentale francophone (Cameroun, Congo, Côte d'Ivoire, Gabon) FO:MISC/78/10; 116 pp. FAO, Rome; 1978

du Saussay, Ch. Projet de code de la protection de la faune. Document de Travail N° 13, 84 pp. CAF/72/010; FAO, Rome; 1979

Battioni, J.C. Informe final sobre el proyecto de asistencia preparatoria para el desarollo del sector forestal - Guinea Ecuatorial. Document de Trabajo; 16 pp. PNUD/FAO/EQG/80/008. Rome; 1980

du Saussay, Ch. La législation sur la faune et les aires protégées en Afrique. Etude législative, N° 25, 153 pp. FAO, Rome; 1981

Service de Législation/Bureau Juridique Compendium de la Législation sur la Pêche; Région du COPACE. FL/CECAF/82/5 (Distribution préliminaire); 346 pp. FAO, Rome, 1982

ANNEXE

SELECTED TEXTS / TEXTES SELECTIONNES

OUTLINE / TABLE DES MATIERES

LOI N° 81/13 DU 27 NOV. 1981
portant Code Forestier

CAMEROUN

CAMEROON

Loi N° 81/13

Loi N° 81/13 du 27 Novembre 1981
portant régime des forêts, de la faune et de la pêche

L'Assemblée nationale a délibéré et adopté :

Le Président de la République promulgue
la loi dont la teneur suit :

TITRE I : DISPOSITIONS GENERALES

Article 1. - Le régime des forêts, de la faune et de la pêche recouvre l'ensemble des règles édictées par la présente loi et les textes pris pour son application, en vue d'assurer la conservation, l'exploitation et la mise en valeur des ressources forestières, fauniques et halieutiques des domaines forestier, fluvial et maritime.

Article 2. - Sont soumises au régime édicté par la présente loi :

- les forêts domaniales
- les forêts des collectivités publiques
- les forêts des particuliers
- les forêts du domaine national
- la faune sauvage
- les ressources halieutiques du domaine public fluvial et du domaine maritime.

Article 3. - Sont qualifiés forêts, les terrains comportant une couverture végétale et susceptibles :

- soit de fournie du bois ou des produits autres qu'agricoles
- soit d'abriter la faune sauvage
- soit d'exercer un effet indirect sur le sol; le climat ou le régime des eaux.

Article 4. - La faune et la flore du domaine public fluvial et du domaine maritime appartiennent à l'Etat.

Article 5. - Le régime de propriété des forêts et des établissements aquacoles est défini par la législation foncière et domaniale et les dispositions de la présente loi.

Article 6. -

1) Les administrations chargées des forêts, de la faune et de la pêche assurent la gestion et la protection des forêts domaniales, de celles des collectivités publiques locales et de celles du domaine national, ainsi que des ressources halieutiques du domaine public fluvial et du domaine maritime.
2) Elles peuvent prendre toutes mesures nécessaires en vue d'assurer la protection des forêts, de la faune et des ressources halieutiques quel que soit leur régime de propriété.

Article 7. - Les forêts doivent être régénérées dans les conditions fixées par les textes réglementaires.

Article 8. - Nul ne peut faire des forêts, de la faune et des ressources halieutiques du domaine public fluvial et du domaine maritime un usage prohibé par les dispositions de la présente loi et les textes pris pour son application.

Article 9. - Les particuliers, les collectivités publiques locales, les organismes et les établissements publics exercent sur leurs forêts et leurs établissements aquacoles tous les droits résultant de la propriété, sous réserve des restrictions spécifiées dans la présente loi et les textes pris pour son application.

Article 10. - L'administration chargée des forêts dispose, pour les opérations de martelage et de saisie, d'un marteau forestier dont l'empreinte est déposée au greffe de la cour suprême.

Article 11. -

1) Le recouvrement des droits et taxes sur les forêts, la faune et les ressources halieutiques s'effectue de la manière suivante :
   a) en ce qui concerne les produits destinés à la consommation locale: les agents des administrations chargées des forêts, de la faune et de la pêche émettent des titres de perception; le recouvrement est assuré par le Trésor;
   b) en ce qui concerne les produits destinés à l'exportation: les agents des douanes émettent les titres de perception après s'être assurés que les éléments de la déclaration d'exportation (D6) sont conformes aux spécifications établies par les agents des administrations chargées des forêts, de la faune et de la pêche; le recouvrement de ces titres est assuré par le Trésor.
2) Les titres de perception prévus à l'alinéa (1) ci-dessus ont force exécutoire.
3) Les agents des administrations chargées des forêts, de la faune et de la pêche perçoivent, au titre des opérations visées à l'alinéa (1) ci-dessus, des indemnités dans les conditions fixées par décret.

Article 12. - Les administrations chargées des forêts et de la faune assurent en ces matières des missions de contrôle et de répression.

A cet effet, les agents de ces administrations sont astreints dans l'exercice de leurs fonctions au port de l'uniforme, d'armes et de munitions, d'insignes de grade et à une organisation et une discipline de type paramilitaires, selon des modalités fixées par décret.

Toutefois, il peuvent dans certaines circonstances particulières, exercer leurs fonctions en civil.

Dans tous les cas, ils doivent se munir de leur carte professionnelle.

TITRE II : DES FORÊTS

Chapitre premier : Des forêts domaniales

Article 13. -

1) Les forêts domaniales sont celles faisant partie du domaine privé de l'Etat.

2) Sont considérés comme tels :
- les réserves naturelles intégrales
- les parcs nationaux
- les sanctuaires à certaines espèces végétales ou animales
- les réserves de faune
- les forêts de production
- les forêts de protection
- les forêts recréatives
- les périmètres de reboisement
- les jardins zoologiques et botaniques
- les game ranches appartenant à l'Etat.

Article 14. -

1) Le classement des forêts dans l'une des catégories visées au paragraphe 2 de l'art. 13 ci-dessus s'effectue suivant une procédure fixée par décret.

2) Le décret portant création d'une forêt domaniale doit préciser dans quelle catégorie elle est placée. Il doit indiquer en outre le mode de gestion des ressources, les restrictions ainsi que les droits d'usage applicables à l'intérieur de cette forêt.

Article 15. - Les forêts domaniales doivent couvrir 20% de la superficie totale du territoire national.

Article 16. - Les administrations chargées des forêts et de la faune établissent pour chaque forêt domaniale et pour chaque parc national, un plan d'aménagement dans des conditions fixées par décret.

Article 17. - La protection du domaine forestier obéit aux règles édictées par la loi N°80-22 du 14 juillet 1980 portant répression des atteintes à la propriété foncière et domaniale.

Chapitre 2 : Des forêts des collectivités publiques et des particuliers

Article 18. - Une forêt appartient à une collectivité publique lorsqu'elle fait l'objet d'un décret de classement pour le compte de cette collectivité ou a été plantée par celle-ci.

Article 19. - Les forêts des particuliers sont des forêts plantées par ceux-ci sur des terrains détenus en vertu de la législation en vigueur.

Article 20. - Les forêts des collectivités publiques et des particuliers sont la propriété de ces derniers. Toutefois, l'utilisation et la jouissance des droits de propriété attachés à ces forêts doivent s'effectuer suivant des règles fixées par des textes réglementaires.

Chapitre 3 : Des forêts du domaine national

Article 21. -
1) Les forêts du domaine national sont celles non visées aux art. 13, 18 & 19 ci-dessus.

2) Les produits forestiers de toute nature s'y trouvant à l'exception de ceux provenant des arbres plantés par des particuliers ou des collectivités publiques, appartiennent à l'Etat.

3) Toutefois des droits d'usage sont reconnus aux populations dans des conditions fixées par décret.

Article 22. - En cas de nécessité, des restrictions concernant les forêts du domaine national, notamment la réglementation des feux de brousse, des défrichements, des pâturages, des pacages, des abattages, des ébranchages et des mutilations des essences protégées, ainsi que la liste de ces essences peuvent être édictées par l'administration chargée des forêts.

Chaptire 4 : De l'inventaire, de l'exploration et de l'exploitation des forêts

Article 23. -
1) L'exploitation de toute zone de forêt est subordonnée à un inventaire préalable de celle-ci.

2) Toute exploration de forêt lorsqu'elle n'est pas faite en régie est subordonnée à l'octroi d'une autorisation délivrée par l'administration chargée des forêts. L'autorisation d'explorer entraîne la perception d'une taxe fixée par la loi de finances.

3) En cas de communication des résultats de l'exploration à toute autre personne physique ou morale, celle-ci doit acquitter la taxe d'exploration prévue au paragraphe (2) ci-dessus.

4) Les modalités d'application du présent art. sont fixées par décret.

Article 24. -
1) La superficie totale pouvant être accordée à un même exploitant est fonction des installations industrielles existantes ou à mettre en place. Elle ne peut excéder 200 000 ha.

2) Toute prise de participation majoritaire ou création d'une société d'exploitation par un exploitant forestier titulaire d'une licence, ayant pour résultat de porter la superficie totale par lui détenue au delà de 200 000 ha est interdite.

Article 25. - L'exploitation des forêts s'effectue soit en régie, soit par licences, soit par ventes de coupe, soit par permis ou autorisations de coupe, accordés aux sociétés ou aux particuliers, dans les conditions fixées par décret.

Article 26. -

1) L'exploitation des forêts domaniales s'effectue en régie, par les soins de l'administration chargée des forêts, ou par ventes de coupe. Toutefois, une forêt domaniale peut être concédée en exploitation à une société d'Etat ou à une société au sein de laquelle l'Etat détient au moins 51% du capital.

2) Dans tous les cas, l'exploitation doit s'effectuer conformément au plan d'aménagement établi pour la forêt concernée.

Article 27. -

1) L'Exploitation des forêts du domaine national s'effectue soit par ventes de coupe, soit par licences accordées aux sociétés d'Etat, d'économie mixte ou aux exploitants privés agréés, soit exceptionnellement en régie.

2) L'attribution de tout titre d'exploitation forestière s'effectue suivant une procédure fixée par décret.

Article 28. - Les licences sont accordées pour une période de cinq ans renouvelable. Leur renouvellement est soumis à une procédure fixée par décret.

Article 29. - Toute licence de superficie inférieure ou égale à 25 000 ha ne peut être attribuée qu'aux nationaux pris individuellement ou regroupés en société.

Toutefois, l'exploitant étranger peut être autorisé à soumissioner en vue d'étendre son exploitation sur une superficie contiguë inférieure ou égale à 25 000 ha.

Article 30. -

1) Toute exploitation par un particulier ou société est assortie d'un cahier des charges comportant des clauses générales et des clauses particulières. Si l'exploitation s'effectue par licence, le cahier des charges comporte une clause de participation à la réalisation d'infrastructures socio-économiques.

2) Les clauses générales concernent toutes les conditions techniques relatives à l'exploitation des produits concernés.

3) Les clauses particulières concernent les charges financières ainsi que les obligations en matière d'installations industrielles incombant aux titulaires des titres d'exploitation.

Article 31. -

1) Les charges financières prévues à l'art. 30 ci-dessus sont constituées par :
   - la redevance de reforestation
   - la redevance territoriale
   - la contribution aux travaux de développement forestier
   - le prix de vente des produits forestiers
   - la participation à la réalisation d'infrastructures socio-économiques.

2) Les taux des taxes et redevances ci-dessus sont fixés par la loi de finances.

Article 32. -

1) La redevance territoriale est reversée en totalité au Fonds d'Equipement Intercommunal (FEICOM).

2) La redevance de reforestation est reversée à l'organisme d'Etat chargé de la régénération forestière.

3) Le prix de vente des produits est réparti de la façon suivante:
- 20% au budget de l'Etat
- 25% à l'organisme d'Etat chargé des inventaires forestiers
- 55% à l'organisme d'Etat chargé de la régénération forestière.

4) La contribution aux travaux de développement forestier dont le taux est fixé par la loi de finances est répartie ainsi qu'il suit :
- 40% pour l'équipement et le contrôle forestier
- 35% pour l'aménagement des forêts
- 25% pour la promotion du bois.

5) La participation à la réalisation d'infrastructures socio-économiques dont le taux est fixé par la loi de finances est reversée en totalité aux communes concernées, aux mêmes fins. Elle ne peut recevoir aucune autre destination.

Article 33. - Aucun exploitant, aucun exportateur ou transformateur de produits forestiers, quel que soit le régime fiscal dont il bénéficie, ne peut être exonéré du paiement du prix de vente des produits forestiers et du versement de tout droit, taxe ou redevance destiné à la régénération forestière.

Article 34. - Toute personne physique ou morale désirant exploiter la forêt par licence ou vente de coupe doit se faire agréer selon une procédure fixée par décret.

Article 35. -

1) Les licences d'exploitation forestière ne peuvent être accordées qu'aux personnes physiques résidant au Cameroun ou aux sociétés y ayant leur siège et dont la composition est connue de l'administration chargée des forêts.

2) L'attribution de chaque licence ou de chaque coupe entraîne la perception de la taxe d'agrément dont le taux est fixé par la loi de finances.

Article 36. - L'attribution, le renouvellement et le transfert de tout titre d'exploitation forestière sont subordonnés à la constitution d'un cautionnement dont le taux est fixé par la loi de finances.

- S'il s'agit d'un national ou d'une société dans laquelle l'Etat ou les nationaux détiennent au moins 51% du capital, le cautionnement peut être bancaire;
- Dans les autres cas, le cautionnement est constitué par un versement au Trésor.

Les modalités d'application du présent art. sont fixées par décret.

Article 37. -

1) La vente ainsi que l'affermage des titres d'exploitation des produits forestiers sont interdits.

2) Le transfert de titres d'exploitation forestière ainsi que toute prise de participation ou cession de parts dans une société d'exploitation forestière sont soumis à l'autorisation préalable de l'administration chargée des forêts.

3) Les modalités d'application du présent article sont fixées par décret.

Article 38. - Le transfert d'une licence donne lieu à la perception d'une taxe dont le montant est fixé par la loi de finances.

Article 39. -

1) La licence d'exploitation forestière, la vente de coupe, le permis et l'autorisation de coupe de perches, de bois de chauffage et de charbon confèrent à leur détenteur, sur la surface concédée, le droit de récolter exclusivement, pendant une période déterminée, les produits désignés dans le titre, mais ne créent aucun droit de propriété sur le terrain y afférent.

En outre, le bénéficiaire ne peut faire obstacle à l'exploitation des produits récoltés traditionnellement.

2) La récolte de graines, de racines, de feuilles, de sève, d'écorces ou de tout autre partie de plante est déterminée par des textes réglementaires.

Article 40. - L'administration chargée des forêts peut marquer en réserve tout arbre qu'elle juge utile, sur une superficie concédée en exploitation.

De même, elle peut marquer les arbres nécessaires à l'exécution de travaux d'utilité publique.

Article 41. - Les titres d'exploitation délivrés jusqu'à la date d'entrée en vigueur de la présente loi demeurent valables, sous réserve des dispositions prévues à l'art. 28 ci-dessus.

Chapitre 5 : De l'utilisation des billes échouées sur la côte atlantique

Article 42. - Les billes sans marques apparentes locales, échouées sur la côte atlantique peuvent être récupérées par toute personne physique ou morale, moyennant paiement d'une taxe dont le taux est fixé par la loi de finances, selon des modalités fixées par décret.

Chapitre 6 : De la promotion et de la commercialisation du bois et des produits forestiers

Article 43. - L'exportation du bois en grumes est réservée, dans des conditions fixées par décret, aux nationaux pris individuellement ou regroupés en société, titulaires d'un titre d'exploitation forestière ou à tout autre exploitant détenteur d'un titre d'exploitation et justifiant d'une industrie de transformation locale.

Article 44. - Les quotas d'exportation des différents produits forestiers bruts ou travaillés sont fixés par l'administration chargée des forêts.

Article 45. - Des mesures particulières peuvent être fixées par décret en vue de la promotion des essences peu ou pas connues et d'autres produits forestiers.

TITRE III : DE LA FAUNE SAUVAGE

Chapitre premier : De l'exercice du droit de chasse

Article 46. - Est considérée comme acte de chasse, toute action visant à poursuivre, tuer, capturer, photographier, cinématographier un animal sauvage ou à guider des expéditions à cet effet.

Il en est de même de la photographie et de la cinématographie à des fins commerciales.

Article 47. - La chasse traditionnelle est autorisée sur toute l'étendue du territoire sauf dans les aires protégées pour la conservation de la faune. Les conditions de son exercice sont fixées par décret.

Article 48. - Tout acte de chasse autre que le cas prévu à l'art. 47 ci-dessus est subordonné à l'octroi d'un permis ou d'une licence.

Article 49. - La délivrance de tout permis de chasse ou licence entraîne la perception de droits dont le taux est fixé par la loi de finances.

Article 50. - Les droits et obligations résultant de l'octroi de permis et licences ainsi que les modalités de leur attribution sont fixés par décret.

Article 51. - Les permis et licences sont personnels et incessibles. Il ne peut être délivré à la même personne qu'un seul permis de chasse au titre de la même saison de chasse.

Article 52. - Le permis de chasse ne peut être délivré qu'aux personnes qui se sont conformées à la réglementation en vigueur sur la détention des armes à feu.

Article 53. - L'abattage et la capture de certains animaux donnent lieu à la perception de taxes dont le taux est fixé par la loi de finances.

La liste de ces animaux est fixée par l'administration chargée de la faune.

Article 54. -

1) Certaines zones spécialement définies peuvent être déclarées zones cynégétiques par l'administration chargée de la faune après avis de celle chargée des forêts. L'exploitation de ces zones s'effectue soit en régie, soit par toute autre personne physique ou morale, selon les modalités fixées par décret, pour une durée de cinq ans renouvelable. Elle est assujettie à un cahier des charges dont les clauses sont définies par l'administration chargée de la faune.

2) L'administration chargée de la faune peut autoriser l'exercice de la profession de guide de chasse dans les zones banales suivant les modalités fixées par décret.

Article 55. - La chasse dans une zone cynégétique donne lieu à la perception d'une taxe journalière dont le taux est fixé par la loi de finances.

Article 56. - Les personnes titulaires d'un permis de chasse disposent librement des dépouilles et des trophées des animaux régulièrement abattus par elles, sous réserve de s'acquitter des taxes y afférentes. Toutefois, elles doivent prendre toutes les dispositions pour éviter l'abandon des dépouilles de ces animaux au lieu d'abattage.

Article 57. -

1) Constituent des trophées : les pointes, carcasses, crânes ou dents des animaux ou de grands carnassiers, les queues d'éléphants ou de girafes, les peaux, les sabots ou pieds, les cornes et les plumes d'oiseaux.

2) La détention et la circulation des trophées d'animaux protégés sont subordonnées à une formalité d'enregistrement et de marquage préalable par l'administration chargée de la faune.

3) Les titulaires de trophées acquis antérieurement à la date de promulgation dela présente loi ont un délai d'un an pour les faire enregistrer et marquer par l'administration chargée de la faune. Passé ce délai, les trophées non conformes aux dispositions du présent alinéa seront saisis pour le compte de l'Etat.

Article 58. -

1) Tout détenteur de dépouilles d'animaux protégés ou de leurs trophées non marqués doit présenter son permis de chasse ou de capture à toute réquisition.

2) La détention et la circulation à l'intérieur du territoire national d'animaux protégés vivants ou morts, de leurs dépouilles ou de leurs trophées, sont subordonnées à l'obtention d'un certificat d'origine délivré parl'administration chargée de la faune.

3) Le certificat d'origine comporte les caractéristiques des animaux et les spécifications des trophées permettant d'identifier les produits en circulation.

4) L'exportation d'animaux sauvages, de leurs dépouilles ou de leurs trophées est subordonnée à l'obtention d'un certificat d'origine et d'une autorisation d'exportation délivrée par l'administration chargée de la faune.

Article 59. - La capture d'animaux sauvages est subordonnée à l'obtention d'un permis suivant les conditions fixées par décret et moyennant paiement des taxes dont les taux sont fixés par la loi de finances.

Article 60. - La gestion des "game ranches" s'effectue en régie.

Toutefois ils peuvent être confiés à des organismes spécialisés ou à des particuliers suivant les modalités fixées par décret.

Article 61. - Des zones tampons sont crées autour des aires de protection dans des conditions fixées par décret.

La chasse est interdite dans ces zones au même titre qu'à l'intérieur de ces aires.

Article 62. - L'exercice de la profession de guide de chasse dans les zones d'intérêt cynégétique ou dans les zones banales est subordonné à l'obtention d'un permis dans les conditions fixées par décret et moyennant paiement des taxes dont les taux sont fixés par la loi de finances.

Chapitre 2 : De la protection des personnes et des biens contre les animaux

Article 63. - Au cas où certains animaux constitueraient un danger ou causeraient des dommages, l'administration chargée de la faune peut faire procéder à des battues contrôlées suivant les modalités fixées par décret.

Article 64. - Aucune infraction ne peut être relevée contre quiconque a fait acte de chasse d'un animal protégé dans la nécessité immédiate de sa défense, de celle d'autrui, de celle de son cheptel domestique ou de celle de sa récolte.

La preuve de la légitime défense doit être fournie dans un délai de 72 heures au responsable de l'administration chargée de la faune le plus proche.

Article 65. - Les trophées résultant des actes prévus à l'art. 64 ci-dessus sont remis à l'administration chargée de la faune qui procède à leur vente aux enchères publiques ou de gré à gré en l'absence d'adjudicataire et en reverse le produit au Trésor.

Chapitre 3 : Des armes de chasse

Article 66. - Est prohibée toute chasse effectuée au moyen :
- d'armes ou munitions de guerre composant ou ayant composé l'armement réglementaire des forces militaires ou de police nationales
- d'armes à feu susceptibles de tirer plus d'une cartouche sous une seule pression de la détente
- de projectiles contenant des détonants.

Article 67. - L'administration chargée de la faune peut réglementer le calibre et le modèle d'arme pour la chasse de certains animaux. Elle peut également interdire l'emploi de certains modèles d'armes ou de munitions en vue de la protection de la faune.

Article 68. - Les entreprises de tourisme cynégétique dûment patentées et déclarées peuvent, dans les conditions fixées par décret, mettre à la disposition de leurs clients des armes de chasse correspondant à des types dont l'utilisation est autorisée par le permis détenu par le client concerné. L'entreprise est dans ce cas civilement responsable des dommages ou infractions imputables au client, sans préjudice des poursuites qui pourraient être exercées contre ce dernier.

Chapitre 4 : De la protection de la faune et de l'environnement

Article 69. -

1) Les espèces animales vivant sur le territoire national sont réparties en trois classes : A, B et C du point de vue de leur protection.

2) Sous réserve des dispositions de l'art. 64, les espèces de la classe A sont intégralement protégées et ne peuvent en aucun cas être abattues.
Toutefois, leur capture ou détention est subordonnée à l'obtention d'un permis de capture délivré par l'Administration chargée de la faune.
Les espèces de la classe B bénéficient d'une protection partielle. Elles peuvent être chassées, capturées ou abattues après obtention d'un permis approprié.
Les espèces de la classe C ne bénéficient d'aucune protection. Cependant leur abattage est réglementé.

3) Les espèces animales se trouvant dans les parcs nationaux, les réserves de faune et les sanctuaires bénéficient du régime de protection de la classe A, sauf pour nécessité d'aménagement.

4) Les modalités d'application du présent art. sont fixées par décret.

Article 70. - La chasse de certains animaux peut être fermée temporairement sur tout ou partie du territoire national par l'administration chargée de la faune.

Article 71. - Quiconque, en tous temps ou en tous lieux, est trouvé en possession d'un animal protégé de la classe A ou B vivant ou mort ou partie de cet animal est réputé l'avoir capturé ou tué.

Article 72. - Sauf autorisation spéciale délivrée par l'administration chargée de la faune, sont interdits :

- la poursuite, l'approche et le tir de gibier en véhicule ou engin à moteur;
- la chasse nocturne, notamment la chasse au phare, à la lampe frontale et en général au moyen de tous engins éclairants, conçus ou non à des fins cynégétiques;
- la chasse à l'aide de drogues, d'appâts empoisonnés, de fusils anesthésiques et d'explosifs;
- la chasse à l'aide d'engins non traditionnels;
- la chasse au feu;
- l'importation, la vente et la circulation des lampes de chasse;
- la chasse au fusil fixe et au fusil de traite.

Article 73. - Tout procédé de chasse même traditionnel de nature à compromettre la conservation de certains animaux rares ou utiles peut être interdit ou réglementé par l'administration chargée de la faune.

Article 74. -

1) L'introduction dans le territoire national de tout végétal ou animal sauvage vivant ou mort est soumise à l'autorisation de l'administration chargée des forêts ou de la faune selon le cas, sur présentation d'un certificat d'origine, d'une autorisation d'exploitation et d'un certificat phytosanitaire ou vétérinaire délivré par un organisme compétent du pays de provenance.

2) La sortie du territoire national de tout végétal ou animal sauvage vivant ou mort est soumise à la présentation des pièces ci-dessus énumérées, délivrées par les autorités compétentes.

Article 75. - Il est interdit d'allumer volontairement ou involontairement un feu susceptible de détruire l'environnement. Tout feu doit être contrôlé afin d'éviter la destruction de l'environnement. Les modalités d'application du présent art. sont fixées par des textes réglementaires.

Article 76. -

1) Toute les actions humaines contribuant à la dégration de l'environnement tel que l'abattage abusif d'arbres dans les zones particulièrement exposées à la désertification ou à l'inondation sont interdites.

2) La circulation et la divagation des animaux domestiques ou des bestiaux dans les périmètres de protection ou dans les zones tampons sont interdites.

Article 77. - La destruction de l'environnement sur une distance de 50 mètres de part et d'autre le long des cours d'eau ou sur un rayon de 100 mètres tout autour de leur source est interdite.

Les droits d'usage le long des cours d'eau sont réglementés par un texte réglementaire.

TITRE IV : DE LA PECHE

Chapitre 1 : Des Définitions

Article 78. - Les "ressources halieutiques" désignent des poissons de toutes sortes, issus de la mer, des eaux somâtres, des eaux douces, y compris les organismes vivants appartenant à des expèces sédentaires, c'est à dire les organismes qui, au moment du ramassage, sont soit immobiles au fond du domaine maritime ou du domaine public fluvial, soit incapables de se déplacer à moins d'être en contact avec le fond de la mer, lac, fleuve ou établissement aquacole.

Article 79. -

1) La "pêche ou pêcherie" désigne la capture ou le ramassage des ressources halieutiques ou tout autre activité dont on peut raisonnablement prévoir qu'elle conduit à la capture, ou au ramassage desdites ressources halieutiques, y compris l'aménagement et la mise en valeur des milieux aquatiques en vue de la protection d'espèces animales par la maîtrise totale ou partielle de leur cycle biologique.

2) Selon les moyens mis en oeuvre pour l'obtention des ressources halieutiques l'on distingue :
   - la pêche traditionnelle ou artisanale;
   - la pêche sportive;
   - la pêche scientitique;
   - la pêche semi-industrielle;
   - la pêche industrielle;
   - la mariculture;
   - la pisciculture.

   Ces différents types de pêche sont définis et réglementés par décret.

Article 80. - Le navire de pêche désigne toute embarcation ou bateau quelle qu'en soit la taille, utilisé pour prendre ou chercher à prendre du poisson ou d'autres produits animaux aquatiques.

Article 81. - Est considérée comme engin de pêche, tout outil ou appareil permettant de capturer, ramasser ou récolter les animaux aquatiques.

Article 82. - La maillage est défini comme étant dans la poche du filet, la mesure moyenne de 50 mailles étirées parallèles à l'axe longitudinal de la poche; ou dans toute autre partie du filet, la mesure moyenne de toute série de 50 mailles étirées consécutives, mesurées à la jauge de pression normale; la mesure étant effectuée sur filet mouillé.

Article 83. - Au sens de la présente loi sont désignées sous les termes :

a. Etablissements de traitement des produits de la pêche
   1. Les installations de mareyage qui se livrent à la préparation (triage, lavage, pesée, glaçage) des produits de la pêche.
   2. Les usines de congélation qui se livrent à la conservation par le froid ou simplement au stockage de produits congelés.
   3. Les ateliers de fumage qui se livrent à la préparation des produits de la pêche en utilisant la combustion du bois ou de ses sous-produits.
   4. Les ateliers de séchage qui assurent la déshydratation par l'action directe de la chaleur (soleil ou autres procécés similaires).
   5. Les ateliers de salage qui se livrent à la préparation des produits de la pêche en utilisant le sel marin ou les produits succédanés, à l'exclusion de tout autre moyen de conservation.

b. Etablissements de stockage et de vente
   1. Les chambres froides ou établissements d'entreposage équipés de façon à pouvoir maintenir les produits préalablement congelés à une température au moins égale à 20°C sous zéro (- 20°C).
   2. Les poissonneries qui se livrent à la vente au détail des produits de la pêche.

c. Moyens de transport
1. Les véhicules isothermes qui regroupent les véhicules (automobiles, wagons, containers, etc..) comportant des parois étanches ne permettant pas d'échange de température avec l'extérieur.
2. Les véhicules réfrigérés qui désignent les véhicules disposant d'un compresseur autonome produisant du froid.

Article 84. - Les normes techniques et les conditions d'hygiène au sein des installations définies à l'art. 83 ci-dessus sont fixées par décret.

Chapitre 2 : De l'exercice du droit de pêche

Article 85.- Le droit de pêche dans le domaine maritime et le domaine public fluvial appartient à l'Etat.

Toutefois la pêche est ouverte dans les conditions fixées par décret.

Article 86. -
1) L'exercice de la pêche est subordonné à l'obtention d'une licence de pêche en ce qui concerne la pêche industrielle et d'un permis de pêche, en ce qui concerne les autres catégories de pêche, à l'exception de la pêche traditionnelle ou artisanale.

2) La pêche à la petite crevette (Palaemon hastatus et Pellonula Vorax) est subordonnée à l'obtention d'une autorisation spéciale de pêche accordée dans les conditions fixées par décret.

Article 87. - Les licences de pêche sont réparties en 3 types :
- la licence d'armement à la pêche aux poissons
- la licence d'armement à la pêche à la crevette et autres crustacés
- la licence d'armement à la pêche thonière.

Article 88. - Les permis de pêche sont répartis en 3 types :
- le permis A pour la pêche semi-industrielle
- le permis B pour la pêche sportive
- le permis C pour la pêche scientifique.

Article 89. -
1) La délivrance d'une licence et d'un permis de pêche donnent lieu à la perception d'une taxe d'exploitation dont le taux est fixé par la loi de finances.

2) Cette taxe est également perçue à l'occasion du renouvellement desdits titres.

Article 90. - Les modalités d'octroi des licences et permis de pêche sont fixées par décret.

Article 91. - Toute licence ou permis de pêche doit être présenté à tout moment aux agents habilités.

Article 92. -
1) Toute personne physique ou morale désirant exploiter les ressources halieutiques à des fins commerciales ou industrielles doit se faire agréer suivant une procédure fixée par décret.

2) Cet agrément donne lieu au paiement d'une taxe dont le taux est fixé par la loi de finances.

Article 93. - Les licences de pêche ne peuvent être accordées qu'aux personnes physiques résidant au Cameroun ou aux sociégés y ayant leur siège et dont la composition est connue de l'administration chargée de la pêche.

Article 94. -

1) La vente ainsi que l'affermage des titres d'exploitation des produits de la pêche sont interdits.
2) Le transfert d'une licence ou d'un permis de pêche est subordonné à l'accord de l'administration chargée de la pêche et au paiement d'une taxe dont le taux est fixé par la loi de finances.
3) Les modalités d'application du présent art. sont fixées par décret.

Article 95. - Aucun exploitant de ressources halieutiques, aucun exportateur ou transformateur des produits de la pêche, quel que soit le régime fiscal dont il bénéficie, ne peut être exonéré du paiement des taxes correspondantes.

Article 96. - Tout exploitant de ressources halieutiques doit déclarer ses captures dans les conditions fixées par l'administration chargée de la pêche.

Chapitre 3 : De la gestion et de la conservation des ressources halieutiques

Article 97. - Des restrictions peuvent être apportées à l'exercice du droit de pêche en vue :

- de la protection de la faune et des milieux aquatiques ainsi que de la pêche traditionnelle
- du maintien de la production à un niveau acceptable.

Article 98. - Sont interdites :

a) l'utilisation sur une largeur de deux milles marins à partir de la ligne de base, d'engins traînants;
b) l'utilisation, pour les types de pêche, de tous moyens ou dispositifs de nature à obstruer les mailles des filets ou ayant pour effet de réduire leur action sélective, ainsi que le montage de tout accessoire à l'intérieur des filets de pêche à l'exception des engins de protection fixés à la partie supérieure du filet, à condition que les mailles aient une dimension au moins double du maillage minimum autorisé et qu'ils ne soient pas fixés à la partie postérieure du filet;
c) l'utilisation dans l'exercice de la pêche sous-marine, fluviale, lagunaire, de tout équipement tel que scaphandre autonome;
d) la présence à bord d'un bateau, d'un engin respiratoire tel qu'un scaphandre, une foëme ou une arme dangereuse de pêche, sauf pour des raisons de sécurité;
e) la pratique de la pêche à l'aide de la dynamite ou de tout autre explosif ou assimilé, de substances chimiques,de poisons, de l'électricité ou de phare, d'armes à feu, de pièges à déclenchement automatique ou tout autre appareil pouvant avoir une action destructrice sur la faune ou le milieu aquatique.
f) le développement d'ouvrages tels que les retenues, les digues, les grands chenaux, ou la mise de portuaires sans avis préalable de l'administration chargée de la pêche;

g) le déversement de matières toxiques et nocives telles que les polluants industriels, agricoles (pesticides, fertilisants, sédiments) et domestiques (principalement les détergents) dans les milieux aquatiques;
h) la destruction de l'environnement sur une distence de 50 mètres le long d'un cours d'eau ou, sur un rayon de 100 mètres tout autour de sa source;
i) la présence à bord d'un bateau armé pour la pêche de chalut, de senne ou de tout autre filet traîné ou halé sur le fond ou près du fond de la mer, fleuve ou lac, non pourvu d'un maillage réglementaire;
j) la présence à bord d'un bateau armé pour la pêche d'engins destructeurs ou de substances pouvant enivrer ou détruire les poissons, ainsi que de tous moyens tendant à diminuer ou à obstruer d'une façon ou d'une autre, le maillage d'une partie quelconque du filet;
k) l'exportation de ressources halieutiques sans autorisation préalable de l'administration chargée de la pêche;
l) l'introduction au Cameroun de ressources halieutiques vivantes étrangères;
m) la capture, la détention et la mise en vente des ressources halieutiques dont la liste est fixée par l'administration chargée de la pêche;
n) la pêche dans toute zone ou secteur interdit par l'administration chargée de la pêche.

Article 99. - Des dérogations aux dispositions de l'art. 98 ci-dessus peuvent être accordées à titre exceptionnel par l'administration chargée de la pêche.

Article 100. - L'utilisation de navires de pêche de plus de 250 tonneaux jauge brute (TJB) est interdite à l'intérieur des eaux territoriales. Dans le domaine public fluvial, les navires de pêche ne doivent pas dépasser 10 tonneaux jauge brute.

Article 101. - L'administration chargée de la pêche détermine pour chaque domaine aquatique les engins de pêche et les caractéristiques des filets utilisables.

Article 102. - La dimension des mailles des différents types de filets est fixée par l'administration chargée de la pêche.

Chapitre 4 : De la mariculture et de la pisciculture

Article 103. - La mise en place de toute installation aquacole est subordonnée à l'obtention d'une autorisation délivrée par l'administration chargée de la pêche, dans les conditions fixées par décret. Cette autorisation donne lieu au paiement d'une taxe dont le taux est fixé par la loi de finances.

Article 104. - L'autorisation d'installation peut édicter des restrictions nécessaires à la conservation, à la gestion et à l'exploitation optimale des ressources halieutiques. Elles peuvent en particulier porter sur :
- l'orientation et la construction
- l'aménagement
- le contrôle de la qualité des produits et des conditions sanitaires.

Article 105. - L'administration chargée de la pêche assure la gestion des stations et des centres aquacoles du domaine public fluvial et du domaine maritime.

Chapitre 5 : De la mise en place des établissements de pêche

Article 106. - La création d'une installation de mareyage, d'une usine de congélation, d'un atelier de traitement (fumage, séchage ou salage), d'une usine de conserverie ou d'une poissonnerie, est subordonnée à l'obtention d'un agrément préalable délivré dans les conditions fixées par décret, sans préjudice des conditions particulières édictées en matière de contrôle des établissements classés. Cet agrément donne lieu au paiement d'une taxe dont le taux est fixé par la loi de finances.

Article 107. - Les établissements d'exploitation des produits de la pêche sont classés suivant leur importance et leur nature, par l'administration chargée de la pêche, et la taxe visée à l'art. 106 ci-dessus calculée en conséquence.

Article 108. - L'ouverture au public des établissements visés à l'art. 106 ci-dessus est subordonnée à l'obtention d'un certificat de conformité délivré dans les conditions fixées par décret.

Chapitre 6 : L'inspection sanitaire et le contrôle des produits de la pêche

Article 109. - Nul ne peut exposer, préparer, distribuer, stocker ou transporter pour la vente, des produits de la pêche non soumis à une inspection sanitaire préalable.

Cette inspection qui peut s'effectuer en tout lieu et à tout moment donne lieu au paiement d'une taxe dont le taux est fixé par la loi de finances.

Article 110. - L'inspection sanitaire des produits de la pêche a pour but de vérifier :

- le respect de la nomenclature officielle des espèces commercialisables;
- le respect de la taille marchande des espèces de consommation courante;
- la provenance des prises;
- l'état sanitaire des produits débarqués et mis en consommation.

Les normes de qualité sont fixées par décret.

Chapitre 7 : Le conditionnement et le transport des produits de la pêche

Article 111. - Les produits de la pêche doivent être conditionnés dans des emballages réglementaires.

Article 112. - Le transport par route ou par rail des produits de la pêche doit être assuré au moyen de véhicules aménagés conformément aux normes fixées par décret.

Article 113. - La mise en service des véhicules destinés au transport des produits de la pêche est subordonnée à un agrément préalable donné dans des conditions fixées par décret. Cet agrément donne lieu au paiement d'une taxe dont le taux est fixé par la loi de finances.

TITRE V : DE LA REPRESSION DES INFRACTIONS (art. 114 - 134 omis)

TITRE VI : DISPOSITIONS DIVERSES

Article 135. - Si dans une instance en répression d'une infraction, le prévenu excipe d'un droit de propriété ou de tout autre droit réel, le tribunal statue sur l'incident aux règles suivantes :
L'exception préjudicielle n'est admise que si elle est fondée sur un titre apparent, ou sur des faits de possession équivalents et si les moyens de droit sont de nature à enlever au fait ayant provoqué la poursuite, son caractère délictuel.

Dans le cas de renvoi à fins civiles, le jugement fixe un délai qui ne peut excéder trois mois, dans lequel la partie doit saisir le juge compétent et justifier de ses diligences, sinon il est passé outre.

Article 136. - La délivrance de duplicata de tout titre, licence, permis ou autorisation spéciale d'exploitation de ressources forestières, fauniques ou halieutiques est subordonnée au paiement d'un droit dont le taux est fixé par la loi de finances.

Article 137. - Le produit de la taxe d'exploitation et de la taxe d'inspection sanitaire visées aux art. 89 et 109 ci-dessus est réparti ainsi qu'il suit :
- 50% au Trésor
- 50% au service ou organisme chargé du développement de la pêche.

Article 138. -
1) Le produit des amendes, transactions, dommages-intérêts, ventes aux enchères publiques ou de gré à gré des produits et objets divers saisis est réparti ainsi qu'il suit :
   1. En ce qui concerne les forêts et la faune 25% aux agents des administrations chargées des forêts; 75% au Trésor.
   2. En ce qui concerne la pêche : 25% aux agents de l'administration chargée de la pêche et aux agents assermentés de la marine marchande ayant aidé à la répression des infractions; 40% au service ou à l'organisme chargé du développement de la pêche; 35% au Trésor.
2) Les modalités de distribution des ristournes aux agents susvisés sont fixées par décret.

Article 139. - Sont abrogées toutes les dispositions antérieures à la présente loi, notamment :
- l'ordonnance n° 73-18 du 22 mai 1973,
- la loi n° 74-12 du 16 juillet 1974
- la loi n° 75-4 du 2 juillet 1975.

Article 140. - La présente loi sera enregistrée, puis publiée au Journal Officiel en français et en anglais.

Yaoundé, le 27 novembre 1981

Le Président de la République

Law Nº 81-13 OF 27 NOV. 1981

CAMEROON

CAMEROUN

Law Nº 81-13

LAW Nº 81-13 of 27 November 1981

to lay down forestry, wildlife and fisheries regulations

The National Assembly has deliberated and adopted :

The President of the Republic hereby enacts the Law set out below :

PART I : GENERAL PROVISIONS

Section 1. - The national forestry, wildlife and fisheries regulations shall comprise all the rules laid down by the present law and subsequent implementing instruments with a view to ensuring the conservation, exploitation and development of the forest, wildlife and fishery resources of the forest estate and waterways.

Section 2. - The following shall be subject to the provisions of this law :

- State forests
- local council forests
- private forests
- communal forests
- wildlife
- fishery resources of public waterways and coastlands

Section 3. - Forests shall mean land covered by vegetation which is capable :

- either of producing wood or other produce which is not agricultural produce
- or of providing a habitat for wildlife
- or of exercising an indirect effect on the soil, climate or water regime.

Section 4. - The fauna and flora of either the public waterways or the public coastlands shall belong to the State.

Section 5. - The system of forest ownership shall be determined by the regulations governing land tenure and State lands and by the provisions of this law.

Section 6. -

1) The forestry, wildlife and fishery services shall be responsible for the management and protection of State, local council and communal forests as well as the fishery resources of the public waterways and coastlands.
2) They may take all necessary measures to ensure the protection of forests, of wildlife and fishery resources irrespective of their ownership.

Section 7. - State forests shall be regenerated in accordance with the rules laid down by regulations.

Section 8. - No person may make use of the forests, wildlife and fishery resources of the public waterways and coastlands in any manner that is prohibited by the provisions of this law and its implementing instruments.

Section 9. - Individuals, local councils, public bodies and establishments may exercise on their forest and water resources all the rights that result from owership, subject to restrictions laid down in this law and its implementing instruments.

Section 10. - For purpose of seizure and marking operations, the Administration in charge of forests shall possess a marking hammer whose end mark shall be lodged with the registry of the Supreme Court.

Section 11. -

1) The collection of forestry, wildlife and fisheries duties and taxes shall be carried out in the following manner :
   a) With regard to produce meant for local consumption: officials of the forestry, wildlife and fisheries services shall issue assessment notices and the fees or taxes shall be paid into the Treasury.
   b) With regard to export produce : customs officials shall issue assessment notices after checking to see that information on the export declaration (D6) agrees with the specifications established by officials of the services in charge of forests, wildlife and fisheries. The export duties and taxes shall be paid into the Treasury.

2) The assessment notices provided for in sub-section 1 above shall be enforceable.

3) Forestry, wildlife and fisheries officials shall receive allowances in respect of the operations cited in sub-section 1 above, under conditions to be determined by decree.

Section 12. -

1) The services in charge of forests and wildlife shall carry out control missions and prosecute in matters concerning forest exploitation and hunting.
   For this purpose, forestry officials, in the exercise of their duties, shall wear uniform with badges showing their ranks, carry firearms and ammunitions and be subject to a paramilitary type of organization and discipline in accordance with the rules laid down by decree.

2) However, under certain especial circumstances they may carry out their duties in plain clothes.

3) In any case, they must carry their professional card.

PART II : FORESTS

Chapter 1 : State Forests

Section 13. -
1) State forests shall be those that form part of the private property of the State
2) The following shall be considered State forests :
- integral nature reserves
- national parks
- sanctuaries for certain wild animals or plant species
- game reserves
- production forests
- protection forests
- recreation forests
- forest plantations
- zoological and botanical gardens
- game ranches belonging to the State

Section 14. -
1) The constitution of a forest into any of the categories referred to in section 13 (2) above shall be done in accordance with a procedure fixed by decree.
2) The decree to constitute a State forest shall specify in which of the categories it has been placed. Furthermore, it shall indicate the manner of resource management, the restrictions and the customary rights applicable within the said forests.

Section 15. - State forests shall cover 20% of the total area of the national territory.

Section 16. - The services in charge of forests and wildlife shall draw up a management plan for each State forest and each national park, under the conditions fixed by decree.

Section 17. - The protection of the forest estate shall be governed by the provisions of-Law N° 80-22 of 4 July 1980 to repress infringements on landed property and State lands.

Chapter 2 : Forests belonging to local councils and private persons

Section 18. - A forest shall belong to a local council if it was constituted by decree for the benefit of the local council or was planted by the council.

Section 19. - Private forests shall be forests planted by individual persons on lands owned in compliance with the regulations in force.

Section 20. - Local council and private forests shall remain the property of their owners : Provided that the utilization and enjoyment of ownership rights attached to such forests shall comply with the rules laid down by regulations.

Chapter 3 : Communal forests

Section 21. -

1) Communal forests shall be such forests as are not referred to in Sections 13,18 and 19 above.
2) Forest produce of all kinds found in them, with the exception of produce from trees planted by private individuals or local councils, shall be the property of the State.
3) However, citizens shall be allowed exploitation rights under conditions laid down by decree.

Section 22. - When necessary, restrictions concerning communal forests, especially the regulating of bush fires, and clearing, grazing, pasturing, felling, lopping and mutilation of protected species as well as the list of the said species may be enacted by order of the Minister in charge of forests.

Chapter 4 : Forest inventory, Forest survey and exploitation

Section 23. -

1) The exploitation of any forest zone shall be subject to a prior inventory of the zone.
2) Where any one other than the Administration wishes to carry out a forest survey, he shall first seek a permit from the Administration in charge of forests. The survey permit shall be subject to the payment of a fee fixed by the Finance Law.
3) If the results of the survey are communicated to any natural person or corporate body, the latter shall pay the fee provided for in sub-section (2) above.
4) The conditions of application of this section shall be determined by decree.

Section 24. -

1) The total forest area that may be granted to any one exploiter shall depend on the existing or planned industrial installations. It shall not exceed 200 000 ha.
2) The acquisition of majority shares or the creation of a forest exploitation company by a forest exploiter holding a licence with the intention of exploiting a total area of more than 200 000 ha shall be forbidden.

Section 25. - Forests shall be exploited either under State management, or under licence, or by the sale of standing volume or under a felling permit or authorization granted to companies or individuals, under conditions to be determined by decree.

Section 26. -

1) State forests shall be exploited under State management, under the supervision of the Administration in charge of forests or by the sale of standing volume : Provided that a State forest may be granted for exploitation to a Governmental corporation, or to a company in which Government has at least 51% of the capital.
2) In any case, the exploitation shall be carried out in conformity with the management plan drawn up for the forest in question.

Section 27. -

1) Communal forests shall be exploited either by the sale of standing volume or by the grant of licences to Government corporations, semi-governmental corporations or approved private forest exploiters or, exceptionally, under State management.
2) The grant of forest exploitation rights shall be subject to the procedure laid down by decree.

Section 28. - Licences shall be granted for a renewable period of 5 years. They shall be renewed according to a simplified procedure laid down by decree.

Section 29. - Any licence covering a surface area which does not exceed 25 000 ha shall be granted only to nationals acting individually or grouped into a company.

However, foreign exploiters may be authorized to apply for adjacent areas less than or equal to 25 000 ha for the purpose of extending their existing concessions.

Section 30. -

1) Any exploitation by a private individual or a company shall be regulated by contract specifications comprising general and specific clauses. If the exploitation is carried out under licence, the contract specifications shall include a clause relating to contributions to the execution of socio-economic infrastructures.

2) The general clauses shall deal with all the technical conditions governing the exploitation of the forest produce in question.

3) The specific clauses shall deal with the finance charges as well as the obligations of the holder of the exploitation rights in respect of industrial installations.

Section 31. -

1) The finance charges referred to in Section 30 shall comprise :
   - the regeneration fees
   - the territorial tax
   - a contribution to forestry development
   - the selling price (received by the State) of the forest produce
   - the contribution to the execution of socio-economic infrastructures.

2) The rates of the above-mentioned taxes and fees shall be fixed by the Finance Law.

Section 32. -

1) The territorial tax shall accrue entirely to the Special Council Support Fund (FEICOM).
2) The regeneration fee shall accrue to the State body responsible for forest regeneration.
3) The selling price of forest produce shall be apportioned in the following manner :

- 20% to the State budget
- 25% to the State body in charge of forest inventory
- 55% to the State body in charge of forest regeneration.

4) The contribution to forestry development whose rate is fixed by Finance Law shall be apportioned in the following manner :
   - 40% for forest equipment and control
   - 35% for the management of forests
   - 25% for the promotion of wood.

5) The contribution to the execution of socio-economic infrastructures whose rate is fixed by the Finance Law shall accrue entirely to the local councils concerned for the same purpose in the area of exploitation. It shall not be used for any other purpose.

Section 33. - Irrespective of the fiscal provisions applicable to him, no forest exploiter, exporter of forest produce or industrialist may be exempted from the payment of the selling price of forest produce and the regeneration fee or tax.

Section 34. - Every natural person or corporate body wishing to exploit a forest under licence or by sale of standing volume shall seek approval according to the procedure laid down by decree.

Section 35. -

1) Forest exploitation licences may only be granted to natural persons resident in Cameroon or to companies whose registered offices are in Cameroon and their composition is known to the Administration in charge of forests.
2) The granting of each licence or authorization to sell standing volume shall entail the payment of an approval fee, the rate of which shall be fixed by the Finance Law.

Section 36. -

1) The grant, renewal or transfer of a forest exploitation right shall be subject to the deposition of a security deposit of an amount to be fixed by the Finance Law.
2) In the case of nationals or companies in which the State or nationals hold at least 51% of the capital, the security deposit could be a bank guarantee.
3) In other cases, the security deposit shall be furnished by a payment to the Treasury.
4) The conditions governing the application of this section shall be determined by decree.

Section 37. -

1) The sale as well as the leasing of forest exploitation rights shall be forbidden.
2) The transfer of forest exploitation rights as well as any acquisitions or transfer of shares in a forest exploitation company shall be subject to the prior authorization of the Administration in charge of forests.
3) The conditions of application of this section shall be determined by decree.

Section 38. - The transfer of a licence shall be subject to the payment of a fee, the rate of which shall be fixed by the Finance Law.

Section 39. -

1) All forest exploitation licences, sales of standing volume, permits or authorizations to cut poles, firewood and wood for charcoal shall confer on their holders over the area conceded the exclusive right to collect the produce described in the exploitation right for a specific period, but confer no right of ownership over the corresponding land.
Furthermore, the holder may not prevent the exploitation of produce that is collected in the traditional manner.

2) The collecting of seeds, roots, leaves, exudate, barks and any other parts of plants shall be governed by regulations.

Section 40. -

1) The Administration in charge of forests may mark as reserved any tree which it considers necessary in an area granted under licence for exploitation.
2) Similarly, it may mark trees necessary for the execution of works in the public interest.

Section 41. - The exploitation rights issued untill entry into force of the present law shall remain valid, subject to the provisions of Section 28 above.

Chapter 5 : Use of drift timber washed ashore on the Atlantic Coast

Section 42. - Drift timber without apparent local marks, found along the Atlantic Coast, may be recovered by any natural person or corporate body, according to the procedure determined by decree and subject to the payment of a fee, the amount of which shall be fixed by the Finance Law.

Chapter 6 : Promotion and marketing of timber and forest produce

Section 43. - The export of logs shall, in accordance with the conditions laid down by decree, be reserved to nationals acting individually or grouped into companies who hold a forest exploitation right or to any other exploiter who holds a forest exploitation right and has a local wood processing industry.

Section 44. - The export quotas for the various types of forest produce whether processed or not shall be fixed by the Administration in charge of forests.

Section 45. - Other measures to promote the use of unknown and lesser known species of timber and other forest produce shall be determined by decree.

PART III : WILDLIFE

Chapter 1 : Exercice of hunting rights

Section 46. -

1) Any attemp to pursue, kill or capture a wild animal or to guide expeditions for that purpose shall constitute acts of hunting.
2) The same shall apply to commercial photography and filming.

Section 47. - Traditional hunting is authorized throughout the national territory, except in areas protected for wildlife conservation. The conditions under which it may be carried out shall be fixed by decree.

Section 48. - All acts of hunting other than the case provided for in Section 47 above shall be subject to the grant of a permit or licence.

Section 49. - Fees shall be payable for permits or licences granted; the rates of such fees shall be fixed by the Finance Law.

Section 50. - The rights and obligations resulting from the grant of permits or licences as well as the conditions for their grant shall be determined by decree.

Section 51. - Permits and licences shall be personal and non-transferable, only one hunting permit may be issued to the same person in the course of one hunting season.

Section 52. - Hunting permits may be issued only to persons who have complied with the regulations in force concerning the possession of firearms.

Section 53. -

1) The killing, capture or keeping in captivity of certain animals shall be subject to the payment of fees, the amount of which shall be fixed by the Finance Law.
2) The list of such animals shall be fixed by the Administration in charge of wildlife.

Section 54. -

1) Certain specially defined zones may be declared as zones of cynegetic interest by the Administration in charge of wildlife after consulting the Administration in charge of forests. Such zones may be exploited either by the Administration or by any other natural person or corporate body in accordance with the conditions fixed by decree, for a renewable period of five years. The exploitation of such a zone shall be subject to specifications, the clauses of which shall be defined by the Administration in charge of wildlife.

2) The Administration in charge of wildlife may authorize the practice of the hunting guide profession in unclassified areas in accordance with the conditions determined by decree.

Section 55. - Hunting within a zone of cynegetic interest shall be subject to the payment of a daily fee, the amount of which shall be fixed by the Finance Law.

Section 56. - Persons who hold hunting permits and who have paid the prescribed taxes may freely dispose of the meat and trophies of animals lawfully killed by them : Provided that they shall take all necessary measures to ensure that no meat is abandoned in the bush.

Section 57. -

1) Trophies shall mean tusks, carcasses, skulls or teeth of animals or of carnivorae, tails of elephants or giraffes, skins, hoofs or paws, horns and feathers.
2) The keeping of and traffic in trophies of protected animals shall be subject to their prior formal registration and marking by the Administration in charge of wildlife.
3) Holders of trophies acquired prior to the date of enactment of this law shall be allowed one year within which to have their trophies registered and marked by the Administration in charge of wildlife. Beyond this time-limit, any trophies that do not comply with the provisions of this sub-section shall be confiscated and shall thereafter become State property.

Section 58. -

1) Any person keeping the meat of a protected animal or its unmarked hides and skins or trophies shall present his hunting or capture permit on demand.
2) The keeping in captivity of and traffic in live or dead protected animals, their hides and skins or trophies within the national territory shall be subject to the obtention of a certificate of origin issued by the Administration in charge of wildlife.
3) The certificate of origin shall specify the characteristics of the animals and the registration number of the trophies to enable the identification of animal produce in circulation.
4) The export of wild animals, their hides and skins or trophies shall be subject to the obtention of a certificate of origin and an export permit issued by the Administration in charge of wildlife.

Section 59. - The capture of wild animals shall be subject to the obtention of a permit in accordance with the conditions fixed by decree and subject to the payment of fees, the rates of which shall be fixed by the Finance Law.

Section 60. - The management of game ranches shall be carried out by the State : Provided that they may be entrusted to specialized bodies or private persons under conditions determined by decree.

Section 61. - Buffer zones shall be created around all protected areas in accordance with the conditions determined by decree. Hunting shall be prohibited in such zones as in the protected areas.

Section 62. - The practice of the hunter guide profession in the zones of cynegetic interest or in the open areas shall be subject to the obtention of a permit in accordance with conditions determined by decree and subject to the payment of fees, the rates of which shall be fixed by the Finance Law.

Chapter 2 : Protection of persons and property against animals

Section 63. - In cases where certain animals constitute a danger or cause damage, the Administration in charge of wildlife may undertake game control under the conditions determined by decree.

Section 64. -

1) No person may be charged with any breach of hunting regulations as concerns protected animals if he was compelled to act in his immediate self-defence or in the defence of another person, his own livestock or his own crops.
2) Proof of legitimate defence shall be supplied within 72 hours to the person in charge of the nearest wildlife service.

Section 65. - The trophies resulting from activities referred to in Section 64 hereabove shall be deposited with the Administration in charge of wildlife. The said Administration shall sell them by public auction or by agreement in the absence of a bidder and the revenue shall be paid into the Treasury.

Chapter 3 : Hunting arms

Section 66. - Hunting carried out using the following weapons shall be prohibited :

- war arms or ammunition which were or are part of the standard of the national armed or police forces
- firearms capable of firing more than one cartridge with one press on the trigger
- projectiles containing explosives.

Section 67. - The Administration in charge of wildlife may regulate the calibre or type of arms for hunting certain animals. It may also prohibit the use of certain types of arms or ammunition if the need to protect wildlife so requires.

Section 68. - Duly licensed and registered cynegetic tourist enterprises may, under conditions determined by decree, issue to their clients hunting arms of the type authorized by their hunting permits. In this case, the enterprise shall be civilly liable for any damage caused or offences committed by its clients, without prejudice to legal proceedings which may be taken against the client himself.

Chapter 4 : Protection of wildlife and the environment

Section 69. -

1) All species of animals living in the national territory shall for the purpose of their protection, be classified into three classes : A, B and C.

2) The species of class A shall be totally protected and may on no occasion be killed except as provided for in section 64 (1): Provided that their exploitation shall be subject to the obtention of a capture permit issued by the Administration in charge of forests and wildlife.
The species of class B shall be partially protected. They may be hunted, captured or killed subject to the obtention of the appropriate permit.
The species of class C may not be accorded any protection. However, their hunting shall be regulated.

3) The species of animals found in national parks, game reserves and sanctuaries shall be protected as class A animals except the needs of management dictate otherwise.

4) The conditions of application of this Section shall be determined by decree.

Section 70. - The hunting of certain animals may be temporarily closed in all or part of the national territory by the Administration in charge of wildlife.

Section 71. - Any person found at any time or anywhere in possession of a live or dead class A or B animal or part thereof shall be presumed to have captured or killed it.

Section 72. - Unless specially authorized by the Administration in charge of wildlife, the following shall be prohibited :

1. the pursuit, approach to or shooting of game in motor vehicles or machines;
2. hunting at night, especially with search lamps, head-lamps or in general with any lighting equipment whether designed for cynegetic purposes or not;
3. hunting with drugs, poisoned bait, tranquilizer guns or explosives;
4. hunting with non-traditional equipment;
5. hunting with the use of fire;
6. the importation, sale and circulation of hunting lamps;
7. hunting with fixed guns and dane guns.

Section 73. - Any hunting practice, whether traditional or not which endangers the conservation of certain rare or useful animals may be prohibited or regulated by the Administration in charge of wildlife.

Section 74. -

1) The introduction into the national territory of any plant or live or dead wild animal shall be subject to an authorization issued by the Administration in charge of forests or wildlife depending on the case, on the presentation of a certificate of origin, an authorization to export, and a phytosanitary or veterinary certificate issued by a competent body of the country of origin.
2) The exportation from the national territory of all plants or of live or dead wild animals shall be subject to the presentation of the certificates mentioned above, issued by the competent authority.

Section 75. - Il shall be forbidden to light a fire voluntarily or involuntarily that could destroy the environment. Any such fire shall be controlled to avoid the destruction of the environment. The conditions of application of the present Section shall be determined by regulations.

Section 76. -

1) All human activities tending to degrade the environment such as the unauthorized felling of trees in zones likely to be invaded by the desert or subject to flooding shall be forbidden.
2) The movement or straying of domestic animals or cattle in protected areas or in buffer zones shall be forbidden.

Section 77. -

1) The destruction of the environment within a distance of 50 metres on either side of a water course or within a radius of 100 metres around a water source shall be forbidden.
2) Exploitation rights along water courses shall be governed by regulations.

PART IV : FISHERIES

Chapter 1 : Definitions

Section 78. - Fishery resources shall mean all piscine forms in the marine, estuarine and fresh water environments, including sedentary animals in these environments which at the time of their capture or harvest are either fixed to the substrate in a public waterway or coastland or are incapable of moving around without coming in contact with the bottom of the aquatic environment in marine, estuarine, fresh water or any aquacultural establishment.

Section 79. -

1) Fishing shall be the act of capturing or of harvesting any fishery resource, or any activity that may reasonably be supposed to lead to the harvest, picking or capturing of fishery resources, including all activities connected with the proper management and use of the aquatic environment with a view to protecting the animal species therein by the total or partial control of their life cycle.

2) There are the following types of fishing operations, depending on the means used to obtain fishery resources :
   - traditional or small-scale fishing
   - sport fishing
   - fishing for scientific purposes
   - semi-industrial fishing
   - industrial fishing
   - sea farming
   - fish farming.

3) These different types of fishing shall be defined and regulated by decree.

Section 80 - A fishing vessel shall be any boat, no matter its size, that is used in activities connected with fisheries.

Section 81. - Fishing gear shall refer to tools, implements or appliances used in fishing operations.

Section 82. - Mesh size shall be defined in relationship to the bag of a net to mean the dimension of one of the open spaces between the cords of a net taken or measured at the 50th space of wet and out-stretched net, or any dimension of a consecutive 50th space of the net measured when the net is under normal pressure, wet and stretched out.

Section 83. - Within the meaning of the present law :

a) Fish processing establishments shall comprise :
   1) Fishmongering installations which prepare fishery products (sorting, washing, weighing, icing);
   2) Freezing establishments which preserve fish by means of freezing or simply store frozen products;
   3) Smoking houses or workshops which smoke fish and fishery products using wood or by-products of wood;
   4) Drying workshops which dehydrate fishery products through the direct action of heat produced by solar energy or some other source;
   5) Salting workshops which process fishery products by using exclusively sea salt or its substitutes.

b) Storage and sales establishments shall comprise :
   1) Cold stores or premises equipped for the storing of products at a temperature of at least minus 20°C (-20°C).
   2) Fish shops where fishery products are stored for sale by retail to the public.

c) Means of transportation shall comprise :
   1) Isothermic vehicles which include cars, wagons or containers, etc.. whose walls are made air tight to prevent any exchange of temperature between the interior of the vehicle and the outside.
   2) Refrigerated vehicles which mean vehicles equipped with an autonomous compressor to maintain a cold environment within the said vehicle.

Section 84. - The technical norms of and the conditions of hygiene in the installations listed in Section 83 shall be determined by decree.

Chapter 2 : Exercice of fishing rights

Section 85. - The right to fish waterways and coastlands shall belong to the State : Provided that fishing shall be carried out under conditions to be determined by decree.

Section 86. -

1) The right to carry out industrial fishing shall be subject to the obtention of a fishing licence, and for the other forms of fishing, except traditional or small-scale fishing the participant shall have a fishing permit.

2) Any person who wished to fish small shrimps or cray-fish (Palaenon hastatus and Pellonula vorax) shall first obtain a special authorization granted under conditions to be determined by decree.

Section 87. - Fishing licences shall be of three (3) types :
- the licence to catch fish
- the licence to fish for shrimps and other crustaceans
- the licence to fish for tunae.

Section 88. - Fishing permits shall be of three (3) types :
Permit A - the permit for semi-industrial fishing
Permit B - the permit for sport fishing
Permit C - the permit for fisheries research.

Section 89. -
1) The issuing of a fishing licence or permit shall be subject to the payment of an exploitation tax, the rate of which shall be fixed by the Finance Law.
2) The said tax shall also be paid on the occasion of the renewal of the said licences or permits.

Section 90. - The conditions under which fishing licences and permits are issued shall be determined by decree.

Section 91. - Any fishing licence or permit shall be presented at any time to the competent authorities on demand.

Section 92. -
1) Any natural person or corporate body wishing to exploit fishery resources for commercial or industrial purposes shall first apply for a licence or permit in accordance with the procedure to be determined by decree.
2) The issuing of the fishing licence or permit shall be subject to the payment of an exploitation tax, the rate of which shall be fixed by the Finance Law.

Section 93. - Fishing licences may only be issued to persons resident in Cameroon orto companies whose head office is located in Cameroon, and whose composition is known by the Administration in charge of fisheries.

Section 94. -
1) The sale or lease of fishery exploitation rights shall be forbidden.
2) The transfer of a fishing licence or permit shall be subject to the approval of the Administration in charge of fisheries and the payment of a tax the rate of which shall be fixed by the Finance law.
3) The conditions of application of this Section shall be determined by decree.

Section 95. - Irrespective of the fiscal provisions applicable to him, no exploiter of fishery resources and no exporter or processor of fishery products shall be exempted from the payment of fishing taxes.

Section 96. - All fishermen and fishing companies shall declare their catches in accordance with the conditions laid down by the Administration in charge of fisheries.

Chapter 3 : Management and conservation of fishery resources

Section 97. - Prohibition may be placed on the right to fish in order :
- to protect aquatic fauna, the aquatic environment and traditional fishery operations,
- to maintain fish production at an acceptable level.

Section 98. - The following shall be forbidden :

a) the use of trawlers or fishing vessels equipped with trawling gear within a 2 nautical mile zone;

b) the use, for any type of fishing, of any material likely to obstruct the mesh of nets or having the effect of reducing their selective action. Furthermore, no accessory equipment may be placed at the interior of fishing nets. Protective devices may be permitted if such devices have a dimension of more than two times that of the authorized mesh, and are placed on the upper part of the net and not behind the net;

c) the use for fishing of any diving suit equipped with a respirator;

d) the presence on board a fishing vessel of respiratory equipment such as a diving suit, of a harpoon or of a dangerous fishing weapon, except as a safety precaution

e) the use for fishing of explosives, chemicals, poisons or other noxious substances, electrical currents or headlamps, firearms, light or automatic traps or any other devices likely to destroy aquatic fauna and the aquatic environment;

f) the construction of dams, embankments, large channels or port facilities without the prior approval of the Administration in charge of fisheries;

g) the pouring or discharging into the aquatic environment of toxic or noxious materials such as industrial, agricultural or domestic wastes and pollutants (pesticides, fertilizers, sediments, detergents, etc);

h) the destruction of the environment within a distance of 50 metres along a water course, or over a radius of 100 metres around its sources;

i) the presence on board a fishing vessel of any fishing nets, whose mesh sizes do not conform to prescribed standards;

j) the presence on board a fishing vessel of any destructive devices or of substances that are capable of stunning or disabling fish, as well as any other materials and devices capable of reducing or obstructing the meshes of fishing nets;

k) the export of any fishery resource without the prior approval of the Administration in charge of fisheries;

l) the introduction into Cameroon of foreign living fishery resources;

m) the capture, sale or possession of any fishery resources appearing on a prohibition list established by the Administration in charge of fisheries;

n) fishing in closed areas forbidden by the Administration in charge of fisheries.

Section 99. - Exemptions may be made to the provisions of Section 98 by the Administration in charge of fisheries.

Section 100.- No fishing vessel whose tonnage exceeds 250 tonnes may fish in Cameroon territorial waters. In public waterways, the total weight of fishing vessels may not exceed 10 tonnes.

Section 101.- The Administration in charge of fisheries shall establish for both marine and inland waters the characteristics of permissible fishing nets.

Section 102.- The permissible mesh size for the different types of nets shall be established by the Administration in charge of fisheries.

Chapter 4 : Sea farming and fish farming

Section 103.- No acquacultural establishment may be constructed without the issue of a permit by the Administration in charge of fisheries. The conditions for issuing any such permit shall be determined by decree. The issue of the permit shall be subject to the payment of a tax, the rate of which shall be fixed by the Finance Law.

Section 104.- The construction permit may lay down restrictions to ensure the conservation, the proper management, and the optimum exploitation of fishery resources. Such restrictions may concern, in particular :

- the layout and the characteristics of construction;
- management;
- the control of the quality of the products and health conditions.

Section 105.- The administration in charge of fisheries shall be responsible for the running of fish stations and fish breeding centres in the public waterways and coastlands.

Chapter 5 : Installation of fish Processing establishments

Section 106.- The creation of a fismonger's store, a frozen products plant, a processing workshop (for smoking, drying or salting) a canning factory or a fish shop shall be subject to the obtention of a certificate issued in accordance with conditions to be determined by decree, without prejudice to other conditions concerning the control of classified establishments. The issue of the said certificate shall be subject to the payment of a tax the rate of which shall be fixed by the Finance Law.

Section 107.- Establishments for the exploitation of fishery products shall be classified according to their size and type by the Administration in charge of fisheries and the tax referred to in Section 106 above shall be calculated accordingly.

Section 108.- The opening to the public of any establishment of the type referred to in Section 106 above shall be subject to the obtention of a certificate of conformity issued in accordance with the conditions to be determined by decree.

Chapter 6 : Sanitary inspection and control of fishery products

Section 109. - It shall be forbidden to expose, prepare, distribute store or transport for sale any fishery products that have not been subject to sanitary inspection.

This inspection which may be carried out anywhere and at any time shall be subject to the payment of a tax, the rate of which shall be fixed by the Finance Law.

Section 110. - Fishery products shall be subject to sanitary inspection
- to ensure that the official nomenclature set out for commercial species is respected;
- to ensure that the minimum marketable size of edible species is respected;
- to ensure that the products are not from prohibited fishing zones;
- to ensure that the products landed are proper for consumption.

Quality norms shall be determined by decree.

Chapter 7 : Packaging and transportation of fishery products

Section 111. - Fishery products shall be packed in prescribed container

Section 112. - The transport by road or by railway of fishery products shall be in vehicles built to conform to norms laid down by decree.

Section 113. - Vehicles for use in transporting fishery products shall be certificated before they are used for such purposes. The certification shall comply with conditions laid down by decree and shall be subject to the payment of a tax, the rate of which shall be fixed by the Finance Law.

PART V : PROSECUTION OF OFFENCES (Section 114 - 134 omitted)

PART VI : MISCELLANEOUS PROVISIONS

Section 135. - If, during a prosecution for an offence, the accused pleads a right of ownership or any other right, the court shall decide the matter according to the following rules :
An interlocutory plea shall only be allowed if it is founded either on an apparent title or on equivalent facts of possession, and if these legal grounds are such as to negative the character of the offence attached to the facts which gave rise to the legal proceedings

If the case is sent to the civil court, the judgment shall fix a perio which shall not exceed three months in which the party must bring the case before the competent judges and justify his action, failing which the plea shall be overruled.

Section 136. - The issue of the duplicate of an exploitation right, licence, permit or special authorization to exploit forestry, wildlife and fishery resources shall be subject to the payment of a fee, the rate of which shall be fixed by the Finance Law.

Section 137. - The proceeds of the exploitation tax and the sanitary inspection tax referred to in Sections 89 and 109 above shall be allotted as follows :

- 50% to the Treasury
- 50% to the service or body in charge of fisheries development.

Section 138. -

1) The proceeds of fines, compounding fees, damages and sale by public auction or private contract of produce and various objects seized shall be allotted as follows :
   1. as regards forests and wildlife :
      - 25% to officials of the Services in charge of forests and wildlife,
      - 75% to the Treasury
   2. in case of fisheries :
      - 25% to the officials of the administration in charge of fisheries and the sworn officials of the Merchant Marine who helped in the punishment of the offences,
      - 40% to the service or body in charge of fisheries development,
      - 35% to the Treasury.

2) The conditions of distribution of the rebates to the officials referred to above shall be determined by decree.

Section 139. - All provisions previous to this law are hereby repealed, in particular :

- Ordinance N° 73-18 of 22 May 1973
- Law N° 74-12 of 16 July 1974
- Law N° 75-4 of 2 July 1975.

Section 140. - This law shall be registered and published in the Official Gazette in French and English.

Yaounde, 27 November 1981

THE PRESIDENT OF THE REPUBLIC

DECRET N° 83/169 DU 12 AVRIL 1983
fixant le Régime des Forêts

STRUCTURE ET MATIERES REGLEMENTEES PAR LE DECRET D'APPLICATION N° 83/169 DE LA LOI PORTANT REGIME DES FORETS, DE LA FAUNE ET DE LA PECHE (LOI N° 81/13)

TITRE I : EXPOSITIONS GENERALES

Chapitre Unique : Régénération des Forêts

Article 1. - Objectifs et modalités des opérations de régénération des forêts.

TITRE II : EXPLOITATION DES FORETS

Chapitre I : Des Forêts Domaniales

Section I : Définitions et Droit d'Usage

Article 2. - Définitions des termes "réserve naturelle intégrale", "forêt de production", "forêt de protection", etc..

Article 3. - Exercice de certains droits d'usage dans les forêts domaniales.

Section II : Classement et Déclassement des Forêts

Article 4. - Préparation du décret et d'un dossier pour le classement d'une forêt.

Article 5. - Avis public du projet de classement et avis d'une commission sur les réclamations formulées.

Article 6. - Affectation à une autre destination ou à une utilisation non forestière seulement par déclassement officiel.

Article 7. - Délimitation et bornage des forêts classées.

Section III : Inventaire et Aménagement

Article 8. - Inventaires en forêts domaniales

Article 9. - Contenu d'un plan d'aménagement.

Section IV : Exploitation des forêts domaniales par vente de coupe

Article 10. - Conditions applicables aux ventes de coupes.

Article 11. - Méthodes d'attribution.

Article 12. - Limite minimum pour mise à prix d'une coupe.

Article 13. - Dossier de demande.

Section V : Exploitation en régie et en concession

Article 14. - Vente des produits exploités en régie.

Article 15. - Dossier à fournir pour demande de concession.

Chapitre II : Des Forêts des Collectivités Publiques autres que l'Etat et de celles appartenant aux Particuliers

Article 16. - Opérations techniques de gestion à être exécutées ou approuvées par l'Administration Forestière

Article 17. - Exploitation des forêts appartenant aux particuliers.

Chapitre III : Des Forêts du Domaine National

Section I : Exploration et Inventaire

Article 18. - Dossier de demande pour autorisation d'exploration d'une forêt.

Article 19. - Conditions à respecter en cas de demande d'un exploitant en activité.

Article 20. - Limitation du droit d'exploration.

Article 21. - Délai de validité et obligations du titulaire de l'autorisation d'exploration.

Article 22. - Préparation d'inventaires par un organisme spécialisé.

Section II : Exploitation des Forêts du Domaine National

Sous-Section I La Commission Technique

Article 23. - Fonctions et composition de la commission technique.

Article 24. - Convocation de la commission par le Ministre.

Article 25. - Délibération et décision de la Commission.

Article 26. - Compte-rendu de réunion de la Commission à soumettre au Ministre.

Article 27. - Dossier de demande pour agrément à la profession forestière avec la présentation de projets d'investissements.

Article 28. - Autorisation du Ministre en cas de remplacement du responsable de l'exploitation forestière.

Article 29. - Critères à observer par la commission technique pour examen des dossiers d'agrément.

Sous-Section II : Exploitation par Licence

A. - Procédure d'attribution de licence

Article 30. - Avis au public sur les zones forestières devant être attribuées et dossier de demande à fournir par les exploitants intéressés.

Article 31. - Information de la population concernée par l'attribution d'une licence et réunion des représentants régionaux et locaux lors de la préparation des projets retenus par la commission technique.

Article 32. - Attribution des licences par le Ministre pour les superficies inférieures ou égales à 10 000 ha et le Premier Ministre pour les superficies supérieures à 10 000 ha.

Article 33. - Justification du paiement de la taxe d'agrément et du dépôt du cautionnement par le titulaire.

B. - Droits et obligations résultant de l'exploitation d'une licence

Article 34. - Droit conféré et droit non conféré par l'attribution d'une licence.

Article 35. - Cahier des charges avec clauses particulières de transformation locale (superficies supérieures à 20 000 ha) et charges financières.

Article 36. - Exploitation de la superficie attribuée par tranche de 2 500 ha (assiette de coupe) et règles d'exécution du travail des assiettes de coupes par phases successives.

Article 37. - Inscription du diamètre des arbre abattus pris à 1,30 m du sol sur les carnets de chantier.

Article 38. - Normes des installations industrielles de transformation locale par rapport aux superficies concédées et pourcentage minimum de bois à être transformé localement.

Article 39. - Autorisation pour implantation de toute unité de transformation industrielle locale.

Article 40. - Utilisation de toutes les grumes commercialisables et paiement du prix de vente pour les arbres brisés et abandonnés.

Article 41. - Marques réglementaires des grumes et contrôle du transport routier.

Article 42. - Autorisation pour ouverture de voies d'évacuation traversant une forêt du domaine national non concédé en licence.

Article 43. - Déclaration des grumes transportées par chemin de fer.

C. - Du renouvellement de licence

Article 44. - Dossier de renouvellement.

Article 45. - Conditions préalables au renouvellement.

Section III : Contrôle de l'Exploitation Forestière

Article 66. - Tenue d'un cahier de chantier et inscriptions à y insérer.

Article 67. - Obligation pour les titulaires de licences de préparer un rapport annuel.

Article 68. - Enregistrement des grumes entrées en usine de transformation.

Article 69. - Contresignature des bulletins et registres de contrôle.

TITRE III : DE L'EXPORTATION DES BOIS EN GRUMES ET DE LA PROMOTION DES ESSENCES ET PRODUITS FORESTIERS

Article 70. - Conditions préalables pour les exportateurs de bois en grumes nationaux et non-nationaux.

Article 71. - Quota d'exportation pour produits forestiers bruts ou transformés.

Article 72. - Rapport annuel obligatoire de chaque exporteur.

Article 73. - Publication annuelle d'une liste d'essences en promotion et taux de taxe préférentiel pour ces essences.

TITRE IV : DISPOSITIONS DIVERSES

Chapitre I : Des Bois échoués sur la Côte d'Atlantique

Article 74. - Conditions de récupération.

Chapitre II : Des Prises de Participation

Article 75. - Autorisation et règles sur les prises de participation et cessions de parts des capitaux des sociétés d'exploitation forestières.

Article 76. - Dossier de demande d'autorisation.

Chapitre III : Des Feux de Brousse

Article 77. - Règles de prévention des feux de brousse.

Chapitre IV : Du Constat des Infractions et des Transactions

Article 78. - Contenu du procès-verbal de constat d'infraction forestière

Article 79. - Possibilités de transaction pour certains responsables de l'administration forestière.

Article 80. - Procédure de transaction.

Chapitre V : Dispositions Transitoires et Finales

Article 81. - Délai d'un an pour les exportateurs de produits forestiers pour se conformer aux nouvelles dispositions du présent Décret.

Article 82. - Réglementation particulière concernant les agents de l'administration forestière.

Article 83. -Agrément à la profession forestière.

Article 84. - Publication du Décret et abrogation des textes antérieurs.

DECRET N°83/169
DU 12 AVRIL 1983
fixant le régime des forêts

CAMEROUN

CAMEROON

Décret N° 83/169

LE PRESIDENT DE LA REPUBLIQUE

VU la Constitution
VU la loi N° 83/13 du 27 novembre 1981, portant régime des forêts, de la faune et de la pêche,

DECRETE :

TITRE I : EXPOSITIONS GENERALES

CHAPITRE UNIQUE : REGENERATION DES FORETS

Art. 1.- (1) Le régénération des forêts a pour but d'assurer la pérennité du patrimoine forestier national.
(2) La régénération des forêts domaniales doit respecter les prescriptions des plans d'aménagement correspondants. Elle est assurée par un organisme spécialisé.
(3) Les modalités de régénération des forêts des collectivités publiques ou des particuliers doivent être approuvées par l'Administration chargée des Forêts au cas où cette régénération est assurée par eux-mêmes ou par un organisme de leur choix.

TITRE II : EXPLOITATION DES FORETS

CHAPITRE I : DES FORETS DOMANIALES

Section I : Définitions et droits d'usage

Art. 2.- Pour l'application du présent décret, constitue :

(1) Une réserve naturelle intégrale : un périmètre dont les ressources bénéficient d'une protection absolue. Y sont notamment interdits : les exploitations forestières agricoles, pastorales ou minières; les fouilles, prospections, sondages, terrassements, constructions ainsi que tous les travaux de nature à modifier l'aspect du terrain ou de la végétation, la pollution des eaux, l'introduction d'expèces botaniques locales ou importées, et d'une manière générale, toute intervention humaine non autorisée par l'Administration forestière, susceptible d'engendrer des perturbations dans l'équilibre de la flore.

(2) Une forêt de production : un périmètre destiné principalement à la production des bois d'oeuvre et de service ou de tout autre produit forestier.

(3) Une forêt de protection : un périmètre dont l'objet principal est la protection du sol, du régime des eaux ou de certains ecosystèmes présentant un intérêt scientifique.

(4) Une forêt récréative : un périmètre dont l'objet est de créer ou de maintenir un cadre de loisir, en raison de son intérêt esthétique, artistique, touristique, sportif ou sanitaire.

(5) Un périmètre de reboisement : un terrain destiné à être régénéré.

(6) Un jardin botanique : un site présentant un intérêt scientifique, esthétique ou culturel et groupant des plantes spontanées ou introduites bénéficiant d'une protection absolue.

Art. 3.- (1) Hormis le cas des réserves naturelles intégrales, des périmètres de reboisement et des jardins botaniques où toute intervention humaine non autorisée par l'Administration Forestière est interdite, les populations locales conservent dans les forêts domaniales des droits d'usage qui consistent pour elles dans l'accomplissement à l'intérieur de ces forêts, d'activités traditionnelles telles que la collecte des produits forestiers secondaires : raphia, palmier, bambou, rotin, bois de chauffage et produits alimentaires.

(2) Le décret portant classement d'une forêt de production ou de récréation fixe, pour chaque cas, les droits d'usage reconnus aux populations locales notamment la liste des produits forestiers susceptibles d'être récoltés ainsi que les possibilités d'utilisation du sol.

(3) L'extraction du sable, du gravier, ou de la latérite à l'intérieur des forêts domaniales doit s'effectuer après avis du Ministre chargé des forêts et conformément à la réglementation sur les carrières.

Section II : Classement et déclassement des forêts

Art. 4.- Le classement d'une forêt au domaine privé de l'Etat est sanctionné par décret du Président de la République sur proposition du Ministre chargé des forêts, sur présentation d'un dossier comprenant :
- un plan de situation;
- une note technique précisant le but visé par ce classement;
- le procès-verbal de la Commission prévue à l'Art. (5) ci-dessous.

Art. 5.- (1) Le classement est précédé d'une période de quatre-vingt-dix (90) jours au cours de laquelle, le Ministre chargé des forêts, par un avis affiché dans les Sous-Préfectures, mairies et services extérieurs et publié à la presse écrite, informe les populations concernées du projet de classement, en vue de leur permettre de faire des oppositions ou des réclamations auprès des Chefs de Circonscriptions Administratives compétents. Passé ce délai, les éventuels opposants sont forclos.

(2) Dans les 30 jours qui suivent ce délai de forclusion, se réunit au Chef-lieu de chaque Préfecture concernée une Commission composée comme suit :

- le Préfet ou son représentant .... Président
- le Responsable Provincial de l'Administration Forestière .... Rapporteur
- le Représentant local du Ministre chargé de l'Urbanisme et de l'Habitat .. Membre
- un Député à l'Assemblée Nationale ..... "
- le Représentant local du Ministre chargé de l'Elevage, des Pêches et des Industries Animales .... "
- le Représentant local du Ministre chargé des Mines et de l'Energie .... "
- le Représentant local du Ministre chargé de l'Agriculture .... "
- les Maires des Communes intéressées .... Membres

(3) la Commission dresse un procès-verbal de la réunion assorti de son avis sur les éventuelles réclamations formulées par la population ou par toute personne intéressée.

(4) L'ensemble du dossier est adressé au Ministre chargé des forêts aux fins de préparer le décret de classement.

Art. 6.- (1) Une forêt domaniale ne peut recevoir une destination différente de celle qui lui est assignée lors de son classement qu'après son déclassement partiel ou intégral.

(2) Elle ne peut recevoir une destination non forestière, qu'après qu'une zone de superficie au moins équivalente aura été classée forêt domaniale.

Art. 7.- Les forêts domaniales doivent être délimitées, bornées et identifiées dans les conditions fixées par arrêté du Ministre chargé des forêts.

Section III : Inventaire et aménagement

Art. 8.- L'inventaire des forêts domaniales est assuré selon les prescriptions fixées par l'Administration forestière.

Il consiste pour l'organisme compétent en un sondage permettant d'apprécier la richesse de la forêt en arbres de diamètre supérieur à 20 cm.

Art. 9.- Sur la base des résultats de l'inventaire, le Ministre chargé des forêts fixe pour chaque forêt domaniale un plan d'aménagement précisant notamment l'objet assigné à la forêt, les infrastructures à y réaliser, les modes et conditions d'exploitation ainsi que les charges y afférentes, les voies d'accès à ouvrir ou à entretenir, les zones à mettre en défens, les parcelles à régénérer ainsi que les méthodes sylvicoles à utiliser.

Section IV : Exploitation des forêts domaniales par vente de coupe

Art.10.- (1) Dans les forêts domaniales, les ventes de coupe se font conformément au plan d'aménagement arrêté pour cette forêt. Cependant pour celles comportant un programme de régénération, les ventes de coupe se font conformément au programme de plantation arrêté au plan d'aménagement. La superficie de la coupe vendue à la fois dans une même forêt ne peut être supérieure à 2 500 ha.

Dans tous les cas, avant le début de l'exploitation, la coupe doit faire l'objet d'un inventaire préalable consistant en une évaluation de 100% de tous les arbres exploitables.

(2) Dans une vente de coupe, les arbres sont vendus sur pied. Seuls ceux préalablement inventoriés, marqués, cubés et désignés à la vente peuvent être abattus.

Plusieurs exploitants forestiers peuvent être autorisés à exercer simultanément dans la même coupe, chacun n'exploitant que les arbres qui lui sont attribués.

(3) La vente porte sur les volumes et les espèces convenus dans l'acte de vente entre l'exploitant et l'Administration forestière.

(4) La durée des opérations d'abattage est fonction du volume des bois vendus et figure dans l'acte de vente. A l'expiration de cette durée, il est interdit à l'exploitant forestier de revenir dans la zone de coupe, sauf s'il est titulaire d'une autre coupe.

Art.11.- (1) La désignation du bénéficiaire d'une vente de coupe se fait par adjudication, ou de gré à gré, en l'absence d'adjudicataire.

(2) En cas d'adjudication, il est établi un cahier-affiche contenant les spécifications concernant les espèces, le volume et le lieu d'exploitation.

Avant la date fixée pour l'adjudication, ce cahier-affiche doit faire l'objet pendant 30 jours d'une information au public par voie de presse et d'affichage dans les unités administratives de la zone d'exploitation.

(3) La vente de coupe ne peut être consentie qu'aux exploitants forestiers agréés, la priorité étant toutefois réservée aux nationaux.

(4) Sauf cas d'exploitation urgente commandée par un programme de régénération, les ventes de gré à gré sont exclusivement réservées aux nationaux, aux sociétés d'Etat ou aux organismes dans lesquels l'Etat détient au moins 50% des parts.

Art.12.- La mise à prix de la coupe ne doit en aucun cas être inférieure à la taxe de récupération des produits inventoriés telle que fixée par la loi des finances.

Art.13.- La vente de coupe est autorisée par arrêté du Ministre chargé des forêts sur la base d'un dossier comportant les pièces suivantes :

A)- pour les particuliers :
- une demande timbrée indiquant les nom, prénom, la nationalité, la profession et la résidence du postulant;
- un extrait de casier judiciaire;
- une copie de l'acte d'agrément à la profession forestière;

B)- pour les personnes morales :
- une demande timbrée précisant la raison sociale ou la dénomination sociale et le siège social;
- une expédition des statuts précisant le montant et la répartition du capital social;
- un extrait du casier judiciaire du Directeur ou du gérant.

C)- pour les particuliers et les personnes morales :
- une fiche indiquant la situation, la superficie de la portion de forêt mise en vente et ses limites définies à partir d'un point topographique immuable;
- une carte en cinq exemplaires du Centre Géographique National indiquant la zone en question avec report de sa superficie;
- une déclaration timbrée spécifiant que le postulant a pris connaissance de la réglementation forestière en vigueur et qu'il s'engage à la respecter.

Section V : Exploitation en régie et en concession

Art.14.- Les produits forestiers exploités en régie par l'Administration Forestière peuvent être vendus de gré à gré ou par adjudication. En cas d'adjudication, la vente fait l'objet d'un cahier-affiche rendu public dans les conditions prévues à l'Art.11 alinéa 2 ci-dessus.

Art.15.- La concession d'une forêt domaniale est sanctionnée par décret du Premier Ministre sur proposition du Ministre compétent au vu d'un dossier comprenant :

- une demande timbrée précisant la raison sociale et le siège de la société;
- les statuts de la société précisant le montant et la répartition du capital social;
- l'extrait du casier judiciaire du Directeur ou du Gérant datant de moins de trois mois;
- le curriculum vitae du Directeur;
- une déclaration timbrée indiquant que la société a pris connaissance de la réglementation et qu'elle s'engage à la respecter;
- une déclaration timbrée indiquant que la société s'engage à respecter le plan d'aménagement arrêté pour la forêt concernée;
- un extrait du dépôt au greffe de la Cour d'Appel compétente de l'empreinte du marteau forestier du postulant. Cet extrait doit porter le fac-similé de l'empreinte;
- une copie certifiée de l'acte d'agrément;
- cinq exemplaires de la carte géographique de la zone concernée.

La demande doit en outre préciser le volume et le programme des investissements à réaliser, le nombre et l'emploi des personnels à recruter, ainsi que leur programme de formation.

CHAPITRE II : DES FORETS DES COLLECTIVITES PUBLIQUES AUTRES QUE L'ETAT ET DE CELLES APPARTENANT AUX PARTICULIERS

Art.16.- En vue de leur exploitation rationnelle, la gestion technique des forêts des collectivités publiques notamment les travaux d'exploitation, de régénération, ou la surveillance de ces forêts doivent être exécutés par l'Administration Forestière, ou approuvés par elle s'ils sont effectués par des tiers.

Art.17.- L'exploitation d'une forêt appartenant à un particulier peut s'effectuer par son propriétaire ou par toute personne de son choix. Toutefois, le propriétaire est tenu d'en aviser au préalable l'Administration chargée des Forêts.

Le Ministre chargé des forêts peut suspendre cette exploitation si elle est de nature à causer un préjudice à l'environnement. Dans ce cas, la procédure d'expropriation peut être engagée conformément à la législation en vigueur.

CHAPITRE III : DES FORETS DU DOMAINE NATIONAL

Section I : Exploration et inventaire

Art.18.- L'exploration d'une forêt du domaine national est subordonnée à l'obtention d'une autorisation du Ministre chargé des forêts sur présentation d'un dossier comprennant les pièces suivantes :

- une demande timbrée indiquant :
  a) s'il s'agit d'un particulier : le nom, prénom, nationalité, profession et résidence;
  b) s'il s'agit d'une personne morale : la raison sociale ou la dénomination, le siège social, le nom du Directeur et du Gérant.
  Dans les deux cas, doivent être indiquées : la situation, la superficie de la portion de forêt à explorer, ainsi que ses limites qui doivent être définies à partir d'un point topographique immuable.
- une carte en cinq exemplaires du Centre Géographique national, sur laquelle est indiquée la zone sollicitée;
- une déclaration sur papier timbré spécifiant que le postulant a pris connaissance de la réglementation forestière en vigueur et qu'il s'engage à la respecter;
- une quittance attestant le paiement des droits d'exploration fixés par la loi des finances.

Art.19.- Tout dossier d'exploration déposé par un exploitant déjà en activité ne peut être instruit que si l'intéressé s'est acquitté de tous les droits et taxes forestiers grevant sa ou ses licences, et s'il a respecté strictement les clauses de son cahier des charges ainsi que son programme d'investissement.

Art.20.- Le titulaire d'une autorisation d'exploration ne peut disposer d'aucun produit forestier dans la zone explorée.

L'autorisation d'exploration ne confère aucun droit particulier quant à concession d'un droit d'exploitation ultérieure sur la zone explorée.

Art.21.- (1) Le délai de validité d'une autorisation d'exploration ne peut excéder six mois.

(2) Au terme de la période de validité, le titulaire de l'autorisation adresse à l'Administration chargée des forêts :
- les résultats de ses prospections;
- les documents topographiques qu'il a pu constituer.

Art.22.- (1) L'inventaire de la richesse en arbres par un organisme spécialisé constitue le préalable à toute exploitation d'une forêt du domaine national.

(2) L'inventaire se fait par chantier de 2 500 ha, chaque chantier constituant une assiette de coupe. Il consiste en une énumération à 100% de tous les arbres ayant atteint le diamètre d'exploitabilité, tel que fixé par les clauses générales du cahier des charges.

Section II : Exploitation des forêts du domaine national

Sous-Section I : La Commission technique

Art.23.- (1) Les demandes d'agrément à la profession forestière, les demandes d'octroi de licence d'exploitation forestière, de renouvellement, de transfert, ou d'abandon de ces titres; les demandes de permis spéciaux d'exploitation de plantes médicinales, sont instruites par l'Administration chargée des Forêts après avis d'une Commission Technique composée ainsi qu'il suit :

- Le Ministre chargé des Forêts ou son Représentant .... Président
- un représentant de l'Assemblée Nationale .. Membre
- un représentant du Ministre de l'Administration Territoriale .... "
- un représentant du Ministre des Finances .. "
- un représentant du Ministre de l'Economie et du Plan .... "
- un représentant du Ministre de l'Urbanisme et de l'Habitat .... "
- un représentant du Ministre des Mines et de l'Energie .... "
- le Délégué Général à la Sûreté Nationale ou son représentant .... "
- le Délégué Général au Tourisme ou son représentant .... "
- le Directeur des Forêts .... "

(2) Le Ministre chargé des forêts peut convoquer toute autre personne de son choix en raison de ses compétences.

Art.24.- La commission technique se réunit sur convocation du Ministre chargé des forêts en tant que de besoin, en tout cas au moins deux fois l'an.

Art.25.- (1) La commission technique ne peut valablement délibérer que si les 2/3 au moins de ses membres sont présents;

(2) Les avis de la commission technique sont émis à la majorité simple des voix, celle du Président est prépondérante.

(3) L'avis de la commisison technique peut être :
-favorable si tous les critères sont réunis;
-favorable sous condition, lorsqu'un complément d'information est nécessaire. Dans ce cas, le postulant dispose d'un délai de trois mois à compter de la date de notification de l'avis pour apporter le complément d'information. Passé ce délai, l'avis de la commission cesse d'être favorable;

- ajourné lorsque l'un des critères majeurs permettant d'étudier valablement le dossier fait défaut. Dans ce cas, le postulant dispose d'un délai de trois mois à compter de la date de notification de l'avis pour compléter son dossier, en vue de son réexamen à la session suivante de la commission technique;
- défavorable lorsque le dossier ne répond pas aux critères réglementaires.

Art.26.- Le compte-rendu de chaque réunion de la commission technique est signé par son président et soumis à l'appréciation du Ministre chargé des Forêts.

Art.27.- L'agrément à la profession forestière est sanctionné par arrêté du Président de la République sur la base d'un dossier comportant les pièces suivantes :

a) s'il s'agit d'un particulier :
- une demande timbrée précisant les nom, prénom, nationalité, profession et résidence;
- un extrait du casier judiciaire datant de moins de trois mois;
- un curriculum vitae.

b) s'il s'agit d'une société :
- une demande timbrée précisant la raison sociale et le siège de la société;
- les statuts;
- l'extrait du casier judiciaire du directeur ou du gérant datant de moins de trois mois;
- le curriculum vitae du directeur ou du gérant.

Dans les deux cas, la demande précise la nature de l'activité postulée, les investissements prévus et leur plan de financement, le nombre et l'emploi des agents à recruter.

Le dossier doit comporter en outre, des pièces justificatives:

- des connaissances techniques du responsable de l'exploitation forestière;
- des investissements réalisés ou les garanties de ceux prévus;
- de la libération du capital qui doit être équivalent à au moins 20% des investissements prévus, conformément aux comptes d'exploitation prévisionnels. Toutefois, pour les nationaux, un capital initial de 5% peut être accepté sous réserve qu'il soit augmenté à au moins 20% deux ans après l'attribution de la licence.

Art.28.- Tout remplacement du responsable de l'exploitation forestière est subordonné à une autorisaiton préalable du Ministre chargé des forêts.

Art.29.- Lors de l'examen des dossiers d'agrément, la commission technique doit tenir compte entre autres, des critères suivants :

a) connaissances techniques en matière forestière;
b) moyens financiers et matériels avec à l'appui toutes les pièces justificatives;
c) capital qui doit être équivalent au moins à 20% des investissements prévus conformément aux comptes d'exploitation prévisionnels.

Sous-Section II : Exploitation par licence

A. Procédure d'attribution de licence

Art.30.- (1) L'octroi de toute licence d'exploitation forestière est précédé d'une période d'information au cours de laquelle, l'Administration chargée des forêts, après avoir choisi la zone forestière à ouvrir à l'exploitation, la déclare libre par un avis au public qui précise la localisation, les limites et la superficie de la forêt concernée.

(2) Les exploitants forestiers intéressés font parvenir au Ministre chargé des forêts un dossier comprenant :

- une demande timbrée précisant :
  - les nom, prénom, nationalité, profession et résidence de l'exploitant, s'il s'agit d'un particulier;
  - la raison sociale, le siège social, le nom du Directeur ou du Gérant et la liste des associés, s'il s'agit d'une société;
  - l'indication de la situation, les limites et la superficie de la portion de forêt sollicitée;
- cinq exemplaires de la carte géographique de la zone sollicitée obtenus auprès du Centre Géographique National;
- un extrait du dépôt au Greffe de la Cour d'Appel compétente de l'empreinte du marteau forestier du postulant. Cet extrait doit porter le fac-similé de l'empreinte;
- le programme d'exploitation, le matériel disponible ou à mettre en oeuvre, la consistance des établissements industriels installés ou envisagés, les productions prévues par année budgétaire et par catégorie de produit, la composition de la main-d'oeuvre et, le cas échéant, le programme de formation de celle-ci;
- un extrait du casier judiciaire ayant moins de trois mois à la date de la demande, si le postulant est une personne physique;
- une expédition authentique des status de la société et les pouvoirs du signataire de la demande et un extrait du casier judiciaire dudit signataire ayant moins de trois mois à la date de la demande, si l'exploitation est sollicitée par une personne morale;
- l'extrait du casier judiciaire du Directeur de l'exploitation forestière ayant moins de trois mois à la date de signature ainsi que son curriculum vitae;

- une déclaration sur l'honneur sur papier timbré spécifiant que le postulant :
  - exploitera lui-même et qu'il n'affermera pas son exploitation;
  - coopérera avec l'Administration chargée des forêts lors du contrôle de ses chantiers d'exploitation et de ses usines notamment en acceptant de signer tous les carnets de contrôle, et en laissant libre accès aux agents commis à cet effet;
  - a pris connaissance de la réglementation forestière en vigueur et qu'il s'engage à la respecter;
  - se conformera strictement au plan d'investissement, au programme de recrutement et de formation de la main-d'oeuvre ainsi qu'aux clauses de ses cahiers des charges;
- une copie certifiée conforme de l'acte d'agrément.

Art.31.- (1) Le dossier retenu par la Commission Technique est transmis au Gouverneur de la Province concernée pour les formalités d'affichage, de publication et la tenue d'une réunion d'information dans un délai ne dépassant pas trois mois.

(2) Dans les 30 jours suivant la date d'affichage, les oppositions et avis des populations sont reçues par le ou les Sous-Préfets ainsi que le Responsable Provincial de l'Administration chargée des forêts. Après ce délai, le Préfet compétent dispose d'une nouvelle période de 30 jours pour organiser une réunion d'information au public.

(3) Cette réunion détermine la route à inscrire au cahier des charges de l'exploitant, et examine les avis des populations. Elle les informe également sur les dispositions réglementaires en matière d'exploitation forestière ainsi que sur les droits et les obligations de l'exploitant forestier.

Participent à cette réunion :
- le Préfet ou son représentant .... Président
- un Député à l'Assemblée Nationale .... Membre
- Le Responsable Provincial de l'Administration chargée des forêts .... Rapporteur
- Le Sous-Préfet ou Chef de District intéressé .... Membre
- le ou les Maires des Communes concernées . "
- les Chefs Traditionnels et les Notabilités concernées .... "
- l'Exploitant Forestier concerné ou son représentant .... "

(4) Si la demande de licence couvre plusieurs départements, une réunion d'information est tenue au niveau de chaque département intéressé.

(5) Le procès-verbal de la réunion est rédigé séance tenante et signé de tous les membres. Il est adressé au Ministre chargé des Forêts pour établissement du cahier des charges qui est signé conjointement avec l'exploitant forestier et enregistré par les soins de ce dernier.

Art.32.- Les licences d'exploitation forestière sont accordées par arrêté :
- du Ministre chargé des forêts si la superficie à exploiter est inférieure ou égale à 10 000 ha;
- du Premier Ministre si la superficie à exploiter est supérieure à 10 000 ha.

Art.33.- Avant la notification, le titulaire de la licence doit justifier du paiement :
- de la taxe d'agrément;
- du dépôt du cautionnement.

B. Droits et obligations résultant de l'exploitation d'une licence

Art.34.- (1) La licence confère à son titulaire le droit d'exploiter exclusivement les bois destinés à l'exportation ou à la transformation locale, sous réserve des restrictions propres à certaines essences résultant de la réglementation forestière en vigueur, ou du cahier des charges.

(2) La licence ne confère notamment à son titulaire aucun droit à l'exploitation d'essences spéciales, de perches, de bois de chauffage et à charbon ou de produit forestier secondaire.

Art.35.- L'octroi de toute licence d'exploitation forestière est assorti :

1- d'un cahier des charges dont les clauses particulières précisent notamment l'importance, le lieu d'implantation ainsi que le délai d'installation des équipements industriels de transformation locale pour les licences de plus de 20 000 ha ; ce délai ne peut être supérieur à 24 mois à compter de la date d'octroi de la licence ;

2- des charges financières suivantes :
  a) le prix de vente des bois, calculé par m³ de bois, selon que les grumes sont exportées ou transformées localement. Le prix de vente des bois destinés à l'exportation est perçu pour partie aux taux frappant les bois destinés à la transformation locale à la sortie du chantier d'exploitation, et pour partie aux taux de sortie du territoire.
  b) la redevance territoriale qui est annuelle et calculée par hectare de forêt concédée;
  c) la redevance de reforestation qui est annuelle et calculée par hectare de forêt concédée;
  d) la contribution aux travaux de développement forestier qui est annuelle et se calcule par hectare de forêt concédée;
  e) la participation à la réalisation d'infrastructures socio-économiques qui se calcule par mètre cube de bois exploité.

Art.36.- (1) L'exploitation d'une forêt par licence se fait par chantier de 2 500 ha, chaque chantier constituant une assiette de coupe.

L'exploitation effective d'une assiette de coupe ne peut commencer qu'après notification de la licence par l'Administration chargée des forêts. Elle donne lieu à la délivrance d'un certificat d'assiette de coupe valable un an et renouvelable à la requête de l'exploitant sur présentation d'une carte du Centre Géographique National en cinq exemplaires.

L'attribution d'une nouvelle assiette de coupe se fait après fermeture par l'Administration chargée des forêts de celles antérieurement ouvertes à l'exploitation. A cet effet, la demande d'attribution d'une nouvelle assiette de coupe est adressée au Ministre chargé des forêts appuyée d'une attestation délivrée par le responsable provincial de l'Administration forestière certifiant que les chantiers antérieurement ouverts sont épuisés et qu'ils ont été exploités dans le respect des prescriptions du cahier des charges.

(2) L'exploitant forestier est tenu de matérialiser les limites des assiettes de coupe par des layons pour faciliter le contrôle permanent de son exploitation par les agents de l'Administration Forestière.

(3) Chaque assiette de coupe ouverte à l'exploitation doit être épuisée dans un délai maximum de 3 ans et fermée par l'Administration Forestière. Après cette fermeture, il est interdit à l'exploitant forestier d'y retourner à nouveau.

Art.37.- L'exploitant forestier est tenu de mentionner sur les carnets de chantier le diamètre pris à 1,30 m du sol ou le diamètre pris juste au-dessus des contreforts de chaque arbre abattu.

Il ne peut abattre que les arbres figurant dans son cahier des charges.

Art.38.- Les normes des installations industrielles de transformation locale du bois à implanter par l'exploitant forestier, compte tenu des superficies de forêt qui lui ont été concédées, sont les suivantes :

Superficie totale de la forêt concédée en ha	Qualité des installations industrielles à mettre en service	% minimum du volume de bois devant être transformé localement
moins de 20 000	pas de normes fixes	60% au moins
de 20 000 à 60 000	au moins une usine de sciage d'une capacité annuelle de transformation supérieure à 25 000m³ grumes	60% au moins
de 60 000 à 100 000	au moins une usine de sciage d'une capacité annuelle de transformation supérieure à 50 000 m³ grumes	60% au moins

de 100 000 à 150 000	au moins une usine de sciage d'une capacité de transformation supérieure à 50 000 m³ de grumes ou toute autre unité de transformation jugée au moins équivalente	60% au moins
de 150 000 à 200 000	un complexe industriel comportant au moins soit une usine de déroulage, soit une usine de fabrication de contre-plaqués ou de panneaux et doublée d'une usine de sciage ou de toute autre unité de transformation	60% au moins

(2) Les exploitants forestiers en activité à la date de signature du présent décret disposent d'un délai de deux ans pour se conformer aux dispositions de l'alinéa ci-dessus. Passé ce délai, les superficies totales détenues seront d'office réduites pour les conformer à ces normes.

(3) Compte tenu de l'incidence économique et sociale de l'implantation industrielle de transformation locale du bois, le lieu de son implantation est déterminée en accord avec les autorités administratives locales.

Art.39.- L'implantation de toute unité de transformation industrielle locale des produits forestiers est subordonnée à l'autorisation pralable des Administrations compétentes. Pour faciliter le contrôle de ses activités, elle doit être enregistrées auprès de l'Administration Forestière.

Art.40.- Les titulaires de licences d'exploitation forestière sont tenus de sortir de la forêt toutes les grumes provenant des arbres abattus, sauf celles jugées inutilisables par les agents de l'Administration chargée des forêts. Les arbres brisés à l'abattage et abandonnés en forêt ne sont pas dispensés de la perception du prix de vente prévu par la loi des finances pour le bois transformé localement. Les agents de l'Administration chargée des forêts tiennent, dans chaque chantier d'exploitation, l'inventaire des arbres abattus et abandonnés.

Art.41.- (1) Tout transport de bois d'oeuvre notamment des grumes non revêtues des marques réglementaires prescrites dans le cahier des charges est interdit.

(2) En cas de transport par route, le conducteur du grumier doit être muni d'une lettre de voiture extraite d'un carnet à souche du modèle réglementaire.

Les agents de l'Administration chargée des forêts peuvent, à tout moment, effectuer des contrôles permanents ou inopinés pour s'assurer de la conformité des documents présentés avec les produits transportés.

(3) La circulation des grumiers par route est réglementée par un arrêté conjoint des Ministres chargés des Transports et des Forêts.

Art.42.- L'ouverture des voies d'évacuation traversant une forêt du domaine national non concédée en licence est subordonnée à l'autorisation préalable du Ministre chargé des forêts. L'exploitant peut être autorisé à récupérer les arbres abattus sur l'emprise de la voie, moyennant paiement du prix de vente de ces bois. Le taux applicable de ce prix de vente est celui fixé pour les grumes récupérées.

Art.43.- Les grumes transportées par chemin de fer font l'objet d'une déclaration spéciale dont une copie est adressée par le Chef de Gare concerné au Responsable Provincial de l'Administration chargée des forêts.

Cette déclaration mentionne le nom de l'expéditeur des grumes, le nom de la gare expéditrice, le nombre des grumes par essence, leur destination, leur volume et leur poids.

C. Du renouvellement de licence

Art.44.- Le renouvellement d'une licence tel que prévu à l'Art.28 de la Loi N°81/13 du 27 novembre 1981 est, sur avis de la Commission Technique, sanctionné par l'autorité l'ayant délivrée. Le dossier de renouvellement qui doit être déposé auprès du Responsable provincial de l'Administration chargée des forêts revêtu de son avis motivé, comprend les pièces suivantes :

- une demande timbrée indiquant les nom, prénom ou raison sociale, nationalité de l'exploitant;
- une copie conforme de l'acte accordant la licence sollicitée en renouvellement;
- une carte en cinq exemplaires de la zone concernée établie par le Centre Géographique National;
- les pièces attestant le paiement de toutes les taxes et redevances grevant la licence;
- un rapport établi par le responsable provincial de l'Administration chargée des forêts justifiant que l'intéressé a transformé au moins 60% de sa production localement;
- une attestation certifiant le dépôt du cautionnement valable pour cinq ans.

Art.45.- (1) Le dossier de renouvellement est déposé au moins six mois avant l'expiration de la licence au Service Provincial de l'Administration chargée des Forêts contre récépissé.

(2) La licence ne peut être renouvelée que si son titulaire s'est conformé à la réglementation forestière et aux clauses du cahier des charges y relatif.

(3) A chaque renouvellement, il est établi un nouveau cahier des charges.

Art.46.- Toute licence dont la demande de renouvellement n'est pas déposée avant la date d'expiration est considérée comme abandonnée. L'exploitation est, à compter de la même date, arrêtée et la procédure de retrait engagée. Ce retrait ne dispense pas l'exploitant forestier du paiement des charges au titre des périodes échues.

Art.47.- Le renouvellement d'une licence porte sur la totalité de sa superficie de forêt qu'elle couvre. Cependant, l'exploitant ne peut revenir dans une assiette de coupe précédemment fermée à l'exploitation que sur autorisation du Ministre chargé des forêts, notamment après que des jeunes arbres de cette zone auront atteint le diamètre d'exploitabilité réglementaire.

D. Du transfert de licence

Art.48.- (1) Toute personne physique ou morale sollicitant le transfert à son profit du droit d'exploitation d'une zone précédemment accordée par licence à un autre exploitant forestier, doit être elle-même préalablement agrée à la profession forestière.

(2) Les demandes de transfert sont simultanément adressées par les deux parties au Ministre chargé des forêts.

(3) En outre, le bénéficiaire du transfert doit introduire le dossier réglementaire d'octroi de licence prévu à l'Art. 30 ci-dessus.

(4) En cas d'accord sur le principe du transfert, le bénéficiaire doit compléter ce dossier par la production d'une quittance attestant le paiement de la taxe de transfert et du cautionnement.

Art.49.- (1) Le transfert est sanctionné par l'autorité ayant accordé la licence. Il porte sur la totalité de la zone couverte par la licence.

(2) Toutes les clauses du cahier des charges non encore exécutées par le précédent détenteur de la licence incombent au bénéficiaire du transfert.

De nouvelles clauses particulières peuvent être établies compte tenu des nouvelles superficies résultant du transfert.

(3) L'exploitation de la zone de forêt transférée ne peut commencer qu'après notification de l'acte de transfert par l'Administration chargée des forêts et acquittement de tous les droits ou taxes afférents à ce transfert.

E. Abandon de licence

Art.50.- (1) L'abandon de licence est sanctionné par arrêté de l'autorité l'ayant accordée.

(2) Le dossier de demande d'abandon est déposé auprès du Ministre chargé des forêts. Il comprend les pièces suivantes :

- une demande timbrée indiquant les nom, prénom, ou raison sociale, nationalité, résidence ou siège social de l'exploitant;
- une copie conforme de l'acte ayant accordé la licence dont l'abandon est sollicité;
- une carte du Centre Géographique National en cinq exemplaires de la zone concernée;
- les pièces attestant le paiement de toutes les taxes et redevances grevant la licence;
- une attestation du responsable provincial des forêts certifiant l'arrêt effectif du chantier.

(3) L'abandon porte sur la totalité de la zone forestière couverte par la licence;

(4) L'abandon ne dispense pas l'exploitant du paiement des charges au titre des périodes échues.

Art.51.- En cas du décès d'une personen titulaire d'une licence d'exploitation forestière, cette licence est retirée sauf si l'ayant-droit a été agréé à la profession forestière dans un délai de 18 mois, auquel cas la licence est transférée à son nom.

Sous-Section III : Exploitation par vente de coupe

Art.52.- (1) L'exploitation des forêts du domaine national par vente de coupe s'effectue :

- dans les zones nécessitant une coupe de sauvetage avant leur mise en valeur ou dans celles ayant déjà fait l'objet d'un inventaire par les soins de l'Administration forestière;
- dans les zones enclavées dont la superficie n'exède pas 2 500 ha;
- dans les cas soit d'ouverture de pistes, ou de layons dans les forêts du domaine national non attribué par licence, soit des travaux publics ou d'installation de sociétés industrielles ou de développement nécessitant l'abattage des arbres. Les programmes des travaux doivent dans ce cas être communiqués au Ministre chargé des forêts au moins six mois à l'avance pour lui permettre d'organiser l'extraction préalable des bois exploitables.

(2) Le rythme des coupes est fixé par le Ministre chargé des forêts.

Art.53.- Toute personne ou société qui, en vertu de la législation foncière obtient un titre de propriété sur une zone de forêt du domaine national, est tenue de soumettre avant toute mise en valeur, cette zone à l'exploitation par coupe organisée par l'Administration forestière.

Si elle est elle-même agréée à la profession forestière, cette coupe est organisée de préférence à son profit.

Sous-Section IV : Exploitation par permis

A. Permis spéciaux

Art.54.- (1) Les produits forestiers secondaires, notamment, le bois, les racines, l'écorce des tiges ou des racines, les feuilles, les fruits, la sève ou tout autre partie de certaines essences présentant soit certaines propriétés à caractère médicinal, soit un intérêt économique particulier pour certains usages, ne peuvent être exploités même à l'intérieur d'une forêt concédée en exploitation qu'avec un permis spécial.

(2) La liste de ces essences, dites spéciales, est fixée par arrêté du Ministre chargé des forêts.

(3) Le permis spécial est personnel et incessible.

Art.55.- (1) Le permis spécial est accordé pour une durée d'un an, par arrêté du Ministre chargé des forêts, après avis de la Commission Technique. Toutefois, des permis de plus longue durée peuvent être accordés dans certains cas particuliers, notamment en cas d'installation d'unité de transformation locale des produits.

(2) Le permis spécial indique notamment les quantités de produits à récolter, la liste des essences spéciales dont l'exploitation est autorisée, la zone d'exploitation, les conditions d'exportation ou d'utilisation locale des produits, ainsi que les conditions de son renouvellement.

(3) Les quotas annuels de chaque type de produit à exploiter par l'ensemble des titulaires de permis spéciaux est fixé en début de chaque campagne par arrêté du Ministre chargé des forêts.

Art.56.- L'obtention d'un permis spécial est subordonnée à la présentation d'un dossier comprenant :

1° une demande timbrée précisant :
- les nom, prénom, nationalité, profession et résidence, s'il s'agit d'un particulier;
- la raison sociale, les statuts, le siège social, le capital social et sa répartition, le nom du Directeur ou du Gérant, s'il s'agit d'une société.

2° les moyens financiers engagés;

3° Les investissements prévus et la garantie de leur financement. Ces investissements doivent indiquer notamment :
- les moyens de transport envisagés;
- les magasins de stockage existants et autres moyens à mettre en oeuvre pour assurer une bonne conservation des produits;
- les dispositions prises en vue de transformer localement une partie de la production;

4° La liste des essences à exploiter, les quantités de produits à récolter ainsi que les lieux de récolte;

5° Un extrait du dépôt de l'empreinte du marteau forestier s'il s'agit de grumes;

6° Une déclaration sur l'honneur spécifiant que le demandeur a pris connaissance de la réglementation en vigueur, qu'il s'y conformera et collaborera avec l'Administration chargée des forêts pour le contrôle de son activité.

Art.57.- En cas de renouvellement du permis, le dossier doit comporter les pièces ci-après :
- une demande timbrée;
- une copie de l'ancien permis;
- une copie de chacun des certificats d'origine si le titulaire s'est livré à l'exportation des produits;
- les quittances attestant le paiement de la taxe de reforestation et du prix de vente des produits;
- un rapport circonstancié sur les activités de la campagne écoulée avec les précisions sur les quantités des produits qui ont été exportées ou transformées localement.

Art.58.- Tout détenteur d'un permis spécial souscrit auprès de l'Administration forestière, un cahier des charges dont les clauses indiquent notamment :
- les conditions d'exploitation des produits;
- les conditions de leur transport;
- les modalités de paiement des taxes.

Art.59.- (1) Les détenteurs des permis spéciaux sont tenus de livrer à la transformation locale au moins 60% de leurs produits.
(2) A la fin de chaque campagne, l'exploitant adresse un rapport de ses activités à l'Administration forestière.

Art.60.- L'exportation des produits forestiers secondaires est subordonnée à l'obtention d'un certificat d'origine délivré par le Ministre chargé des forêts. Ce dossier comprend :
- les pièces attestant le paiement des taxes;
- une copie du permis spécial;
- une copie de l'acte d'agrément à la qualité d'exportateur;
- des attestations délivrées par une ou plusieurs unités de transformation, permettant de certifier que le titulaire du permis a livré au moins 60% de sa production à la transformation locale.

Art.61.- (1) Pour faciliter les contrôles de l'Administration forestière les produits récoltés sont soumis, le cas échéant, à l'obligation de marquage ou de numérotage notamment lorsqu'il s'agit des billes.

(2) L'exploitant doit en outre tenir un cahier de chantier.

(3) La circulation des produits est accompagné d'une lettre de voiture.

B. Permis et autorisation de coupe d'arbres

Art.62.- (1) Dans les forêts du domaine national non concédés en licence, tout abattage d'arbres protégés est interdit.

(2) Toutefois, en vue de satisfaire leurs besoins domestiques notamment en bois de chauffage et de construction, des nationaux résidant dans les zones concernées peuvent abattre un nombre limité d'arbres.

Art.63.- (1) Lorsque les nationaux se livrent à l'exploitation artisanale des forêts du domaine national dans le but de commercialiser les perches, le bois de construction de chauffage ou de charbon de bois, ils doivent être titulaires d'un permis de coupe d'arbres.

(2) A cet effet, ils doivent fournir un dossier comprennant les pièces suivantes :
- une demande timbrée précisant les motifs de la demande du permis;
- une copie de la carte nationale d'identité;
- la liste des essences sollicitées ainsi que leur localisation;
- le cas échéant, une copie du certificat d'enregistrement comme transformateur artisanal de bois.

Ce dossier est transmis au Ministre chargé des forêts par le responsable provincial de l'Administration forestière avec son avis motivé.

(3) Les permis de coupe d'arbres est délivré par le Ministre chargé des forêts après étude du dossier et paiement au taux fixé par la loi des finances par le postulant du prix de vente du bois dont la coupe est sollicitée.

(4) Les arbres dont l'abattage est autorisé sont préalablement marqués par les Agents de l'Administration forestière.

(5) Le permis ordinaire de coupe d'arbres est accordé pour une durée d'un an. Il peut être renouvellé par l'autorité l'ayant délivré après vérification que le titulaire a respecté les règles d'exploitation qui lui avaient été fixées.

Art.64.- Le permis de coupe d'arbres peut être accordé à l'intérieur d'une zone concédée en licence pour les essences ne faisant pas l'objet d'une exploitation commercialle ou pour celles que le titulaire de la licence n'est pas disposé à exploiter.
Toutefois, ce permis ne peut être accordé sur une assiette de coupe fermée à l'exploitation forestière.

Art.65.- Toute exploitation par permis de coupe doit respecter les clauses du cahier des charges y afférentes, notamment les diamètres d'exploitabilité fixés par le Ministre chargé des forêts.

Section III : Contrôle de l'exploitation forestière

Art.66.- (1) Tout titulaire de licence doit tenir, par assiette de coupe, un carnet de chantier dont le modèle est établi par l'Administration forestière. Ce carnet est visé et paraphé par le Responsable Départemental de cette administration. Dans le carnet de chantier sont inscrits chaque jour les arbres abattus avec indication du diamètre pris à 1,30 m du sol ou au-dessus des contreforts, le numéro d'abattage figurant sur la souche de l'arbre, la longueur des grumes, leur diamètres aux gros et fins bouts, et leur volume. Le carnet de chantier est signé conjointement par l'exploitant et l'agent de l'Administration forestière affecté au chantier.

(2) L'agent de l'Administration forestière affecté au chantier procède au martelage de toutes les billes avant leur sortie de forêt.

Si l'arbre est abandonné en forêt après l'abattage, le motif de l'abandon est mentionné dans le carnet de chantier.

(3) A la fin de chaque semaine, les feuillets du carnet de chantier sont transmis au responsable départemental de l'Administration forestière qui, après vérification et compilation, transmet mensuellement les résultats et les spécifications au responsable provincial de l'Administration forestière pour le calcul du prix de vente des produits et établissement des sommes dues à recouvrer par les soins des services du Trésor.

(4) Pour le cubage, le volume de chaque bille est calculé d'après le barême confectionné à partir de la formule suivante :

$$V = \frac{Pi}{4} \times D2 \times L$$

ou V = volume de la bille
L = longueur de la bille
D = diamètre de la bille sous écorce

$$\frac{Pi}{4} = 0,785$$

Le volume est exprimé en mètres cubes suivis de trois décimales.

La longueur est exprimée en mètres, décimètres et centimètres couverts.

Le diamètre est la zone moyenne arithmétique des diamètres des deux bouts.

Art.67.- Durant la période de validité de la licence, le titulaire doit adresser au Ministre chargé des forêts, un rapport indiquant :
- la destination des produits transformés localement;
- le programme d'activité envisagé pour l'année suivante;
- la main-d'oeuvre actuelle et envisagée ainsi que sa composition.

Art.68.- Pour les grumes transformées localement, il est tenu dans chaque usine de transformation un registre à souches paraphé qui enregistre toutes les entrées. Ce registre doit être présenté pour vérification à toute réquisition à l'agent de l'Administration forestière commis au contrôle.

Chronologiquement, ce registre mentionne, par essence, le numéro de l'arbre, le diamètre, le volume et le nombre de grumes entrées à l'usine ainsi que l'indication de leur chantier d'origine.

Aucune grume ne doit être admise à l'usine si elle ne porte de manière visible, les marques réglementaires dont elle doit être revêtue avant sa sortie du chantier d'exploitation.

Art.69.- Les exploitants forestiers, les exportateurs de produits forestiers, les transformateurs locaux de grumes sont tenus de contresigner suivant le cas, les bulletins de contrôle, les registres d'entrée à souche ainsi que les rapports de contrôle établis par les agents de l'Administration forestière qui visitent leur exploitation.

TITRE III : DE L'EXPORTATION DES BOIS EN GRUMES ET DE LA PROMOTION DES ESSENCES ET PRODUITS FORESTIERS

Art.70.- (1) A l'effet d'exporter du bois en grumes, les nationaux pris individuellement ou regroupés en société doivent être :
- agréés à la profession forestière;
- titulaires d'un titre d'exploitation forestière,
- enregistrés comme exportateur auprès de l'Administration forestière.

(2) Les non-nationaux qui veulent se livrer à l'exportation des bois en grumes doivent en plus des conditions prévues au paragraphe précédent, justifier de l'installation préalable d'une industrie de transformation locale.

Art.71.- Le Ministre chargé des forêts fixe pour chaque exportateur le quota des produits forestiers bruts ou transformés exportables compte tenu des besoins du marché national.

Art.72.- Les exportateurs des produits forestiers bruts ou transformés sont tenus d'adresser au Ministre chargé des forêts un rapport annuel indiquant notamment :
- la nature des produits forestiers exportés;
- les quantités exportées par essence et par destination;
- la provenance des produits.

Art.73.- En plus des actions entreprises par l'organisme compétent en matière de promotion du bois, il est publié chaque année par l'Administration des forêts et tenu à la disposition des exploitants forestiers une liste d'essences en promotion.

Par essence en promotion, il faut entendre les essences peu ou pas connues pour lesquelles des propriétés technologiques satisfaisantes pour leur utilisation ont été mises en évidence par des instituts spécialisés et qui se trouvent en quantité économiquement exploitables dans les forêts.

La loi des finances fixe chaque année un taux préférentiel pour ces essences en promotion.

TITRE IV : DISPOSITIONS DIVERSES

CHAPITRE I : DES BOIS ECHOUES SUR LA COTE ATLANTIQUE

Art.74.- (1) On entend par billes échouées, celles des essences, sans marques locales apparentes, exploitées hors du territoire national et parvenues au hasard dans les eaux territoriales de la Côte Atlantique du Cameroun.

(2) La récupération de ces billes telle que prévue à l'Art.42 de la Loi 81/13 du 27 novembre 1981 est sujette à l'obtention préalable d'une autorisation écrite du responsable provincial de l'Administration des forêts. Cette autorisation ne peut être délivrée qu'après martelage et cubage des bois et paiement par l'intéressé d'une taxe de récupération dont le taux est fixé par la loi des finances.

CHAPITRE II : DES PRISES DE PARTICIPATION

Art.75.- (1) Les prises de participation et les cessions de parts des capitaux des sociétés d'exploitation forestière doivent être autorisées par le Ministre chargé des forêts.

(2) Ces transactions doivent obéir aux règles suivantes :

- Lorsqu'il s'agit d'une société constituée par des nationaux, la part de capital détenue par des non-nationaux, soit du fait des cessions, soit à la suite des augmentations de capital ne doit pas être supérieure à 30% du capital social.
- Lorsqu'il s'agit d'une société constituée par des nationaux et des non-nationaux, les modifications ultérieures du capital de la société, soit du fait des cessions des parts, soit à la suite des augmentations de capital ne doivent pas avoir pour effet de baisser le pourcentage des parts, détenues par les nationaux tel que fixé dans le capital social initial.

- Lorsqu'il s'agit d'une société constituée par des non-nationaux, les modifications ultérieures du capital de la société au profit des non-nationaux non agréés à la profession forestière pris individuellement ou en société, soit du fait des cessions de parts, soit à la suite des augmentations de capital, ne doivent pas porter sur plus de 30% du capital social initial.

Art.76.- (1) Dans tous les cas, toute prise de participation ou cession de parts est subordonnée à l'autorisation préalable du Ministre chargé des forêts sur présentation d'un dossier comprenant les pièces suivantes :

- une demande timbrée précisant les motifs de la prise de participation;
- une fiche de renseignements du cédant et du cessionnaire;
- les statuts actuels de la société ainsi que la répartition actuelle et prévue du capital social;
- le procès-verbal de l'Assemblée Générale au cours de laquelle les nouvelles prises de participation ont été agréées.

(2) L'acte authentique des changements intervenus est communiqué au Ministre chargé des forêts.

CHAPITRE III : DES FEUX DE BROUSSE

Art.77.- (1) Il est interdit de provoquer un feu susceptible de se propager dans la brousse et de détruire la végétation. Tout feu provoqué doit être maîtrisé par son auteur.

(2) Les Gouverneurs de Provinces, par arrêté pris sur proposition des responsables provinciaux de l'Administration forestière, réglementent les feux de brousse et fixent notamment les dates et les conditions d'allumage des feux précoces.

CHAPITRE IV : DU CONSTAT DES INFRACTIONS ET DES TRANSACTIONS

Art.78.- Omis
Art.79.- Omis
Art.80.- Omis

CHAPITRE V : DISPOSITIONS TRANSITOIRES ET FINALES

Art.81.- Les personnes physiques ou morales enregistrées en qualité d'exportateurs des produits forestiers et en activité à la date de signature du présent décret disposent d'un délai d'un an pour se conformer à ses dispositions.

Art.82.- Un texte particulier détermine les modalités de port d'armes et de l'uniforme ainsi que les règles particulières de discipline auxquelles sont astreints les personnels de l'Administration forestière.

Art.83.- (1) Les personnes physiques ou morales agréées à la profession forestière à la date de signature du présent décret conservent cette qualité.

(2) Les autres exploitants forestiers dont les dossiers d'agrément sont en cours ne pourront bénéficier de l'agrément que s'ils sont en règle vis à vis de la législation forestière.

Art.84.- Le présent décret qui abroge toutes dispositions antérieures notamment le décret N°74/357 du 17 avril 1974, sera enregistré puis publié au Journal Officiel en français et en anglais.

YAOUNDE, le 12 avril 1983

LE PRESIDENT DE LA REPUBLIQUE

DECRET N° 82-636 DU 8 DECEMBRE 1982

CAMEROUN

Portant création de l'Office National de Régénération des Forêts

CAMEROON

Décret N° 82-636

LE PRESIDENT DE LA REPUBLIQUE

VU la Constitution
VU la Loi N° 81-13 du 27 novembre 1981 portant régime des Forêts, de la Faune et de la Pêche

DECRETE

Chapitre I : DISPOSITIONS GENERALES

Article 1. - Il est créé un établissement public à caractère industriel et commercial, doté de la personnalité juridique et de l'autonomie financière dénommé Office National de régénération des Forêts, en abrégé "ONAREF".

Article 2. - 1) L'office national de régénération des Forêts est placé sous la tutelle du Ministère chargé des Forêts.

2) Il est classé à la 2è Catégorie des établissements publics.

3) Le siège de l'ONAREF est fixé à Yaoundé. Des agences et succursales peuvent être créées à l'intérieur du territoire national en tant que besoin, par le conseil d'administration.

Article 3. - L'Office National de régénération des Forêts a pour objet, la mise en oeuvre de la politique du gouvernement en matière de régénération des forêts, de reboisement, de conservation et de restauration des sols.

A ce titre il est notamment chargé :
- de la régénération des forêts domaniales en vue de l'accroissement de leur productivité, dans le cadre des plans d'aménagement approuvés par le ministre de tutelle;
- de l'exécution des projets de reboisement, de protection et restauration des sols, de lutte contre les effets de sécheresse;
- de l'étude et de l'exécution, sur leur demande et sur leur financement, des projets de reboisement des particuliers et des collectivités locales;
- de la réalisation de toutes opérations financières, commerciales, industrielles, mobilières et immobilières se rattachant à son objet.

Article 4. - A l'exception de ceux effectués à la demande des collectivités locales ou des particuliers, les travaux de l'ONAREF s'exécutent dans les forêts domaniales et les périmètres faisant l'objet d'une procédure régulière de classement.

Chapitre II : ADMINISTRATION

Article 5. - L'administration de l'office national de régénération des Forêts est assurée par les organes suivants :
- un conseil d'administration
- une direction générale.

Du Conseil d'Administration

Article 6. - 1) Le conseil d'administration de l'ONAREF est composé ainsi qu'il suit :
Président : - une personnalité nommée par décret;
Membres :
- un représentant des Services du Premier ministre
- " du Ministère de l'Agriculture
- " " de l'Urbanisme et de l'Habitat
- " " de l'Economie et du Plan
- " " des Finances
- " " de l'Equipement
- " de la Chambre d'Agriculture, de l'Elevage et des Forêts
- " de la Délégation générale à la Recherche Scientifique et Technique
- le Directeur Général du CENADEFOR
- une personnalité désignée par le Chef de l'Etat en raison de sa compétence.

2) Le président du conseil d'administration peut inviter toute personne à prendre part aux délibérations du conseil d'administration avec voix consultative, en raison de sa compétence.

3) Le Directeur Général de l'ONAREF assure le secrétariat du Conseil d'Administration.

Article 7. - 1) Les fonctions d'administrateur sont gratuites. Toutefois, les administrateurs et les personnes appelées en consultation perçoivent une indemnité de session, conformément aux textes en vigueur.

2) Il est alloué au Président du Conseil d'Administration, une indemnité mensuelle, conformément aux textes en vigueur.
A l'occasion des réunions ou des missions spéciales, les frais de transport et de séjour des administrateurs et des personnes invitées en application de l'art. 6 paragraphe (2) ci-dessus, sont à la charge de l'ONAREF.

Article 8. - Le conseil d'administration se réunit deux fois par an, sur convocation de son Président.
Il peut se réunir en session extraordinaire, sur autorisation écrite de l'autorité de tutelle.

Article 9. - Le conseil d'administration ne peut délibérer valablement qu'en présence des deux tiers au moins de ses membres.

Tout membre empêché peut déléguer, par écrit, ses pouvoirs à un membre. Toutefois, le mandataire ne peut avoir plus de deux voix, y compris la sienne.
Les décisions sont prises à la majorité simple des membres présents ou représentés. En cas de partage des voix, celle du Président est prépondérante.

Les délibérations du conseil d'administration font l'objet d'un procès-verbal signé par le Président et par le Secrétaire de séance, et consigné dans un registre spécial tenu au siège de l'ONAREF.

Les décisions du conseil d'administration sont rendues exécutoires après approbation de l'autorité de tutelle.

Article 10. - 1) Le conseil d'administration est investi des pouvoirs les plus étendus pour la gestion et l'administration de l'ONAREF, à ce titre :
- il approuve l'organigramme, le statut du personnel, le réglement intérieur, les programmes et les comptes-rendus d'activités;
- il vote le budget;
- il approuve les comptes et bilans;
- il recrute, nomme et révoque le personnel cadre;
- il autorise les emprunts;
- il passe tous contrats et conventions;
- il accepte les dons et legs;
- il autorise la passation des marchés de fournitures et de travaux, conformément à la réglementation.

2) Le conseil d'administration peut déléguer une partie de ses pouvoirs au Directeur Général .
Toutefois, ne peuvent faire l'objet de délégation :
- le vote du budget, et
- l'approbation des comptes et du bilan.

Article 11. - 1) L'autorité de tutelle peut, en tant que de besoin, autoriser le conseil d'administration à désigner en son sein un comité ad hoc chargé d'émettre un avis technique sur le fonctionnement et la réalisation des programmes de l'ONAREF.

2) Les membres du comité jouissent des mêmes avantages que les membres du conseil d'administration.

De la Direction Générale

Article 12. - La Direction de l'Office National de régénération des des Forêts est assurée par un Directeur Général, assisté d'un Directeur Général-Adjoint, tous deux nommés par décret.

Article 13. - 1) Le Directeur Général prépare les réunions du conseil d'administration; il exécute les décisions du conseil d'administration dont il reçoit à cette fin, les délégations de pouvoirs nécessaires.

2) Le Directeur Général assure la gestion administrative, financière et technique de l'ONAREF sous le contrôle du conseil d'administration.
En particulier, le Directeur Général :
- soumet à l'approbation du conseil d'administration, le statut du personnel, l'organigramme et le règlement intérieur;
- prépare et exécute le budget de l'ONAREF dont il est l'ordonnateur;
- recrute, nomme et révoque le personnel non cadre;
- gère le patrimoine de la société;
- représente l'ONAREF en justice et dans tous les actes de la vie civile.

Le Directeur Général peut déléguer une partie de ses pouvoirs au Directeur Général-Adjoint.
Le Directeur Général-Adjoint remplace le Directeur Général en cas d'empêchement ou d'absence de celui-ci.

Chapitre III : DISPOSITIONS FINANCIERES

A. Régime comptable et financier

Article 14. - 1) La gestion financière de l'ONAREF s'effectue dans le cadre d'un budget annuel approuvé par le conseil d'administration.

L'exercice budgétaire commence le 1er juillet de chaque année et se termine le 30 juin de l'année suivante.

2) Dans les trois mois qui suivent la clôture de l'exercice budgétaire, le Directeur Général établit le bilan au 30 juin, ainsi que le rapport d'activités, le rapport financier et les comptes d'emploi des subventions.

Article 15. - Les ressources de l'ONAREF sont constituées par :
- le produit des prélèvements fiscaux, conformément aux textes en vigueur;
- les subventions de l'Etat;
- le produit de ses activités;
- les emprunts;
- les dons et legs de toute nature.

B. Commission Financière

Article 16. - Il est créé auprès de l'ONAREF, une commission financière composée ainsi qu'il suit :
Président : un représentant du Ministère chargé de l'Inspection Générale de l'Etat.
Membres : un représentant du Ministère des Finances;
un représentant du Ministère de tutelle.

Article 17. - 1) La commission dispose de tous pouvoirs d'investigation et de contrôle sur les documents financiers et comptables de l'ONAREF.
En particulier, elle est chargée de vérifier livres, caisses et comptes bancaires; de contrôler la régularité et la sincérité des inventaires ainsi que l'exactitude des informations sur les comptes de l'ONAREF.
La commission peut, à tout moment, effectuer les vérifications ou les contrôles qu'elle juge opportuns. En cas d'urgence, elle peut demander la convocation du conseil d'administration.

2) La commission apure les comptes et rédige un rapport annuel sur la gestion de l'ONAREF; ce rapport est adressé au conseil d'administration.
Le Ministre de tutelle en reçoit le double.

Article 18. - Il est alloué aux membres de la commission financière une indemnité dont le montant est fixé par le conseil d'administration.

Chapitre IV : DISPOSITIONS DIVERSES

Article 19. - Les biens meubles et immeubles du Fonds National Forestier et piscicole sont dévolus à l'ONAREF, à l'exception de ceux relatifs à des activités piscicoles qui reviennent au Ministère de l'Elevage, des Pêches et des Industires Animales.

Article 20. - Le présent décret abroge toutes dispositions antérieures contraires, notamment celles du décret N° 74-935 du 15 novembre 1974 portant application de l'Ordonnance N° 73-18 du 22 mai 1973 fixant le Régime Forestier National.

Article 21. - Le Ministre chargé des Forêts est chargé de l'exécution du présent décret qui sera enregistré puis publié au Journal Officiel en français et en anglais.

Yaoundé, le 8 décembre 1982

LE PRESIDENT DE LA REPUBLIQUE

DECRET N° 81/223 DU 9 JUIN 1981

Portant création du Centre National de Développement Forestier

CAMEROUN

CAMEROON

Décret N° 81/223

LE PRESIDENT DE LA REPUBLIQUE

VU la Constitution de la République Unie du Cameroun;

VU le Décret N° 76/256 du ler juillet 1976 portant réorganisation du Ministère de l'Agriculture;

DECRETE

Chapitre I : DISPOSITIONS GENERALES

Article 1. - Il est créé, sous la dénomination "CENTRE NATIONAL DE DEVELOPPEMENT DES FORETS" en abrégé "CENADEFOR", un établissement public à caractère industriel et commercial doté de la personnalité juridique et de l'autonomie financière.

Article 2. - 1) Le CENTRE NATIONAL DE DEVELOPPEMENT DES FORETS est placé sous la tutelle du ministre chargé des Eaux et Forêts et son siège social est fixé à Yaoundé.

2) Il est classé à la 2è catégorie des Etablissements Publics.

3) Des Agences du Centre National de Développement des Forêts peuvent être créées par arrêté à l'intérieur du Territoire National, sur proposition du Conseil d'Administration.

Article 3. - Le Centre National de Développement des Forêts est chargé de la mise en valeur des forêts et de la promotion des Bois Camerounais tant à l'intérieur qu'à l'extérieur du Territoire.

Il a notamment pour objet de :
- réaliser une reconnaissance forestière générale;
- réaliser l'inventaire et l'aménagement du domaine forestier permanent;
- faire toutes expériences en vue de promouvoir la transformation et l'utilisation des essences encore peu connues;
- concevoir, à des fins commerciales, divers articles en bois;
- établir un mécanisme de coopération entre Gouvernement, exploitants, transformateurs et architectes, en vue d'accroître la proportion du matériau bois et des articles manufacturés en bois, notamment dans construction, l'ameublement et les travaux publics;
- assurer le perfectionnement technique des nationaux dans le domaine d'inventaire forestier, du travail et de l'industrie du bois;
- organiser des conférences et participer aux expositions au Cameroun ou à l'étranger;
- mettre au point une normalisation des produits transformés;
- assurer le conditionnement et établir les règles de classement des bois;
- faire toutes opérations commerciales, financières, industrielles, mobilières et immobilières se rattachant à son objet.

Article 4. - Le Centre National de Développement des Forêts est chargé de l'exécution des conventions ayant trait à la réalisation des objectifs fixés à l'art. 3 ci-dessus, passés entre la République Unie du Cameroun et les Pays et Organismes Etrangers.

Article 5. - Pour lui permettre de réaliser les objectifs qui lui sont assignés en matière de promotion du Bois, le Centre National de Développement des Forêts peut être autorisé à exploiter en régie certains massifs forestiers.

Chapitre II : ADMINISTRATION DU CENADEFOR

Article 6. - L'administration du CENADEFOR est assurée par les organes suivants :
- Un Conseil d'Administration;
- Une Direction.

(Art. 7 à 14 Omis)

Chapitre III : REGIME COMPTABLE ET FINANCIER

(Art. 15 à 18 Omis)

Chapitre IV : COMMISSION FINANCIERE

(Art. 19 à 22 Omis)

"Les dispositions des articles 7 à 22 sont analogues mutatis mutandis aux dispositions du Décret N° 82/636"

Chapitre V : DISPOSITIONS DIVERSES

Article 23. - Le Centre national de Développement des Forêts hérite de l'actif et du passif de l'ancien Projet Forestier à compter de la date de signature du présent décret.

Article 24. - Le présent décret sera enregistré et publié au Journal Officiel en français et en anglais.

Yaoundé, le 9 juin 1981

LE PRESIDENT DE LA REPUBLIQUE

REPUBLIQUE CENTRAFRICAINE

CENTRAL AFRICAN REPUBLIC

Loi N° 61/273

LOI N° 61/273 portant création d'un Code forestier centrafricain (version consolidée d'après la Loi N° 62/341 du 7.12.1962 et l'Ordonnance N° 71/045 du 27.3.1971)

TITRE I : GENERALITES

Article 1. - Dans ce qui suit, l'on appellera forêt par abréviation de terrain à vocation forestière, toutes superficies qui supportent des formations végétales sauvages et qui hébergent du cheptel sauvage.

Ces formations végétales sont les forêts sensu stricto, les savanes et les steppes dont les fruits exclusifs sont : les bois d'ébénisterie, d'industrie ou de service, les bois de feu et à charbon ou des produits accessoires tels que : les écorces et fruits à tanin, les écorces textiles et tinctoriales, le kapok, le caoutchouc, la glu, les résines, les gommes, les bambous, les palmiers spontanés, les drogues et tous autres végétaux ne constituant pas un produit agricole.

Le présent code concerne le régime propre aux forêts en général et aux formations végétales.

Le régime de la faune sauvage fait l'objet de textes spéciaux sur la chasse et la pêche.

TITRE II : DES DIFFERENTS DOMAINES FORESTIERS

Chapitre I : Définitions

Article 2. - En dehors des forêts du domaine public on distingue :

1) Les forêts du domaine classé qui comprennent :
   - les forêts du domaine privé de l'Etat (forêts domaniales, réserves de faune et de flore, les réserves intégrales, les parcs nationaux).
   - les forêts soumises qui sont celles appartenant aux collectivités ou établissements publics (forêts communales, rurales, etc.).

2) Les forêts du domaine coutumier sur lesquelles les collectivités coutumières ou les particuliers exercent librement des droits coutumiers non immatriculés.

3) Les forêts privées particulières qui comprennent celles sises sur les titres fonciers et celles sur l'étendue desquelles les droits coutumiers ont été immatriculés, ainsi qu'il est prévu au Code foncier, Section I.

Chapitre II : Gestion des forêts du domaine classé

Section 1 : Constitution

Article 3. - Sont constituées en forêts classées :

1) Toutes les forêts domaniales, les réserves de faune et de flore, les réserves intégrales et les parcs nationaux existant à la date de promulgation du présent code.
2) les forêts domaniales, ou soumises, qui seront classées à partir des forêts coutumières ou des forêts vacantes et sans maître après application de la procédure suivante.

Article 4. - La direction des Eaux et Forêts et Chasses, de sa propre initiative pour les forêts domaniales ou sur sollicitation de la part des collectivités ou établissements publics, procède à l'établissement d'un dossier qui fait ressortir :

- les données relatives à la situation et à l'étendue de la superficie
- son régime actuel
- les intérêts en cause et en particulier les droits coutumiers
- les buts d'intérêt général ou d'intérêt particulier, les buts économiques, sociaux, etc.

Ce projet est soumis au ministre chargé des Forêts qui, s'il l'approuve :

1) Désigne une commission de classement représentant tous les intérêts en cause, ayants-droit coutumiers, chefs de villages, etc. Elle est présidée par le sous-préfet du ressort ou le maire de la commune ou leur représentant et comportant un député comme observateur. Le secrétaire rapporteur est le directeur des Eaux, Forêts et Chasses ou son représentant.

2) Le fait insérer au "J.O.".

3) Le transmet au sous-préfet. Celui-ci l'affiche et le porte à la connaissance des intéressés par tous les moyens conformes aux réglements et usages locaux.

Deux mois après l'affichage, le sous-préfet réunit la commission qui :

1) S'assurera des limites, de la régularité de l'affichage et de la publicité, de la représentativité des membres de la commission.

2) Se prononcera sur la suite à donner au classement.

3) Procédera à l'examen des oppositions, s'il y a lieu, et au réglement des droits coutumiers.
   Pour ce dernier point, les titulaires des droits pourront :
   - Soit en faire l'abandon gracieux,
   - Soit en demander le rachat. Le paiement de ce rachat sera accordé obligatoirement sous forme d'un prélèvement sur le revenu en argent ou en produits matériels lorsque la forêt considérée sera rendue au stade de l'exploitation.

Il sera établi un procès-verbal des opérations qui sera transmis au ministre chargé des Forêts.
Celui-ci fera sanctionner le classement par une loi qui explicitera :

- la situation et l'étendue de la forêt
- le but du classement
- le réglement des droits coutumiers.

Article 5. - Au cas où la commission n'aurait pas donné une suite favorable ou si le réglement des usages s'avérait impossible, la procédure d'expropriation pour cause d'utilité publique est la seule applicable. Elle le sera aussi si le projet porte en tout ou en partie sur une superficie immatriculée en ce qui concerne le titre foncier considéré.

L'expropriation ne pourra toutefois être prononcée que si le projet a pour but :

- le maintien des terres sur les pentes,
- la défense du sol contre les érosions, les envahissements des fleuves et des rivières ou des torrents,
- d'assurer la défense des sources,
- pour la fixation des dunes,
- pour la salubrité publique,
- pour la défense militaire.

Article 6. - Les forêts du domaine classé ne pourront être aliénées ou ne pourront changer de régime que par l'intervention d'une loi prise sur proposition du ministre chargé des Forêts, après avis du directeur des Eaux, Forêts et Chasses.

Article 7. - Dans les forêts du domaine forestier classé, la prescription acquisitive ne jouera ni en ce qui concerne la propriété du sol ni en ce qui concerne les usages sous réserve que les actes de gestion aient été régulièrement effectués.

Section 2 : Règles générales de gestion

Article 8. - Les usages à caractère commerciaux ou non (récolte de fruits, bois gisants, feuilles, latex, sèves, résines, rotin, raphias, lianes, écorces, herbes, racines, etc.) reconnus par le réglement des droits coutumiers fixé par la loi de classement, sont strictement personnels et inaliénables et ne peuvent être exercés que par le titulaire ou les membres de sa famille.

Les récoltes doivent être effectuées de façon à ne pas détruire les végétaux producteurs.

Article 9. - Les cultures par essartage ou toutes autres méthodes et les déboisements sont formellement interdits en forêt classée sauf dans les réserves de faune où les cultures peuvent exceptionnellement donner lieu à cantonnement.

Article 10. - Le pacage et de parcours des bovins, chèvres, moutons, porcs, animaux de bât, sont strictement interdits en forêt classée.

Article 11. - Tout émondage ou ébranchage est interdit en forêt classée.

De nuit, le port de la matchette, coupe-coupe, crochet, scie et tout instrument analogue est interdit hors des voies de circulation et des campements autorisés dans toutes les forêts classées. La sagaie de défense est autorisée.

La même interdiction est valable de jour comme de nuit dans les forêts classées de la zone soudanienne.

Article 12. - Dans les forêts classées situées en zone de savane, il est interdit d'allumer du feu le long des routes et chemins qui les traversent.

Cependant, des charbonnières et fours à charbon ou des fours pour la distillation du bois pourront être établis en forêt par les exploitants dûment autorisés, sous leur responsabilité et après désherbage du sol, dans un rayon de 50 mètres autour de chaque installation.

De même, les feux pour les besoins domestiques y seront tolérés sous la responsabilité entière de leurs auteurs dans les campements réguliers prévus pour les usagers et les passagers.

Article 13. - Les agents forestiers procèderont d'office, en saison favorable, à l'incinération des herbages aux environs des forêts classées afin de les préserver des effets possibles des mises à feu inconsidérées. Les populations bordurières seront averties au préalable.

Article 14. - Les agents forestiers, dans le cadre des missions qui leur sont imparties, sont habilités à effectuer dans les forêts classées autres que les réserves intégrales toutes opérations sylvicoles ou sanitaires ou de réalisation touchant la faune ou la flore nécessitées par le service.

Ils ont à charge d'établir et de maintenir en bon état le bornage et le panneautage.

Article 15. - Sauf dans les réserves intégrales, des autorisations exceptionnelles de déboisement peuvent être accordées dans les forêts classées par décret uniquement dans le but de recherches minières. Le décret se prononcera sur l'indemnité à accorder. Elle devra couvrir les frais de remise en état et le préjudice et ne saura être inférieure à une somme fixée chaque année par une loi spéciale.

Section 3 : Exploitation

Article 16. - Dans les forêts domaniales et soumises, l'exploitation ne peut avoir lieu qu'en régie ou par vente par voie d'adjudication publique de coupe ou de lots définis. La création des régies et le programme des ventes seront décidés par décret.

Article 17. - Le bénéfice des régies et des ventes par adjudication est versé, après défalcation, s'il y a lieu, du rachat des droits coutumiers prévu par le classement comme indiqué à l'art.4, soit à l'Etat s'il s'agit de forêt domaniale, soit à leur propriétaire : commune, établissement public etc., s'il s'agit de forêt soumise.

Sous-section a) Régies

Article 18. - Les coupes en régie pourront être effectuées soit par le service des Eaux, Forêts et Chasses, soit par des services publics pour leurs besoins propres, soit par tout organisme créé à cet effet par décret.

Sous-section b) Ventes par adjudication

Article 19. - Ne pourront prendre part à l'adjudication que les personnes n'ayant pas fait l'objet d'une interdiction d'obtenir un permis de coupe.

Les personnes désirant prendre part à l'adjudication en adresseront, par lettre recommandée, la demande au ministre chargé des Forêts.

Chaque demande indiquera les nom, prénoms, nationalité et adresse du demandeur et la superficie qu'il désire.

Elle devra en outre être accompagnée :

a) d'un certificat de l'autorité administrative du lieu de résidence établissant que le demandeur réunit les deux premières conditions;
b) d'un extrait du casier judiciaire n'ayant pas plus de trois mois de date;
c) d'une déclaration d'élection de domicile dans un centre administratif;
d) d'un récépissé constatant le versement du cautionnement;
e) d'une procuration légalisée, si le demandeur a l'intention de se faire représenter par un tiers.

Les titulaires d'une autorisation d'exploiter en cours de validité sont dispensés de fournir les pièces énumérées aux alinéas a. b. c.

Toute demande devra parvenir au ministre chargé des Forêts au moins un mois avant la date prévue pour les adjudications.

A l'expiration de ce délai qui sera décompté de quantième en quantième sans qu'aucune prolongation pour cas de force majeure puisse être admise, le ministre chargé des Forêts adressera la liste des demandes jugées recevables, avec les dossiers, au président de la commission d'adjudication.

Il retournera aux intéressés, avec son avis motivé, les demandes jugées irrecevables parce que non conformes aux prescriptions du présent code ou parvenues hors délais.

Le montant des cautionnements sera fixé chaque année par une loi spéciale.
Ce cautionnement pourra être bancaire.

Article 20. - La commission d'adjudication sera composée comme suit:

Le conservateur de la Propriété foncière, président
Le directeur des Eaux, Forêts et Chasses, membre,
Un fonctionnaire désigné par le ministre chargé des Forêts, secrétaire.

Un procès-verbal sera dressé à la fin d' l'adjudication. Celle-ci ne sera définitive qu'après parution de l'arrêté l'approuvant. Cet arrêté prononcera le remboursement des cautions des personnes qui n'auront pas été déclarées adjudicataires.

Article 21. - L'adjudication se fera au plus offrant et dernier enchérisseur. L'enchère minimale ne devra pas être inférieure au vingtième de la mise à prix.

Article 22. - Les adjudicataires seront tenus de verser à la caisse du receveur des Domaines, dans les cinq jours qui suivront l'adjudication, le quart du montant de leur offre et la totalité des frais accessoires.

Il leur sera remis une copie du procès-verbal d'adjudication.

Le cinquième du cautionnement des adjudicataires qui, dans le délai prévu au premier paragraphe de cet article n'auront pas effectué ce versement, restera acquis au Trésor/

Le solde de l'offre devra être acquitté dans les délais impartis par le cahier des charges. A l'exception des délais, la totalité du montant de l'offre est exigible. Les sommes versées sont définitivement acquises au Trésor.

Article 23. - Le titre d'exploitation sera constitué par une copie du cahier des charges remis après que les adjudications auront été déclarées définitives et après versement des sommes fixées pour la délivrance par le cahier des charges. Ce titre sera dûment enregistré.

Chapitre III : Gestion des forêts du domaine coutumier

Section 1 : Constitution

Article 24. - Les forêts coutumières existent et se transmettent par la coutume.

En cas de déclassement d'une forêt du domaine forestier classé, la loi pourra prononcer le retour au domaine forestier coutumier.

Les forêts coutumières ne peuvent changer de régime que par la procédure fixée à l'art.4 ci-dessus, les instituant en forêts classées ou par celles prévues au Code foncier dans sa deuxième section.

Section 2 : Règles générales de gestion

Article 25. - Les collectivités coutumières ou les particuliers titulaires de droits coutumiers jouiront de la plénitude de leurs droits sur les forêts qui leur appartiennent sous les réserves figurant aux art. 26 et 28 ci-après et celles fixées à la section 3 du présent chapitre.

Article 26. - Les feux de brousse ayant pour but le renouvellement des pâturages ou la préparation des terrains de culture ou l'assainissement des lieux habités et des pistes sont autorisés du 15 novembre au 1er mars. Passé cette date ils sont strictement prohibés.

L'incinération des rémanents sur les terrains défrichés et nettoyés en vue des cultures est autorisée en tout temps.

Les mises à feux sont soumises aux prescriptions suivantes et à toutes celles qui pourront être fixées par la suite par voie d'arrêté du ministre chargé des Forêts.

La mise à feu ne peut être effectuée que de jour et par temps calme.

Elle est faite avec l'autorisation du chef de village. Tous les hommes valides doivent se tenir prêts à intervenir pour combattre l'incendie qui se propagerait hors des limites prévues.

Article 27. - En cas d'incendie, la direction des secours appartient à l'agent des Ea x, Forêts et Chasses présent s'il est agent technique ou du grade supérieur; à défaut elle appartient à un représentant de l'administration la plus proche (chef de village, fonctionnaire etc).

Ces personnes pourront requérir quiconque en vue de combattre un incendie que ce soit en forêt coutumière. La réquisition sera réputée valable lorsqu'elle sera faite verbalement.

La réquisition peut intéresser collectivement les habitants d'un village et sera réputée valablement faite lorsqu'elle aura été adréssée au chef de village.

Seuls, les agents du service forestier, les préfets et sous-préfets, les maires et les conseillers municipaux et les chefs de village que ces derniers auront habilités, sont autorisés à ordonner et à diriger les contre-feux.

Article 28. - Si ce n'est pour l'établissement d'une culture pérenne, sont interdits l'arrachage, l'abattage, la mutilation grave des karités, kolatiers, arbres et lianes à caoutchouc, roniers, palmiers à huile, sauf pour ce dernier, nécessité d'éclaircie lorsqu'il est jeune.

Section 3 : Exploitation

Article 29. - Les forêts coutumières sont exploitées sous le contrôle de l'Etat et avec son entremise entre les titulaires des droits coutumiers et les exploitants.

Les profits de gestion sont attribués aux titulaires des droits coutumiers représentés par leurs communes et répartis conformément aux dispositions de l'art.66 du titre III, dans le cas des règ des permis temporaires d'exploitation et des permis spéciaux de pieds en bordure qui figurent respectivement aux paragraphes 1°, 2° et 3° b de l'art.30;

Dans le cas des permis spéciaux qui figurent aux paragraphes 3° a) et 3° c) de l'art.30, le rachat des droits coutumiers est débattu directement entre celui qui les détient et la personne qui demande un permis.

L'Etat garantit l'exécution du contrat de rachat passé, conformément aux dispositions de lart.67 du titre III.

Article 30. - L'exploitation des forêts coutumières par les particuliers ou les services publics ne peut être faite que :

1) soit en régie
2) soit par permis temporaire d'exploitation de bois d'oeuvre obtenu par voie d'adjudication de "droit de coupe"
3) soit enfin par permis spéciaux de coupe d'un nombre limité d'arbres, de pièces, mètres cubes, stères, kilogrammes ou d'hectares, lorsqu'elle a pour but :
   a) la production commerciale de bois de feu et à charbon, des bois de mines, perches de construction, poteaux, pirogues, des produits accessoires énumérés à l'art.1 ainsi que pour satisfaire aux besoins purement locaux en bois d'oeuvre, ne se trouvant pas sur le marché.
   b) l'exploitation des peuplements en bordure des permis temporaires et des régies économiquement exploitables par le titulaire du titre d'exploitation principal.
   c) la réalisation des déboisements nécessaires aux cultures industrielles, aux exploitations minières et autres.

Les titulaires de droits coutumiers ne sont pas astreints à se faire délivrer un permis spécial lorsqu'ils exploitent dans les buts définis au paragraphe 3 a) ci-dessus les parcelles de forêt sur lesquelles portent leurs droits coutumiers.

Article 31. - Des arrêtés pourront, pour une période donnée, limiter soit le volume des bois à abattre, soit, sous réserve des droits acquis, ouvrir ou fermer à l'exploitation forestière certaines zones déterminées, soit protéger certaines espèces, soit fixer le diamètre minimum d'exploitation.

Sous-section a) Régies

Article 32. - Des coupes en régie peuvent être effectuées soit par les Eaux, Forêts et Chasses, soit par des services publics pour leurs besoins propres, soit par tout organisme créé à cet effet par décret.

Les redevances à verser aux titulaires des droits coutumiers sont les mêmes que celles auxquelles sont soumises les exploitations privées.

Toutefois, aucune redevance ne sera due si l'exploitation est faite au bénéfice des titulaires des droits coutumiers ou de leur collectivité ou pour la viabilité.

Ces régies seront réglées par voie de décret.

Sous-section b) Permis temporaires d'exploitation

Article 33. - Les permis temporaires d'exploitation sont accordés par arrêté. Ils peuvent avoir une superficie comprise entre 500 et 20 000 hectares.

Ils sont strictement personnels et ne peuvent donner lieu à affermage. Aucune mutation ne peut intervenir sans une autorisation administrative sanctionnée par arrêté du ministre chargé des Forêts.

Les transferts sont subordonnés au paiement d'une redevance spéciale.

Article 34. - Les permis temporaires d'exploitation sont accordés aux personnes ayant acquis un "droit de coupe" de même superficie. Ces droits sont attribués par voie d'adjudication.

Toute personne ayant obtenu un "droit de coupe" aura le droit de déposer sur une partie disponible de la forêt ouverte à l'exploitation, à l'endroit de son choix, le permis correspondant.

Article 35. - Les adjudications auront lieu aux dates et dans les centres prévus par arrêté du ministre chargé des Forêts, sur proposition du directeur des Eaux, Forêts et Chasses.

Ces mêmes arrêtés fixeront le programme des adjudications et le cahier des charges. Ceux-ci pourront prévoir des droits réservés à certaines catégories d'exploitants et des clauses de préférence.

Les mises à prix sont fixées par la loi des finances sur la base du prix à l'hectare.

Article 36. - Les conditions de participation et la procédure des adjudications seront les mêmes que celles fixées aux art. 19,20, 21 et 22 pour les adjudications de lots dans les forêts domaniales.

Article 37. - Les demandes de permis issus de "droits de coupe" seront établis suivant la réglementation en vigueur.

Elles devront être déposées au siège de l'Inspection forestière dont dépend la superficie demandée, au plus tard un an après la date de parution au "Journal officiel" de l'arrêté approuvant l'adjudication.

Elles pourront être déposées sous la responsabilité du demandeur dès le versement de la totalité de l'offre, sous plis cachetés et scellés. Ceux-ci ne seront ouverts qu'à la parution de l'arrêté rendant définitives les adjudications.

Article 38. - Les adjudicataires qui, dans les délais prévus à l'art. précédent, n'auraient pas déposé une demande recevable de permis temporaire d'exploitation seront déchus de leurs droits.

Si le permis temporaire d'exploitation correspondant a été prévu par le titulaire en plusieurs lots, l'adjudicataire sera censé avoir renoncé aux lots qui n'auront pas fait, dans les délais prescrits, l'objet d'une demande recevable.

Article 39. - Le cautionnement versé au moment de l'adjudication, ainsi qu'il est prévu à l'art.19, restera consigné pendant la durée de validité du permis temporaire d'exploitation qui lui correspond, pour servir à garantir le versement des redevances domaniales et l'exécution des obligations contractées par l'intéressé vis-à-vis de sa main-d'oeuvre et des titulaires de droits coutumiers. Il pourra être remboursé par anticipation si l'exploitant justifie de l'introduction, sur son exploitation, d'un matériel forestier d'une valeur au moins égale à dix fois ce cautionnement.

Article 40. - Le titulaire d'un permis temporaire d'exploitation, déposé à la suite d'une adjudication de droit, dont le permis arrive à expiration avant d'avoir été épuisé, pourra, après enquête du service forestier, acquérir un nouveau titre correspondant à un permis suffisant pour permettre de terminer son exploitation.

La redevance à verser sera calculée sur la moyenne des dernières adjudications pour des permis de même ordre que le nouveau permis demandé.

La durée et la surface de ce nouveau permis ne pourront excéder celle du permis expiré.

La forme de la demande et la procédure applicable sont celles fixées par les textes en vigueur.

Sous-section c) Permis spéciaux de coupe

Article 41. - Les demandes de permis spéciaux sont faites dans les formes prévues par la réglementation en vigueur.

Dans le cas des permis prévus à l'art.30, paragraphes 3 a), 3 c), le demandeur devra fournir à l'appui de sa demande une expédition d'une convention de règlement des droits coutumiers passée avec les ayant-droit et homologuée devant le Tribunal civil du second degré.

Cette convention comportera accord formel des titulaires des droits coutumiers et le montant et les échéances du paiement du prix du rachat convenu.

Dans le cas des permis prévus à l'art.30, paragraphe 3 b), la convention sera passée avec le maire de la commune rurale.

Article 42. - Les demandes sont transmises à l'Inspection forestière du ressort qui les instruit et les transmet au sous-préfet qui délivre le titre d'exploitation sous forme d'une décision.

Article 43. - Le commerce du bois de chauffe et des bois de service, lorsqu'il est exercé à l'échelon coutumier pour les Centrafricains, ne donne pas lieu à la délivrance d'un permis Spécial de coupe.

Il est seulement soumis aux dispositions prévues par la réglementation fiscale en vigueur.

Chapitre IV : Forêts des particuliers

Article 44. - Les particuliers propriétaires de terrains boisés et les titulaires de droits coutumiers immatriculés sur les terrains boisés y exercent tous les droits résultant de la propriété, mais ne pourront en pratiquer le défrichement qu'en vertu d'une autorisation administrative, délivrée par le ministre chargé des Forêts, lorsque l'étendue du défrichement dépassera 50 hectares.

Les bois et forêts sis sur des concessions rurales accordées à titre provisoire ne sont pas considérés comme bois particuliers. Pour l'exploitation de ces bois, le concessionaire est astreint aux conditions imposées par la présente réglementation aux forêts coutumières.

Article 45. - L'autorisation ne peut être refusée que si le défrichement est susceptible de compromettre :

1) Le maintien des terres sur les montagnes et les pentes.
2) La défense du sol contre les érosions et les envahissements des fleuves, rivières ou torrents.
3) L'existence des sources ou cours d'eau.
4) La fixation des dunes.
5) La salubrité publique.
6) La défense militaire.

Article 46. - En cas de contravention à l'art.44, indépendamment des amendes encourues, le propriétaire pourra être mis en demeure de rétablir les lieux défrichés en nature de bois.

Article 47. - Si dans un délai d'un an après la mise en demeure, tout ou partie de la superficie à reboiser n'est pas replantée par le propriétaire, il y sera pourvu à ses frais par les services des Eaux, Forêts et Chasses.

Article 48. - Sont exceptés des dispositions de l'art. 44 :

1) Les jeunes bois, pendant les trente premières années après leur semis ou leur plantation, sauf le cas prévu à l'art. précédent.
2) Les parcs et jardins clos ou attenant aux habitations.

Chapitre V : Forêts du domaine public

Article 49. - Les forêts du domaine public fluvial suivent, en ce qui concerne la réglementation forestière, le régime de la forêt contigue, sauf s'il en est disposé autrement par un texte particulier.

Chapitre VI : Dispositions diverses concernant l'exploitation

Article 50. - Marquage des bois .
Les bois en grume provenant des exploitations quelles qu'elles soient, y compris celles de forêts particulières, ne peuvent circuler sans être revêtus de l'empreinte du marteau portant la marque de l'exploitant, marque triangulaire qui doit être déposée au Greffe du Tribunal de grande instance ou d'instance et au Service forestier. Ces bois doivent, en outre, être accompagnés d'une feuille de route.

Article 51. - Droits des tiers .
Les titres d'exploitation forestière sont accordés sous réserve des droits des tiers et des droits coutumiers reconnus.

Article 52. - Droit d'occuper pour les installations nécessaires .
Le titre d'exploitation ne donne aucun droit sur le sol que celui d'y établir, à titre précaire, des logements magasins, cultures, chantiers nécessaires à l'organisation et au fonctionnement de l'exploitation.

L'administration conserve le droit d'accorder des concessions dans le périmètre des permis de coupe, à charge par elle ou le concessionnaire d'indemniser à titre d'expert le titulaire du permis de coupe, soit du manque à gagner pour les arbres qui lui seraient enlevés, soit du préjudice que lui causerait l'obligation de modifier ses installations.

Conformément à l'art.93 du Code foncier, le droit d'occuper pourra être transformé en un titre définitif s'il y a une mise en valeur suffisante au moment de l'expiration du titre d'exploitation.

Article 53. - Coupe pour les besoins d'exploitation .
Les exploitants d'un titre d'exploitation forestière sont autorisés à faire, pour les besoins stricts de leur exploitation, les abattages nécessaires à l'établissement des pistes, voies d'évacuation, campements, etc.

Les bois abattus dans ces conditions ne seront pas portés au carnet de chantier et ne pourront sortir du chantier. Ils n'acquitteront aucune taxe.

Article 54. - Troubles de jouissance .
Les exploitants ne pourront formuler aucune réclamation, ni prétendre à aucune indemnité, restitution ou compensation quelconque du fait :

1) Soit des travaux d'installation, d'occupation de terrains, provisoire ou définitive, effectuées par l'administration dans le périmètre des massifs forestiers concédés, pour un motif d'intérêt général ou pour les besoins de ses services.
2) Soit du chevauchement de permis consécutif à des plans inéxacts ou incomplets, présentés à l'appui des demandes, l'administration laissant au demandeur la responsabilité entière du plan fourni, dont une ampliation sera jointe à l'arrêté lui-même. Toutefois, en cas de chevauchement, l'exploitation de la partie commune appartiendra toujours au premier exploitant en date.
3) Soit de la coupe des arbres servant à la viabilité.
4) Ils devront, en outre, faciliter les déplacements des agents de l'administration, de passage sur leurs permis, en leur fournissant en location la main-d'oeuvre, les moyens de transport qui leur seraient nécessaires et l'usage des voies d'évacuation et de débardage.

Article 55. - Chevauchement des exploitations non forestières .
Dans le cas de chevauchement de permis d'exploitation forestière et de permis non forestiers, le titulaire du permis forestier ne pourra refuser à l'autre partie les abattages et l'exploitation des bois nécessaires à son activité.

Toutefois, aucune coupe ne pourra être exécutée avant l'accord du titulaire du permis d'exploitation forestière et le versement préalable d'une indemnité. Le titulaire du permis forestier reste responsable de toutes les infractions à la réglementation forestière relevées sur son permis, comme prévu à l'art.104.

Article 56. - Besoins des exploitations non forestières .
Les personnes non titulaires de permis temporaires, non titulaires de droits coutumiers d'exploitation qui, en raison de leur activité, utilisent le bois sous une forme quelconque, devront, lorsqu'elles ne peuvent se fournir dans le commerce, se munir de permis de coupe spéciaux. Ainsi en est-il des exploitations minières, pour les bois indispensables à leurs travaux que l'intérêt économique commande de couper à proximité immédiate des chantiers.

Article 57. - Servitude de passage .
Tout exploitant aura le droit d'accéder par des routes, pistes, chemins de tirage ou voies férrées et sans qu'aucune entrave puisse être apportée par l'occupant du fonds traversé, à une voie d'évacuation publique (rivière ou fleuve, voie férrée ou route).

Toutefois, au moment de l'établissement du tracé du réseau d'évacuation, l'occupant du fonds traversé qui estimerait subir un préjudice, pourra demander qu'une enquête soit effectuée par le chef de l'Inspection forestière du ressort qui jouera le rôle d'arbitre.

Si le différend persiste, il sera réglé par une commission composée du sous-préfet ou son délégué, président, ayant voix prépondérante, du chef de l'Inspection forestière, d'un représentant de chacune des deux parties, pris autant que possible parmi les représentants des organismes professionnels.

Cette commission pourra, soit confirmer la nécessité du tracé, soit prescrire qu'il en soit recherché un autre, ou encore provoquer un réglement d'exploitation du réseau d'évacuation en cause, ou fixer l'indemnité due à l'occupant du fonds traversé. Sa décision, prise à la majorité, sera sans appel.

Sous réserve des dispositions de l'art.54, 4), l'exploitant possède l'exclusivité d'utilisation, pour le charroi du réseau d'évacuation qu'il a établi, aussi bien sur la partie de ce réseau située sur le fonds d'autrui que sur celle située sur le domaine non concédé.

Il pourra, toutefois, en autoriser l'utilisation par les personnes de son choix.

Aucune entrave ne doit être apportée par quiconque à cette utilisation ou à celle du réseau d'évacuation public.

De même, les exploitants doivent laisser continuellement le libre usage des sentiers et pistes traversant leur permis.

TITRE III : TAXES ET REDEVANCES

Chapitre I : Généralités

Article 58. - Toutes les recettes forestières sont considérées comme des recettes domaniales.

Les poursuites pour le recouvrement des taxes dues sont engagées et les instances introduites comme en matière d'enregistrement.

Chapitre II : Des différentes taxes

Section 1 : Taxes de superficie

Article 59. - Les titulaires de permis temporaires d'exploitation, les régies, les acheteurs de coupe en forêts domaniales ou soumises, sont astreints au paiement d'une taxe annuelle de superficie. Cette redevance est exclusive de toute autre taxe d'occupation pour les installations nécessaires à l'exploitation.

Les permis spéciaux de coupe définis à l'art.30, paragraphe 3, ne sont pas soumis à cette taxe.

Article 60. - Le montant de la taxe de superficie est exigible à terme échu et doit être versé à la caisse du receveur des Domaines ou, à défaut, à celle de l'agent spécial au début de l'année suivante.

Tout retard dans le paiement de la taxe de superficie entraînera une augmentation de 10% de la somme due par trimestre de retard. Le non paiement pendant une année pourra entraîner la fermeture du chantier.

Article 61. - Le taux de la taxe de superficie sera fixé chaque année par une loi spéciale.

Section 2 : Taxes de transferts

Article 62. - Le transfert d'un permis de coupe donne lieu au paiement d'une redevance égale à cinq fois la redevance de superficie annuelle.

L'échange de permis ou de parcelle de ces permis entre exploitants est considéré comme un transfert unique et donne lieu au paiement de la même redevance par moitié entre les parties.

La redevance est due même si le transfert du permis a pour résultat de regrouper sous une même raison sociale des permis précédemment attribués à des titulaires différents.

La redevance de transfert pourra être réduite à la moitié lorsqu'un ascendant, conjoint ou descendant en ligne directe du titulaire demande le transfert à son profit de permis accordé au défunt dont il est appelé à recueillir la succession.

Section 3 : Redevance de déboisement

Article 63. - Les personnes titulaires de permis d'occuper à titre provisoire qui, en vertu de leur activité, sont dans l'obligation de détruire la forêt comme il advient en matière d'exploitation minière ou agricole, devront obtenir le permis spécial de déboisement prévu à l'art.30, paragraphe 3 c).

Les demandes de permis spéciaux de déboisement devront être adressés avant que le défrichement ne soit intervenu. Cependant, en ce qui concerne les exploitations minières, il sera toléré que les déboisements soient régularisés à la fin de chaque année.

L'aboutissement de la procédure foncière d'octroi des superficies détenues à titre provisoire ne dispense pas du paiement de la redevance de déboisement.

Cette redevance sera fixée chaque année par une loi spéciale.

Les titulaires de droits coutumiers ne sont pas soumis à cette taxe lorsque le déboisement qu'ils effectuent porte sur leur terrain coutumier.

Section 4 : Taxes de production (dite taxe d'abattage)

Article 64. - La perception de la taxe sur les permis spéciaux pour l'exploitation commerciale prévue à l'art.30, paragraphe 3 a) et 3 b) se fera par avance, au moment du dépôt de la demande qui devra être accompagnée du récépissé de versement.

Les taxes seront fixées chaque année par une loi spéciale.

Article 65. - Il est institué une taxe dite taxe de production frappant les grumes vendues hors de l'Union Douanière Equatoriale, les sciages et autres produits de transformation quelle que soit leur destination.

Une Loi spéciale fixera les modalités d'assiette et de recouvrement de la taxe de production.

Tout intéressé est tenu, avant le 15 de chaque mois, de présenter à l'agent forestier habilité ou au fonctionnaire en tenant lieu, un état indiquant par essence et cubage, pour le mois écoulé :

a) le stock de bois en grumes au premier jour du mois,
b) le stock de bois débité au premier jour du mois,
c) le total des débits livrés à l'exportation pendant le mois écoulé,
d) le total des débits livrés à la consommation locale pendant la même période.

Cet état sera accompagné d'une ampliation certifiée exacte des bordereaux de livraisons et de feuilles de route concernant les grumes.

Au vu de cet état, et après vérification, le fonctionnaire en question dressera le "bon à percevoir" qui permet à l'intéressé de verser la redevance à la caisse du receveur des Domaines ou à l'agent spécial du ressort.

Chapitre III : Du réglement des droits coutumiers

Article 66. - Les recettes provenant des adjudications de droit de coupe, des attributions de permis temporaires d'exploitation, des bénéfices sur les règles prévues à l'art.30, paragraphe 1, des taxes sur les permis spéciaux de coupe prévus à l'art.30, paragraphe 3 b) et des taxes de superficie, seront réparties de la manière suivante :

- 60% du montant total au profit du budget de l'Etat,
- 40% du montant total au profit de l'Office National des Forêts.

Dans ce but, le ministère des Eaux, Forêts, Chasses et Pêches établira chaque trimestre un état récapitulatif des recettes visé par la Direction de la Conservation Foncière et des Domaines et adressé au ministère des Finances qui fera effectuer les virements nécessaires au compte de l'Office National des Forêts ouvert au Trésor public.

Article 67. - Dans le cas des permis spéciaux prévus à l'art.30, paragraphe 3 a) et c), le non paiement par l'exploitant dans les délais impartis des sommes fixées par le contrat d'abandon des droits coutumiers sera assimilé au non paiement des taxes et redevances forestières et poursuivi comme tel au bénéfice de l'ayant droit.

TITRE IV : REPRESSION DES INFRACTIONS

Chapitre I : Procédure
Section 1 : Recherche et constation des délits (art. 68 à 81 omis)
Section 2 : Confiscation et saisie (art. 82 à 86 omis)
Section 3 : Actions et poursuites (art. 87 à 91 omis)
Section 4 : Transactions (art. 92 à 93 omis)

Chapitre II : Infractions et pénalités
Section 1 : Coupes et exploitation non autorisées - Mutilation d'arbres (art. 94 à 98 omis)
Section 2 : Marteaux forestiers - Marques (art. 99 omis)
Section 3 : Exploitation (art.100 à 105 omis)
Section 4 : Cultures en forêts classées, feux de brousse, Incendies de forêt (art.106 à 108 omis)
Section 5 : Pâturages (art.109 à 110 omis)
Section 6 : Infractions diverses (art.111 à 123 omis)

TITRE V : DISPOSITIONS GENERALES

Article 124. - Le dixième du produit des amendes, transactions, confiscations, restitutions, dommages-intérêts et contraintes, sera attribué aux agents du service forestier et, le cas échéant, aux agents des autres services habilités conformément aux dispositions de l'Art.69, qui auraient verbalisé en matière forestière.

La répartition en sera fixée par les textes en vigueur.

Article 125. - Sont abrogées toutes dispositions antérieures contraires à la présente loi.

Article 126. - La présente loi sera promulguée et publiée au "Journal officiel" de la République Centrafricaine.

Elle sera exécutée comme loi de l'Etat.

Bangui, le 5 février 1962

REPUBLIQUE CENTRAFRICAINE

CENTRAL AFRICAN REPUBLIC

LOI N° 61/282
DU 5 FEVRIER 1962

Loi N° 61/282

Loi N° 61/282 Fixant le Taux de Redevances Forestières

Article 1. - En cas d'octroi d'autorisation de déboisement du domaine Forestier classé prévu à l'Art.15 du Code Forestier,l'indemnité compensatrice du préjudice subi est fixée en 1962 à 15 000 F par Hectare.

Article 2. - Le Cautionnement exigé pour présenter une demande de participation aux adjudications forestières (Art.19 et 39 du Code Forestier) est fixé à 20 F par hectare.

Article 3. - La mise à prix pour les adjudications prévue à l'Art.35 du Code Forestier de "droit de coupe" est fixée à 1 200 F par Hectare.

Article 4. - La taxe de superficie prévue à l'Art.61 du Code Forestier est fixée à 20 F par Hectare et par an.

Article 5. - La taxe de déboisement prévue à l'Art.63 du Code Forestier est fixée à 5 000 F par Hectare.
Elle sera réduite à 250 F si le déboisement a pour but l'installation de cultures pérennes.

Article 6. - Les taxes sur les permis spéciaux prévues à l'Art.64 du Code Forestier sont les suivantes :

- Bois d'oeuvre - Diamètre supérieur à 0,80m : 1 000 F par pied.
- Bois d'oeuvre - Diamètre compris entre 0,30 et 0,60 m : 200 F par pied.
- Poteaux - Diamètre compris entre 0,15 et 0,30m : 100 F par pied.
- Perches - Diamètre compris entre 0,05 et 0,15m : 25 F par pied.
- Gaulettes - Diamètre jusqu'à 0,05 m : 250 F le cent.
- Bois de chauffe ou à carboniser ou de râperie : 15 F le stère.
- Autres produits : 10 F le stère.

Article 7. - Les taxes à la production prévues à l'Art.65 du Code Forestier sont ainsi fixées :

- Grumes vendues hors de l'Union Douanière Equatoriale : 105 F/ m^3
- Grumes vendues hors de la République Centrafricaine, mais à l'intérieur de l'Union Douanière Equatoriale : 105 F/ m^3
- Sciages et autres produits de transformation quelle que soit leur destination : 210 F le m^3 plus 200 F financement Centre pilote (J.O. du 1er juin 1964, page 413)

Cette taxe est réduite à 150F pour les sciages destinés à l'Exportation et dont les dimensions sont inférieures à 1,80m en longueur ou 0,15m en largeur (sciages dits "short and narrow" et frises à parquet).

Article 8. - Le tarif prévu par la présente loi sera applicable du 1er Janvier 1962 au 31 Décembre 1962.

Article 9. - La présente loi sera promulguée et publiée au J.O. de la République Centrafricaine. Elle sera exécutée comme Loi de l'Etat.

Bangui, le 5 Février 1962

ORDONNANCE N° 69/49
DU 23 SEPTEMBRE 1969

REPUBLIQUE CENTRAFRICAINE

CENTRAL AFRICAN REPUBLIC

Ordonnance N° 69/49

Ordonnance N° 69/49 Portant création de l'Office National des Forêts

Article 1. - Il est créé l'Office National des Forêts (ONF)

Article 2. - L'ONF est appelé à mettre en oeuvre certains programmes de développement du secteur qui lui est confié. Il est spécialement chargé :

Dans le Domaine Forestier

1° de la promotion de l'Economie et des Industries Forestières
2° des Aménagements Forestiers
3° de l'aménagement des bassins versants
4° de l'Etude de l'influence des forêts sur les milieux
5° de sélectionner et développer la plantation d'essence de plus grande valeur
6° de protéger les régions non boisées, propres à l'implantation des peuplements forestiers pour la production, la protection ou les loisirs
7° d'encourager l'utilisation du bois pour la construction de l'habitation à bon marché
8° de la lutte contre les incendies en général et les incendies de forêts en particulier
9° de l'institution de l'enseignement forestier pour la formation des cadres Techniciens à tous les échelons en vue de satisfaire les besoins des industries forestières.

Dans le Domaine de la Pêche

10° de développer la pêche fluviale
11° d'étudier la biologie du milieu aquatique
12° de développer les Industries des bateaux de pêches et en vulgariser l'utilisation
13° de procéder à l'extention des bassins et pisciculture
14° de promouvoir la commercialisation des produits de pêche.

Article 3. - Un décret, pris en conseil des Ministres, portera approbation du Statut dudit Office.

Article 4. - La présente Ordonnance sera publiée au J.O. Elle sera exécutée comme loi de l'Etat.

Fait à Bangui, le 23 Septembre 1969

ORDONNANCE N° 74/014
DU 24 JANVIER 1974

REPUBLIQUE CENTRAFRIQUE

CENTRAL AFRICAN REPUBLIC

Ordonnance N° 74/014

Ordonnance N° 74/014 Portant modification de la taxe à la production applicable aux bois Centrafricains exportés

Article 1. - Les dispositions de l'Art.3 B-G de la Loi 62/382 du 4 Janvier 1963 sont abrogées.

Article 2. - Les Taux de la taxe à la production prévue à l'Art.65 du Code Forestier sont fixés ainsi qu'il suit :

1) Grumes Exportées hors de l'UDEAC
Ayaus, Limba = 500 Francs le mètre cube
Autres bois = 1 000 Francs le mètre cube

2) Sciages et autres produits de transformation exportés hors de l'Union Douanière Economique de l'Afrique Centrale (UDEAC) = 500 Francs le mètre cube

3) Sciages et autres produits de transformation vendus à l'intérieur de l'UDEAC
= 210 Francs le mètre cube

Article 3. - La présente Ordonnance ne s'applique pas aux sociétés pour lesquelles les textes d'agrément auraient prévu un régime fiscal particulier. Elles restent à la fiscalité qui leur a été accordée.

Article 4. - La présente Ordonnance sera publiée au J.O. selon la procédure d'urgence. Elle sera exécutée comme loi de l'Etat.

Bangui le 24 Janvier 1974

ORDONNANCE N° 79/025
DU 8 MAI 1979

REPUBLIQUE CENTRAFRIQUE

CENTRAL AFRICAN REPUBLIC

Ordonnance N° 79/025

Ordonnance N°79/025 Portant création d'une taxe dite "Taxe de reboisement et de formation"

Article 1. - Il est créé une taxe forestière dite "Taxe de Reboisement et de Formation" frappant les grumes de sapelli et de sipo vendues à l'exportation y compris les pays de l'U.D.E.A.C.

Article 2. - Le montant de cette taxe est fixée à 1 000 F le m³ grume Exporté. *

Article 3. - Toutes Sociétés Forestières sont soumises au paiement de cette taxe sans exception et quels que soit les termes de leurs conventions.

Article 4. - Cette Taxe de Reboisement et de Formation sera perçu sur les Ordres de Recettes émis par la Direction des Forêts et sera reversée à l'Office National des Forêts.

Article 5. - Le Produit de cette taxe sera utilisé pour les travaux Forestiers de Reboisement de la Forêt et pour couvrir les frais de Formation. Le Ministre des Eaux et Forêts Chasses et Pêches présentera chaque Année un programme d'utilisation de ces fonds.

Article 6. - La Présente Ordonnance sera enregistrée, publiée au J.O. selon la procédure d'urgence. Elle sera exécutée comme Loi de l'Etat.

Fait à Bangui, le 8 Mai 1979

* L'Ordonnance N° 80/80 du 26 Septembre 1980 relève la Taxe de Reboisement et de Formation à 2 000 F le m³ dans le cas des essences Sapelli et Sipo.

Loi N° 004/74 Portant Code Forestier

(Version consolidée d'après la Loi N° 32.82
du 7 Juillet 1982)

L'Assemblée Nationale populaire a délibéré et adopté :

Le Président de la République promulgue la Loi dont la teneur suit :

Vu la Constitution,

Vu la Loi N° 34-61 du 20 Juin 1961 fixant le régime forestier dans la République Populaire du Congo,

Vu la Loi N° 31-61 du 3 Juin 1961 fixant les redevances en matière forestière,

Vu la Loi N° 37-63 du 4 Juin 1963 modifiant la Loi N° 31-61 du 3 Juin 1961,

Vu la Loi N° 38-63 du 4 Juillet 1963 sur l'organisation et le fonctionnement du Fonds Forestier.

TITRE PREMIER : LE DOMAINE FORESTIER ET LES DROITS D'USAGE

Chapitre Premier : Le Domaine Forestier

Article 1 (nouveau). - Les forêts et les périmètres de reboisement définis aux art. 2 et 10 ci-après appartiennent au domaine privé de l'Etat et sont constitués en domaine forestier.

Article 2 (nouveau). - Sont considérées comme forêts au sens de la présente loi, les terrains dont les fruits exclusifs ou principaux sont les bois d'ébénisterie, d'industrie ou de service, les bois de chauffage ou à charbon, ou des produits accessoires tels que les écorces et fruits à tanin, les écorces textiles et tinctoriales, le kapok, le caoutchouc, la glue, les résines, les gommes, les bambous, les palmiers spontanés et tous autres végétaux ne constituant pas un produit agricole.

Article 3 (nouveau). - Les parcelles du domaine forestier sont classées en forêts de production, forêts de développement communautaire et forêts de protection. Toute parcelle du domaine forestier n'ayant pas fait l'objet d'un classement conformément à la procédure instituée par les art. 5, 10 et 11 ci-après est considérée comme forêt protégée.

1) Les forêts de production ont pour vocation la production de bois. Elles peuvent faire l'objet des permis et contrats prévus par la présente Loi.

L'exercice des droits d'usage peut être réglementé dans les parcelles ouvertes à l'exploitation forestière par le Ministre des Eaux et Forêts en vue de rendre ceux-ci compatibles avec les exigences de l'exploitation forestière.

2) Les forêts de protection ont pour vocation la conservation ou la restauration des peuplements forestiers, de la flore, de la faune, des sols et systèmes hydrauliques.
L'exercice des droits d'usage, les permis et contrats d'exploitation forestière, le droit d'y résider ou d'y accéder peuvent y être supprimés, interdits ou réglementés conformément aux intérêts ayant motivé le classement des parcelles forestières concernées.
Le régime juridique des forêts de protection s'applique de droit aux réserves et parcs nationaux ainsi qu'aux périmètres de reboisement.

3) Dans les parcs nationaux nul n'est admis de façon permanente. Aucune activité autre que celles nécessaires à la conservation ou de la restauration des richesses naturelles objets de la création de la réserve ne peut être entreprise. L'accès du public peut y être interdit.

4) Dans les réserves naturelles, le Ministre des Eaux et Forêts réglemente les activités et le droit de résider en fonction de la conservation ou de la restauration des richesses naturelles objets de la création de la réserve. L'accès du public peut y être interdit.
Les réserves naturelles pourront être à vocation générale ou spécialisées dans la protection d'une ou plusieurs espèces de la faune ou de la flore conformément aux indications de leur acte constitutif.
Les réserves naturelles créées pour la conservation d'espèces forestières resteront gérées directement par le ministère des Eaux et Forêts.

5) Les réserves naturelles intégrales sont soustraites dans toute la mesure du possible à l'influence humaine. Leur accès est interdit à toute autre personne qu'aux agents chargés de leur surveillance.

6) Les forêts de développement communautaires sont affectées à la subsistance des populations y résidant. Selon les besoins de celles-ci et après qu'elles aient été consultées dans les formes que réglementera le ministre des Eaux et Forêts, l'Etat pourra garantir le maintien des espaces forestiers nécessaires aux populations et entreprendre des programmes de développement économique à leur bénéfice notamment de types sylvi-agricoles ou sylvi-pastoraux. Le défrichement de ces forêts sera subordonné à la condition d'un reboisement préalable équivalent.
Dans les forêts protégées il ne pourra être attribué aucun droit de coupe intéressant des superficies telles que le classement ou l'aménagement futur des parcelles concernées soit compromis.

Article 4 (nouveau). - Le classement d'une forêt est prononcé par décret du Premier ministre publié au Journal Officiel et porté par les soins de l'autorité administrative régionale compétente à la connaissance des villages intéressés.

Ce décret ne pourra être pris qu'à la condition que les actes constitutifs de ces forêts aient déterminé leurs limites d'une façon précise, qu'elles soient reconnues libres de tout droit ou que les droits d'usage aient fait l'objet d'un réglement d'aménagement.

Article 5 (nouveau). - Après avoir entendu l'autorité administrative régionale et les représentants des villages voisins, l'Administration des Eaux et Forêts procède à la reconnaissance du périmètre à classer et des droits d'usage ou autres s'exerçant sur la forêt.

Le projet de classement, comportant les coordonnées exactes et une description précise des limites du périmètre dont le classement est projeté, est remis à l'autorité administrative régionale qui le porte à la connaissance des intéressés par tous les moyens de publicité conformément aux règlements ou aux usages locaux.

Article 6 (nouveau). - Dans les trente jours qui suivent le dépôt du projet de classement au Chef lieu de région, le Ministre des Eaux et Forêts convoque la réunion de la Commission de classement qui comprend, sous la présidence du Ministre, les députés de la circonscription où est située la forêt à classer, le Commissaire Politique Régional, le Président du Comité du District ou de la Commune concernée, le Directeur Régional des Eaux et Forêts ou son représentant, les Présidents et Membres de Comités de chaque village intéressé.

La Commission de classement se réunit dans la région où se trouve la forêt à classer.

Elle examine le bien-fondé des réclamations formulées, détermine les limites de la forêt à classer et constate l'absence ou l'existence des droits d'usage grèvant cette forêt.

S'il existe de tels droits, la Commission constate la possibilité de leur exercice à titre exceptionnel à l'intérieur du périmètre classé.

Un procès-verbal relatant les opérations accomplies par la Commission de classement est transmis au Chef de Gouvernement.

Article 7 (nouveau). - Sans préjudice des recours légaux postérieurement à la prise du Décret de classement, les habitants qui auraient des droits autres que ceux d'usage ordinaires à faire valoir sur la partie de la forêt à classer pourront former opposition au projet de classement pendant un délai d'un mois à compter de la date de sa communication effective aux intéressés par l'autorité administrative régionale conformément à l'art. 6 ci-dessus.

L'opposition et les réclamations formulées à cette occasion sont enregistrées au Chef lieu de la Région et portées devant la Commission de classement qui en tentera le réglement amiable.

En cas d'échec, le litige est porté devant le Tribunal de Grande Instance territorialement compétent.

Article 8. - Les forêts domaniales classées sont gérées directement par le Service des Eaux et Forêts.

Elles ne pourront être aliénées en totalité ou en partie qu'après déclassement par décret pris sur l'avis d'une commission de déclassement comprenant, sous la présidence du Ministre chargé des Eaux et Forêts ou de son représentant, le Chef du Service des Eaux et Forêts, le Receveur des Domaines et l'autorité administrative régionale dont dépendent les forêts concernées.

Article 9. - Abrogé par la loi N° 32/82

Article 10 (nouveau). - Les parties de terrain nu ou insuffisamment boisé, dont le reboisement ou la restauration est reconnue nécessaire, sont classées par Décret comme périmètre de reboisement sur l'initiative du ministre des Eaux et Forêts en vue :

- du maintien des terres sur les montagnes ou les pentes;
- de la défense du sol contre les érosions, contre les inondations et les envahissements des fleuves, rivières ou torrents;
- d'assurer l'existence des sources et cours d'eau;
- de la fixation des dunes maritimes et pour la protection contre les érosions de la mer et l'envahissement des sables;
- de la salubrité publique;
- de la défense militaire;
- de la réalisation des projets d'intérêt économique ou social.

A cet effet, le ministre des Eaux et Forêts établit un projet de classement faisant apparaître les droits des tiers dont le rachat ou l'expropriation seraient nécessaires.

Le projet est notifié individuellement aux intéressés par l'autorité administrative régionale préalablement à l'adoption du décret de classement.

Les intéressés ont droit de recours devant le Tribunal de Grande Instance territorialement compétent dans les dix (10) jours qui suivent cette notification.

Article 11 (nouveau). - Les parcs nationaux et les réserves naturelles sont créées en vue de la conservation de la faune, de la flore, des formations géologiques remarquables et des sites présentant un intérêt touristique, scientifique ou historique.

Le classement d'un terrain ou d'un site comme parc national ou réserve naturelle est prononcé par Décret du Premier Ministre sur proposition du ministre des Eaux et Forêts ou du Département ministériel intéressé. La procédure prévue par l'art.10 ci-dessus est applicable en ce qui concerne l'indemnisation et la reprise des droits concédés et en ce qui concerne le recours éventuel.

Chapitre Deuxième : Les droits d'usages

Article 12 (nouveau). - Les populations et les individus les composant, quelque soit le lieu de leur résidence, continuent d'exercer leurs droits d'usage sur le domaine forestier en se conformant strictement aux dispositions de la présente loi, à la réglementation prise pour son application et aux règles coutumières compatibles avec la loi écrite.

Ces droits d'usage s'exercent même sur les chantiers forestiers sans que les exploitants forestiers puissent prétendre, à ce titre, à aucune indemnité ou compensation.

L'exercice des droits d'usage, strictement limité à la satisfaction des besoins personnels individuels ou collectifs des usagers, est réservé aux seuls nationaux.

Article 12-1 (nouveau). - Il est interdit sur tout le domaine forestier d'abattre, brûler, écorcer ou porter préjudice aux arbres et jeunes plantes des espèces suivantes :
- de valeur commerciale
  - Limba (Terminalia superba)
  - Okoumé (Aucouméa Klaineana)
  - Sapelli (Entendrophragma cylindricum)
  - Sipo (Entendrophragma utile)
  - Kossipo (Entendrophragma candollei)
  - Acajou (Khaya Ivorensis ou anthoteca)
  - Tiama (Entendrophragma angolense ou congolense)
  - Ayous (Triplochiton scleroxylon)
  - Doussié (Afzelia Bipindensis)
  - Iroko (Chlorophora excelsa)
  - Kokrodua (Afrormosia, Pericopsis élata)
- et toutes les espèces d'arbres et arbustres pouvant porter fruits ou noix servant de nourriture pour la faune sauvage.

Au sens de cet article toute plante ligneuse de plus de 60 cm de hauteur est considéré arbre ou arbustre.

Les activités en matière de chasse et de pêche dans le domaine forestier sont soumises aux dispositions relatives à la conservation de la faune et l'exploitation des eaux.

Article 13. - Dans les forêts protégées, l'exercice des droits d'usage continue d'être libre et ne donne lieu au paiement d'aucune redevance forestière. Les bénéficiaires de ces droits pourront se livrer à l'exploitation même commerciale des palmiers, kapokier, rotins et autres plantes dont les récoltes leur appartiennent traditionnellement sous réserve que les récoltes soient faites de manière à ne pas détruire les végétaux producteurs. Ils pourront également se livrer à l'exploitation même commerciale des menus produits forestiers tels que gaulettes, perches, poteaux, bambous, planches éclatées et bois de chauffe à usage domestique, - et faire des cultures sur sol forestier après défrichement et incinération des arbres.

Toutefois des arrêtés du ministre chargé des Eaux et Forêts, réglementeront les saignées ou les interdiront dans certaines zones où la conservation des végétaux concernés serait en péril. L'abattage, la mulitation ou la détérioration des peuplements d'Okoumé, de Limba ou autres essences désignées par un arrêté du ministre des Eaux et Forêts sont interdits pour la préparation des terrains de culture. Les cultures pourront être interdites là où la rareté ou la dégradation des boisements nécessitera cette mesure.

Article 14. - Dans les forêts classées, l'exercice des droits d'usage est, sauf exception expresse, limité au ramassage du bois mort gisant, à la récolte des fruits et des plantes alimentaires, médicinales ou à usage religieux et au parcours des animaux domestiques.

L'exercice des droits d'usage ainsi reconnus est toujours subordonné à l'état et à la possibilité des forêts. Le Ministre chargé des Eaux et Forêts peut restreindre ou interdire l'exercice de tous ou quelques-uns de ces droits en fonction de l'état de boisement.

Article 15 (nouveau). - Les droits d'usage sur les forêts classées pourront être rachetés par voie de cantonnement ou moyennant une indemnité en argent. Les conditions de ce rachat seront déterminées de gré à gré ou à défaut d'accord entre les intéressés et le Service des Eaux et Forêts, fixées par décret.

Par contre le droit de parcours des moutons et chèvres peut être interdit ou retiré sans compensation dans tous les cas où l'intérêt public l'exige. Il est spécialement interdit d'exercer le droit de parcours dans les forêts aménagées, dans les terrains repeuplés artificiellement ou reboisés, dans les parcelles incendiées des forêts classées durant les dix années après l'incendie, ainsi que dans les périmètres de reboisement.

Article 16. - L'abattage et la mutilation des kapokiers, arbres ou lianes à latex, roniers et palmiers à huile sont interdites dans les forêts classées, sauf autorisation préalable du Chef régional du Service des Eaux et Forêts.

Article 17. - L'exploitation commerciale, dans les forêts classées, des palmiers, kapokiers, rotins et autres plantes dont les récoltes appartiennent traditionnellement aux usagers des droits d'usage, est subordonnée à la délivrance par le Service des Eaux et Forêts d'un permis spécial indiquant où peut être effectuée l'exploitation ou la récolte dans un but commercial.

Ce permis spécial, qui est délivré gratuitement à la demande des titulaires des droits d'usage, peut être retiré par le Chef régional du Service des Eaux et Forêts si le bénéficiaire du permis ne se conforme pas à la réglementation en vigueur et notamment ne fait pas les récoltes de manière à ne pas détruire les végétaux producteurs.

Ce permis spécial peut également être accordé à un particulier si les groupements humains concernés par les forêts classées en question ont renoncé à l'exploitation commerciale précisée ci-dessus. Le permis, accordé alors pour une durée déterminée, comportera un cahier des charges dont les clauses tendront à préserver l'avenir de la population locale.

Article 18. - L'exploitation commerciale, dans les forêts classées, des menus produits forestiers tels que gaulettes, perches, poteaux, bambous, planches éclatées, bois de chauffe à usage domestique, est interdite.

Il y est également interdit de faire des cultures après défrichement et incinération des arbres.

Toutefois le Ministère chargé des Eaux et Forêts pourra autoriser sous réserve des prescriptions de l'alinéa 1er de l'art.21 les cultures temporaires sur brûlis placées sous la surveillance du Service des Eaux et Forêts qui en déterminera les emplacements et les modalités d'exécution.

Article 19. - Les périmètres de reboisement sont affranchis de tous droits d'usage.

Il est notamment interdit d'y introduire du bétail ou d'y faire des cultures après défrichement et incinération de la couverture végétale.

Article 20. - Les parcs nationaux sont également affranchis de tous droits d'usage.

Cependant le Ministre chargé des Eaux et Forêts, peut pour chacun des parcs à proximité desquels ne subsisteraient pas des superficies suffisantes pour l'exercice des droits d'usage, prendre un arrêté réglementant l'exercice de certains droits d'usage de manière à ce que la protection et la conservation de la faune, de la flore ou du site soient assurées et l'aménagement du parc national respecté.

Article 21. - Dans toutes les dépendances du domaine forestier, sont interdits l'abattage, la mutilation, la détérioration ou l'incinération des peuplements d'Okoumé, de Limbas, et d'autres essences comprises dans une liste dressée par arrêté du ministre chargé des Eaux et Forêts en vue ou à la suite de la préparation de terrains de culture.

Cependant le Chef du Service régional des Eaux et Forêts peut autoriser, dans les dépendances du domaine forestier où s'appliquent les droits d'usage, un abattage d'okoumés limités au nombre d'unités strictement nécessaires pour la fabrication des pirogues destinées à la satisfaction des besoins personnels des bénéficiaires des droits d'usage, à leur demande.

Article 22. - Dans toutes les dépendances du domaine forestier, il est interdit d'abandonner un feu non éteint.

Il est défendu de porter ou d'allumer du feu en cas d'établissement d'une exploitation, en dehors des habitations et des bâtiments d'exploitation à l'intérieur des forêts classées ou protégées, ou à une distance de 500 mètres de telles forêts situées en bordure de savane ou bien à l'intérieur ou à la même distance des périmètres de reboisement et des parcs nationaux.

Cependant, des charbonnières, des fours à charbon et des fours pour l'extraction du goudron et de la résine pourront être établis dans les forêts protégées et les forêts classées et dans la zone de 500 mètres autour de telles forêts situées près de savanes, par les exploitants forestiers après autorisation du Chef de Service des Eaux et Forêts. Ces installations seront faites, sous la responsabilité des exploitants, sur un sol complètement désherbé dans un rayon d'au moins 50 mètres autour de chaque installation.

Pour prévenir les incendies de forêts, les autorités administratives et forestières locales pourront organiser et diriger l'allumage de feux précoces en bordure des dépendances du domaine forestier et le long des voies qui les traversent.

L'ordre d'allumer ces feux précoces ne pourra être donné par l'autorité administrative locale qu'après qu'une publicité suffisante aura été faite afin que les villages riverains des dépendances du domaine forestier prennent les mesures de sécurité appropriées. La responsabilité de l'administration ou de ses agents sera dégagée en cas de dommages causés par ces feux précoces si la publicité préalablement faite était suffisante.

Pour combattre un incendie d'une dépendance du domaine forestier ou un incendie menaçant une telle dépendance, l'autorité administrative locale ou, à défaut, le Chef local du Service des Eaux et Forêts peut requérir, même verbalement, les habitants des villages riverains de la dépendance du domaine forestier incendiée ou menacée et toute personne se trouvant à proximité.

L'opération sera organisée et dirigée par les autorités locales administrative et forestière. Leur responsabilité n'est pas engagée à l'occasion de l'organisation et de la direction de lutte contre l'incendie.

Les requis pourront par tous moyens faire la preuve de leur réquisition.

TITRE II : L'UTILISATION DU DOMAINE FORESTIER

Chapitre Premier : Les principes fondamentaux de la gestion, de la conservation, de la reconstitution, de l'aménagement et de l'exploitation économique du domaine forestier

Article 23. - Il appartient à l'administration de veiller strictement sur le plan régional et national à ce que les activités autorisées dans le domaine forestier se fassent de manière à éviter la destruction du domaine, à assurer sa permanence, son extension et son exploitation dans les conditions rationnelles.

Article 24. - Les produits forestiers exploités devront, dans toute la mesure du possible, être transformés au Congo, de manière que les exportations portent en définitive non sur des matières premières, mais sur des produits finis.

La première transformation de bois sera effectuée à proximité des coupes.

Article 25. - Tout en sauvegardant les droits des investisseurs étrangers l'économie forestière devra progressivement passer aux mains des nationaux.

A cette fin seront prises des mesures tendant à promouvoir des entreprises para-étatiques, d'économie mixte ou privées dont le capital, comme les cadres seront congolais.

Les entreprises étrangères ou gérées par des étrangers passeront à long-terme sous contrôle congolais; les formalités de passage sous contrôle congolais seront régies par des dispositions contractuelles, arrêtées avec ces entreprises, au moment de leur installation.

Les entreprises privées congolaises pourront bénéficier d'aide technique et financière.

Article 26. - Les taxes domaniales et forestières seront fondées uniquement sur des critères économiques de manière à épouser la valeur des produits sans interrompre ni même freiner l'expansion et la permanence de l'économie forestière dans les régions.

Certaines taxes forestières seront obligatoirement affectées à des comptes spéciaux hors budget pour financer uniquement la conservation, la reconstitution et l'aménagement du domaine forestier et le développement de la pisciculture.

La fiscalité forestière demeurera stable par période quinquénale et ne pourra être revisée que tous les cinq ans.

Chapitre Deuxième : La Gestion, la Conservation, la Reconstitution et l'Aménagement du Domaine Forestier

Article 27 (nouveau). - L'administration forestière prépare le plan d'aménagement, qui comporte les opérations concernant l'évaluation des richesses forestières, les modalités d'exploitation de ces richesses, les mesures et travaux de conservation et l'aménagement du domaine forestier.

Ce plan pourra en outre dénombrer les sites remarquables du point de vue touristique cynégétique. Il précisera les possibilités d'établissement d'entreprises de pisciculture, de parcs nationaux et de réserves naturelles.

Article 28. - Le Service des Eaux et Forêts prépare par ailleurs un inventaire forestier national.

Les normes techniques, les données à relever et les méthodes applicables pour la confection de cet inventaire, ainsi que le programme annuel de ces travaux, doivent être approuvés par le Ministre des Eaux et Forêts.

L'exécution des travaux incombe au Service des Eaux et Forêts qui peut cependant, sous sa direction et sa responsabilité, la sous-traiter à des organismes spécialisés et présentant une qualification suffisante.

Excepté dans les zones du domaine forestier où un inventaire assimilable à l'inventaire déterminé ci-dessus a déjà été effectué, l'installation et l'ouverture des chantiers forestiers sont subordonnées à la confection préalable de l'inventaire forestier national dans la partie concernée du domaine forestier.

Article 29. - Le domaine forestier est divisé en circonscriptions forestières de base pour l'exécution des tâches de gestion, conservation, reconstitution et exploitation du domaine forestier : ce sont les "unités forestières d'aménagement".

Le découpage effectif du domaine en unité d'aménagement est fait par arrêté du Ministre chargé des Eaux et Forêts en fonction des caractéristiques forestières propres à chaque zone et sur la proposition du Service des Eaux et Forêts.

Le plan d'aménagement, établi par le Service des Eaux et Forêts et approuvé par le Ministre des Eaux et Forêts avant son exécution, comporte pour chaque unité d'aménagement : une liste des essences les plus recherchées, la détermination d'un volume maximal annuel de coupe de ces essences et la fixation de la durée de la période d'exploitation de l'unité d'aménagement; cette durée est égale au temps nécessaire aux jeunes arbres subsistant après la coupe, pour atteindre un diamètre nettement supérieur au diamètre minimum d'exploitabilité fixé par arrêté du Ministre chargé des Eaux et Forêts.

L'exploitation globale d'une unité d'aménagement ne peut dépasser pour chaque essence le volume maximal annuel de coupe. Un contingent annuel limitatif concernant chaque essence des plus recherchées est assigné à chaque exploitant forestier en fonction du volume maximal annuel de coupe.

Le plan d'aménagement doit être remis à jour tous les cinq ans; le nouveau plan est également soumis à l'approbation du Ministre chargé des Eaux et Forêts avant son application.

L'application du plan d'aménagement est confiée au Service des Eaux et Forêts.

Le programme de reboisement est exécuté par l'organisme chargé de reboisement.

Article 30. - La gestion, la protection et l'aménagement du domaine forestier, la protection et l'aménagement de la faune et le développement de la pisciculture sont confiés au Service des Eaux et Forêts, et financés par la taxe forestière d'aménagement affectée à un compte spécial de dépôt ouvert à la B.N.D.C. sous le nom de "Fonds d'aménagmeent et des ressources naturelles".

Le financement des travaux de reboisement confié au Service autonome d'Etat à créer, sera assuré par la taxe forestière de reboisement affectée à un compte spécial ouvert à la B.N.D.C. sous le nom de "Fonds de Reboisement".

Chapitre Troisième : L'exploitation économique du Domaine Forestier

Article 31. - L'attribution des droits d'exploitation du domaine forestier excepté dans les cas prévus par la Loi au bénéfice des titulaires des droits d'usage, n'est jamais gratuite.

Toutes les entreprises forestières, même les sociétés d'Etat ou à participation étatique sont également assujetties aux redevances fixées par la Loi. Les procédures de recouvrement forcé, à défaut de paiement aux échéances normales, leur sont également applicables.

Article 32. - La concession des droits d'exploitation du domaine forestier est faite par contrat d'exploitation forestière, par contrat de transformation industrielle de bois, par l'attribution de permis de bois d'oeuvre et permis spéciaux.

Article 33. - Peuvent solliciter le bénéfice des contrats et permis prévus par l'art.32 ci-dessus les Sociétés d'Etat, les sociétés à participation d'Etat, les sociétés privées à capitaux purement congolais ou exclusivement étrangers, les sociétés où les capitaux congolais et étrangers sont associés et les particuliers de nationalité congolaise.

Article 34. - Les contrats d'exploitation forestière et les contrats de transformation industrielle de bois garantissent à leur titulaire le droit de prélever sur une unité d'aménagement, découpée ou non en unité d'exploitation, des contingents annuels limitatif des essences les plus recherchées conformément à l'art.29 ci-dessus.

Pour permettre la fixation sur le terrain de l'assiette de ces contingents annuels limitatifs, les titulaires de ces contrats doivent en dehors de l'évaluation de la richesse forestière faite par le plan d'aménagement, exécuter des comptages préalables des essences les plus recherchées.

Article 35. - Le permis de bois d'oeuvre confère à son titulaire le droit d'exploiter dans les zones forestières déterminées par arrêté du ministre des Eaux et Forêts un nombre limité d'arbres.

Article 36 (nouveau). - Le permis spécial confère à son titulaire le droit d'exploitation en quantité limité des produits forestiers accessoires, destinés exclusivement à la consommation domestique.

Article 37. - Les contrats et permis prévus par l'art.32 ci-dessus sont strictement personnels et ne peuvent être ni cédés, ni sous-traités.

Les personnes appelées à recueillir par voie d'héritage des biens mobiliers se trouvant sur une exploitation en activité, sont autorisées à poursuivre l'exploitation dans les mêmes conditions que leur auteur jusqu'à l'échéance du contrat ou du permis à moins qu'elles ne présentent pas les aptitudes nécessaires pour poursuivre efficacement l'exploitation.

Si une entreprise est judiciairement déclarée en état de cessation de paiements, le Tribunal qui aura constaté cet état pourra, après avis du Chef de Service des Eaux et Forêts, nommer même parmi les fonctionnaires des Eaux et Forêts un liquidateur chargé de poursuivre l'exploitation pendant les opérations de liquidation. Le contrat ou le permis dont cette entreprise est titulaire ne peut être cédé à aucun créancier en compensation des dettes de l'entreprise.

Article 38. - Peuvent bénéficier d'un contrat, les pétitionnaires visés à l'art.33 ci-dessus qui auront été sélectionnés en raison de l'impact économique de leur programme d'action concernant l'exploitation d'une surface forestière déterminée. Au terme du contrat le Gouvernement décidera compte tenu de la gestion du titulaire et de ses propositions pour l'avenir s'il signe le contrat suivant avec l'ancien titulaire ou au contraire avec un nouveau pétitionnaire. Le nouveau contractant est tenu de racheter l'entreprise à son prédécesseur, suivant les conditions prévues par le présent décret d'application.

Il sera rédigé un "contrat d'exploitation" lorsque les activités se limiteront à l'exploitation des arbres : sa durée ne pourra excéder sept ans.

Il sera rédigé un "contrat de transformation" lorsque les activités comporteront en outre l'implantation d'une usine de traitement de grumes. Sa durée sera fonction de volume des investissements auxquels il se rapporte.

L'utilisation de scies ou dérouleuses mobiles ne donneront pas lieu à l'obtention d'un contrat de transformation.

Article 39 (nouveau). - Les candidatures sont suscitées par un arrêté du Ministre des Eaux et Forêts, qui lance un appel d'offre.

L'offre porte sur des surfaces bien définies. L'arrêté précise les conditions auxquelles doivent satisfaire les dossiers des pétitionnaires

Les candidatures et dossiers sont examinés par la Commission Forestière visée à l'art.40 qui émet un avis à l'adresse de l'autorité dont relève l'approbation du contrat.

Article 40 (nouveau). - Les contrats d'exploitation après avis de la Commission Forestière, sont préparés et visés par le Secrétaire Général aux Eaux et Forêts, approuvés et signés par le Ministre des Eaux et Forêts, qui confirmera cette approbation par un arrêté.

La composition et le fonctionnement de la Commission Forestière, présidée par le Ministre des Eaux et Forêts, sont fixés par décret pris en Conseil des Ministres.

Article 41. - Les contrats de transformation relatifs à des entreprises qui ne peuvent pas prétendre au bénéfice des régimes privilégiés prévu au Code des Investissements sont instruits et approuvés dans les mêmes conditions que les contrats d'exploitation.

Article 42. - Les contrats de transformation relatifs à des entreprises qui peuvent prétendre au bénéfice des régimes privilégiés prévu au Code des Investissements sont, après avis de la Commission des Investissements, préparés et visés par le Directeur des Eaux et Forêts et approuvés et signés par le Ministre des Eaux et Forêts qui confirmera cette approbation par un arrêté.

La Commission des Investissements est celle qui est prévue au Code des Investissements.

Article 43. - Les contrats comporteront deux parties :

- Le contrat proprement dit qui a un caractère synallagmatique et détermine les droits et obligations des parties;
- Le cahier de charge particulier qui précise les charges de l'entreprise, autres que celles prévues dans le cahier des charges général, notamment en ce qui concerne : le plan d'exploitation, les installations, la formation professionnelle à l'intérieur de l'entreprise et les infrastructures sociales ou d'exploitation, et dont les différentes clauses s'imposent au bénéficiaire du contrat.

Article 44. - Les permis de bois d'oeuvre sont attribués par décision du Directeur des Eaux et Forêts. Les permis spéciaux sont attribués par décision du Chef d'Inspection Forestière.

Article 45. - Un décret édictera un cahier des charges général concernant les contrats et permis. Il se rapportera à l'organisation, aux modalités et au contrôle de l'exploitation, de la circulation et de la commercialisation des produits forestiers.

Article 46. - Ce décret fixera également les conditions d'exercice des activités du bois et la procédure d'attribution des contrats et permis.

TITRE III : REPRESSION DES INFRACTIONS

Chapitre Premier : De l'exercice des poursuites

Section 1 : Recherches et constatation des délits (art.47 à 57 omis)
Section 2 : Confiscation et saisie (art.58 à 62 omis)
Section 3 : Actions et poursuites (art.63 à 69 omis)
Section 4 : Transactions (art.70 omis)

Chapitre II : Infractions et pénalités

Section 1 : Coupes et exploitations non autorisées, mutilations et autres actions préjudiciables aux arbres (art.71 à 73 omis)
Section 2 : Marteaux forestiers, Marques (art.74 omis)
Section 3 : Exploitation (art.75 à 80 omis)
Section 4 : Cultures en forêts - Feux de brousse - incendie de forêts (art.81 à 83 omis)
Section 5 : Pâturages (art.84 omis)
Section 6 : Infractions diverses (art.93 abrogé) (art.85 à 92 omis)
Section 7 : Dispositions diverses (art.94 à 102 omis)

TITRE V (IV) : DISPOSITIONS TRANSITOIRES
(art. 103 à 108 abrogés par Loi N° 32/82)

Article 109. - Toutes les dispositions réglementaires antérieures à la présente Loi notamment la Loi 34/61 du 20 Juin 1961, les décrets 62/211 et 62/212 du ler Août 1962 et tous les décrets et arrêtés pris postérieurement à ces textes sont abrogés.

Sont abrogées également les dispositions des Lois N° 31/61 du 3 Juin 1961 et 37/63 du 4 Juillet 1963.

Article 110. - La présente loi sera exécutée comme Loi de l'Etat.

Fait à Brazzaville le 4 Janvier 1974

DECRET N°____ DU ____

PORTANT APPLICATION DU CODE FORESTIER

CONGO

Décret d'Application

STRUCTURES ET MATIERES REGLEMENTEES PAR LE DECRET D'APPLICATION (PROJET) DU CODE FORESTIER (LOI N° 004/74)

Le Premier Ministre, Chef du Gouvernement,

Vu la Constitution du 8 Juillet 1979;
Vu la Loi 25/80 du 13 Novembre 1980 portant amendement de l'article 47 de la Constitution du 8 Juillet 1979;
Vu la Loi 004/74 du 4 Juillet 1974 portant Code Forestier;
Vu la Loi 32/82 du 7 Juillet 1982 portant modification du Code Forestier;

Sur proposition du Ministre des Eaux et Forêts;
Le Conseil des Ministres entendu.

DECRETE

TITRE PREMIER : CAHIER GENERAL DES CHARGES DE L'EXPLOITATION FORESTIERE

Chapitre I : Professions du bois

Article 1. - Agrément pour les professions relatives aux activités forestières.

Article 2. - Délivrance d'une carte d'identité professionnelle.

Article 3. - Catégories des professions relatives aux activités forestières.

Article 4. - Activités cumulées.

Article 5. - Application des dispositions aux entreprises publiques et privées.

Article 6. - Publication d'informations statistiques et tarifaires.

Article 7. - Corporations représentant la profession.

Chapitre II : Les modalites de l'exploitation

Article 8. - Subdivisions des surfaces forestières.

Article 9. - Régénération des peuplements.

Article 10.- Catégories d'exploitation.

Article 11.- Les Permis.

Article 12.- Les Contrats.

Fait à Brazzaville, le____ 1982

Le Ministre
des Eaux et Forêts

Le Premier Ministre,
Chef du Gouvernement

ANNEXES AU DECRET D'APPLICATION :

Annexe I : Etat à fournir trimestriellement et annuellement par les exploitants.
Annexe II : Registre d'usine.
Annexe III : Etat à fournir annuellement par tout exportateur même s'il est usinier ou exploitant.
Annexe IV : Etat à fournir mensuellement par les usiniers.
Annexe V : Etat à fournir par les usiniers, trimestriellement et annuellement.

LOI N° 005/74 DU 4 JANVIER 1974

fixant les redevances dues
au titre de l'exploitation des
ressources forestières

CONGO

Loi N° 005/74

L'Assemblée Nationale populaire a délibéré et adopté :

Le Président de la République promulgue la Loi
dont la teneur suit :

VU la Constitution

VU la Loi N° 31/61 du 3 Juin 1961 fixant les redevances en matière forestière modifiée par la Loi 37/63 du 4 Juillet 1963

VU la Loi N° 004/74 du 4 Janvier 1974 portant Code Forestier dans la République Populaire du Congo

TITRE PREMIER : DISPOSITIONS GENERALES

Article 1. - En dehors des dispositions concernant l'exercice des droits d'usage coutumiers, définis et réglementés par les art.11 et suivants de la Loi N° 004 du 4 Janvier 1974 tous les produits de la forêt appartiennent à l'Etat, qui en confie l'exploitation à des entreprises, moyennant le paiement par celles-ci de redevances qui correspondent à la valeur du produit en son état naturel.

Toutes les entreprises paient ces redevances, quel que soit leur statut.

L'Etat reste copropriétaire des produits exploités, quel que soit leur degré de transformation, pour une part correspondant à la valeur des redevances, tant que celles-ci ne sont pas payées.

Les redevances non payées à l'échéance sont automatiquement pénalisées d'une augmentation de 1% par mois de retard.

En cas de défaillance du débiteur, l'Etat représenté par le Ministre des Eaux et Forêts peut procéder au recouvrement de la dette par vente directe des produits qui portent la marque du débiteur ou par prélèvement sur le montant des ventes. L'application de ces dispositions est obligatoire si le retard dépasse six (6) mois. Le Ministre des Eaux et Forêts informe alors les négociants ou usiniers que les produits du débiteur seront assujettis au prélèvement de la dette et qu'ils seront tenus de la verser aux conditions fixées par le Ministre des Eaux et Forêts, à charge pour eux de la déduire des sommes qu'ils ont eux-mêmes à payer au débiteur.

Si par suite de cessation d'activité, la dette ne peut être recouvrée, le Ministre des Eaux et Forêts saisira le Receveur des Domaines.

Article 2. - Ces redevances sont :

- les "redevances sur les bois en grumes".
  Elles sont versées au Trésor et alimentent le budget de l'Etat.

- les "taxes forestières".
  Elles alimentent à part égale le "Fonds de reboisement" et le "Fonds d'aménagement des ressources naturelles".
  Elles sont versées au compte spécial B.N.D.C. de chacun de ces Fonds.

Article 3. - Les redevances sur les bois en grumes sont perçues au moment où les produits sont commercialisés :

- soit auprès des exportateurs : le recouvrement est alors assuré par le Service des Douanes;

- soit auprès des usines de transformation : le recouvrement est alors assuré par le Service des Eaux et Forêts.

Article 4. - Les taxes forestières sont exigibles au moment de la délivrance des autorisations d'exploitation, c'est à dire à la remise de la décision d'attribution d'un permis ou de l'autorisation de coupe annuelle relative à un contrat.

Elles sont payées :

- soit en espèces, en une seule fraction et d'avance;
- soit par "prélèvement d'office" sur le compte bancaire de l'exploitant, en douze mensualités. Dans ce cas, l'exploitant remet au service des Eaux et Forêts un "ordre de prélèvement d'office" à l'adresse de sa banque. Cet ordre stipule les sommes qui doivent être versées, les dates de paiement, les numéros de compte du Trésor auxquels les sommes doivent être virées. Il précise également que cet ordre est valable un an sans possibilité de résiliation et indique les pénalités à payer en cas de retard.

Article 5. - Les "redevances sur les bois" et les "taxes forestières" sont toujours exprimées en pourcentage des valeurs FOB.

La valeur FOB est :

- soit celle qui est pratiquée par un Office National de commercialisation;
- soit la valeur FOB moyenne estimée par cet Office compte tenu du marché et des feuilles de spécification produites par les exportateurs.

Les offices sont tenus de faire connaître au moins 15 jours à l'avance à la Direction des Douanes et à la Direction des Eaux et Forêts le changement de cours des bois dont ils ont le monopole d'exportation.

Pour les autres bois, ils envoient semestriellement avant le 15 Décembre et le 15 Juin un état des valeurs FOB estimées ou calculées sur les six (6) mois précédents qui serviront de base au calcul des redevances pour le semestre suivant.

Les valeurs FOB prises en considération sont exclusivement celles des qualités LM et BC pour les bois divers et QST et sciage pour l'Okoumé.

Article 6. - Le décompte des redevances sur les lots de bois en grumes destinés à l'exportation figurera sur les factures de réception que le négociant ou l'usinier remet au producteur.

Article 7. - Le décompte des redevances sur les lots de bois en grumes exportés figurera au bas de la feuille de spécification.

TITRE II : LES REDEVANCES SUR LES BOIS EN GRUMES

Article 8. - Les redevances sur les bois en grumes sont payées au moment où ils entrent en usine ou au moment où ils sont exportés.

Les bois exportés après transformation en usine ne paient pas de redevance.

Les produits partiellement transformés sur les chantiers par des scies ou dérouleuses mobiles sont assimilés à des grumes et paient la redevance à l'usine ou à l'exportation du bois en grume de la qualité la plus basse.

Article 9. - Les bois exportés en grumes paient une "redevance à la sortie" qui est fonction de leur qualité et de leur origine.

Les forêts sont classées en huit (8) catégories, selon les frais de transport qu'elles doivent supporter pour évacuer leurs produits. A chacune de ces catégories correspond un tarif de redevance.

Les limites des zones d'application des différents tarifs sont précisées à l'art.11 de la présente Loi.

Le marteau triangulaire de l'exploitant porte le numéro correspondant à la catégorie de la région.

Les taux de la redevance à la sortie, exprimés en % des valeurs FOB sont les suivants :

Redevances sur les grumes exportées

Essence	Groupe	Assiette Valeur FOB		Taux de Redevance par Région							
				1	2	3	4	5	6	7	8
Okoumé	1	qualité	standard	33	30	27	24	21	15	6	-
	2	"	sciage	20	15	11	4	4	4	-	
	3	"		4	4	4	4	4	4	-	
Limba	1	"	LM	25	23	21	19	17	14	11	9
	2	"	BC	18	15	12	10	7	3	3	3
	3	"	BC	10	8	5	3	3	3	3	3
Sapelli	1	"	LM	17	15	14	12	11	8	7	5
	2	"	BC	9	7	4	3	3	3	3	3
Sipo	1	"	LM	21	19	18	17	15	14	12	11
	2	"	BC	11	10	8	7	6	3	3	3
Tiama	1	"	LM	18	15	11	9	5	3	3	3
	2	"	BC	4	4	4	4	4	4	4	4

Kossipo	1	qualité	LM	18	15	11	9	5	3	3
	2	"	BC	4	4	4	4	4	4	4
Acajou	1	"	LM	19	16	13	11	7	3	3
	2	"	BC	16	12	7	4	4	4	4
Dibétou	1	"	LM	19	18	16	14	12	10	8
	2	"	BC	11	8	6	4	4	3	3
Douka	1	"	LM	16	13	11	10	7	4	3
	2	"	BC	8	5	3	3	3	3	3
Moabi	1	"	LM	16	13	11	10	7	4	3
	2	"	BC	8	5	3	3	3	3	3
Longui-mukali	1	"	LM	14	12	11	10	9	8	7
	2	"	BC	9	7	6	4	3	3	3
Tchitola	1	"	LM	14	12	11	10	8	6	4
	2	"	BC	9	7	5	4	3	3	3
Agba	1	"	LM	17	14	11	8	6	3	3
	2	"	BC	12	9	5	3	3	3	3
Mutenyé	1	"	LM	15	13	12	11	9	6	3
	2	"	BC	7	5	3	3	3	3	3
Doussié	1	"	LM	14	11	9	8	6	4	3
	2	"	BC	6	4	3	3	3	3	3
Pao-Rose	1	"	LM	20	17	14	12	9	5	3
	2	"	BC	15	12	7	5	3	3	3
Kokrodua *			LM	-	-	-	-	-	14	10
Wengué *			LM	-	-	-	-	-	16	12
Iroko *			LM	17	12	6	3	3	3	3
Padouk *			LM	17	12	6	3	3	3	3
Autres bois*		Okoumé sciage		10	8	6	3	3	3	3

* quelle que soit la qualité

Article 10. - Les bois entrant en usine paient une "redevance à l'usine" qui est fonction de leur qualité.

Les taux de la redevance à l'usine exprimés en des valeurs FOB sont les suivants :

Redevance sur les grumes entrées en usine

Essence et qualités			Assiette Valeur FOB	TAUX
Okoumé	Lot comprenant moins de 40% QST		Qualité QST	14,5
	excédent QST		"	10,5
	sciage et déclassé		"	3,5
Limba	qualités du groupe	1	Qualité LM	4
	"	2	"	2,5
Sapelli	"	1	"	5,5
	"	2	"	3
Sipo	"	1	"	4,5
	"	2	"	3
Tiama	"	1	"	6
	"	2	"	3,5
Kossipo	"	1	"	5,5
	"	2	"	3
Acajou	"	1	"	5,5
	"	2	"	3,5
Dibétou	"	1	"	4,5
	"	2	"	2,5
Douka	"	1	"	5
	"	2	"	2,5
Moabi	"	1	"	5
	"	2	"	2,5
Mutényé	"	1	"	4
	"	2	"	3
Kokrodua	"	1	"	6
	"	2	"	4,5
Wengué	"	1	"	5,5
	"	2	"	4
Tchitola	"	1	"	3
	"	2	"	4
Autres bois quelque soit la qualité			Okoumé qualité sciage	3

Article 11. - Tarif N° 1 - Zone forestière de Pointe-Noire ....
(suit délimitation de la zone)

Tarif N° 2 - Zone forestière de M'Vouti ....
(suit délimitation de la zone)

Tarif N° 3 - Zone forestière de Dolisie ....
(suit délimitation de la zone)

- Zone forestière de Youbi ....
(suit délimitation de la zone)

Tarif N° 4 - Zone forestière de Koukouati ....
(suit délimitation de la zone)

- Zone forestière de Mossendjo - Sibiti-Madingou ....
(suit délimitation de la zone)

Tarif N° 5 - Zone forestière de Kibangou ....
(suit délimitation de la zone)

- Zone forestière de Mayoko ....
(suit délimitation de la zone)

- Zone forestière de Kindamba - Kinkala ..
(suit délimitation de la zone)

Tarif N° 6 - Zone forestière de Divenie ....
(suit délimitation de la zone)

- Zone forestière de Zanaga ....
(suit délimitation de la zone)

Tarif N° 7 - Zone forestière du Nord distante de 50 Km au plus à vol d'oiseau, d'une voie d'évacuation par flottage.

Tarif N° 8 - Zone forestière du Nord distante de plus de 50 Km à vol d'oiseau, d'une voie d'évacuation par flottage.

Article 12. - Afin qu'il n'y ait aucune ambiguïté sur la nature des essences taxées, elles seront toujours désignées sur les feuilles de de spécification par le nom- pilote adopté par l'Association technique internationale des bois tropicaux.

Cependant, pour ne pas gêner les coutumes commerciales, l'expéditeur pourra faire suivre ce terme par le nom, placé entre parenthèse, auquel est habitué son client.

Article 13. - Pour les bois en grumes, les qualités sont désignées suivant les dispositions ci-après :

Le nom de l'essence, inscrit selon les termes de l'art. précédent, est suivi de la qualité commerciale du lot, elle-même suivie de l'indication (placée entre parenthèse), du groupe dans lequel entre cette qualité pour la taxation. Par exemple : Mutényé (Benzi) LM (premier groupe).

Les qualités sont groupées de la façon suivante :

1° - Okoumé :	premier groupe	: QST ou LM, QS exédent 2è, 3è choix
	deuxième groupe	: Sc petit Ø 3è choix 1/
	troisième groupe	: Déclassé
2° - Limba :	premier groupe	: LM-A-B-AB déroulage, tranchage
	deuxième groupe	: BC-A/3 noir, petit diamètre
	troisième groupe	: ½ noir, déclassé, sciage ...
3° - Autres essences	premier groupe	: LM-A-B-AB déroulage, tranchage
	deuxième groupe	: BC - déclassés + sciage

1/ Suivant classement adopté par l'office.

TITRE III : LES TAXES FORESTIERES

Article 14. - Les taxes forestières sont :

- la "taxe d'aménagement" qui est versés au compte de dépôt de la BNDC du Fonds d'Aménagement des ressources naturelles;
- la "taxe de reboisement" qui est versée au compte de dépôt de la BNDC du fonds de reboisement.

Article 15. - Ces taxes sont calculées sur le volume annuel des essences principales que les exploitations s'engagent à produire par contrat ou sur le volume de ces essences qui est porté sur la décision d'attribution du permis.

La valeur FOB à prendre en considération pour l'assiette de ces taxes est celle de ma meilleure qualité : QST pour l'Okoumé, LM pour les autres bois.

La liste des essences principales est spécifiée dans les contrats. Pour les permis, cette liste est fixée au niveau de chaque inspection forestière par le Directeur des Eaux et Forêts ainsi que l'équivalence de l'arbre en volume.

Article 16. - Les deux taxes sont égales.
Le taux global des deux taxes réunies est compris entre 2,5 et 3,5 de la valeur FOB considérée; il est fixé par le Ministre des Eaux et Forêts en fonction des difficultés d'exploitation.

Ce taux est mentionné dans les contrats et décisions relatives aux permis.

Article 17. - Elles sont perçues selon le tarif ci-après et réparties par moitié entre la taxe d'aménagement et la taxe de reboisement.

P R O D U I T S	Unités	Taxation
Poteaux (diamètre supérieur à 0,15 m)	par pieds	50
Perches (" " à 0,10 m)	"	20
(" inférieur à 0,10 m)	"	10
Gaulettes ou bambous	par pièce	2
Bois de chauffe	par stère	100
Ebène	par Kg	50

TITRE IV : DISPOSITIONS TRANSITOIRES

Article 18. - Les dispositions des titres un, deux et trois sont applicables dans un délai de deux mois après la publication de la présente loi. Cependant des dispositions transitoires sont prévues pour certains permis qui sont en cours de validité, mais qui en aucun das ne pourront être prorogés.

Article 19. - Le montant des "taxes forestières" payées par les titulaires de permis temporaires d'exploitation sera égal à la taxe territoriale relative à ce permis dont le tarif en vigueur est de 30 Francs par hectare. Ce tarif sera doublé au 1er janvier 1975 et triplé au 1er janvier 1976.

Article 20. - Les montants des "taxes forestières" et les "redevances à l'usine" devront être aménagées dans chaque cas particulier de façon à ne pas faire supporter à l'entreprise des charges plus importantes que celles qui lui incombaient antérieurement.

Article 21. - La première échéance annuelle des taxes forestières est calculée à partir de la date d'application de la présente loi, proportionnellement au nombre de mois entiers qui restent à courir jusqu'à la fin de l'année, déduction faite de la part excédentaire de la taxe territoriale qui aurait été payée d'avance sur les mois soumis aux taxes forestières.

Article 22. - Le Ministre des Eaux et Forêts établira à la date de mise en application de la présente Loi, un état des sommes dues au titre des taxes et redevances antérieures. Un échéancier de liquidation sera établi en accord avec les entreprises intéressées faute de quoi les dispositions de l'art. premier de la présente Loi seront appliquées six mois après la date de mise en application dudit texte.

Les produits de cette liquidation seront versés au Trésor, suivant la procédure habituelle.

La taxe de fermage, perçue sur les permis affermés sera maintenue jusqu'à ce que le fermage ait pris fin.

Article 23. - Les permis spéciaux et de bois d'oeuvre en cours de validité ne sont pas soumis aux dispositions relatives aux taxes forestières.

Article 24. - Les grumes d'Okoumé en provenance de permis situés dans la région du NIARI en cours de validité à la date de la mise en application de la présente Loi seront tous soumis au tarif N° 5 de la redevance à la sortie.

Article 25. - Afin de permettre aux entreprises de s'adapter progressivement aux nouveaux principes de taxation, le barème, objet de l'art.9 précédent, a été établi de façon à redresser seulement partiellement les iniquités existantes. En conséquence, en dehors de la révision quinquennalle prévue à l'art.26 du Code Forestier, une révision exceptionnelle sera effectuée 30 mois après la date de mise en application de la présente loi, afin de procéder à une nouvelle réduction des inégalités régionales.

TITRE V : FONDS D'AMENAGEMENT - FONDS DE REBOISEMENT

Article 26. - Conformément à l'art.30 du Code Forestier, il est créé un "Fonds d'aménagement des ressources naturelles". Ce Fonds est géré par le ministre des Eaux et Forêts qui en est l'ordonnateur et en dispose dans le cadre d'un programme annuel :

- pour le fonctionnement des services de gestion
- pour le financement de certains travaux d'amélioration, d'inventaire et autres tâches qui lui incombent.

Ce fonds est alimenté :
- par la taxe d'aménagement prévue à l'art. 2 ci-dessus
- par les subventions diverses, emprunts ou avances
- par les produits éventuels des activités du service forestier.

Un compte de dépôt B.N.D.C. au Trésor hors budget sera ouvert au Trésor, à l'intitulé de ce fonds.

Un décret précisera :
- les modalités de la gestion de ce fonds
- les modalités d'établissement et d'approbation du programme par l'autorité de tutelle.

Article 27. - Conformément à l'art.30 du Code Forestier, il est créé un "Fonds de reboisement". Ce fonds est géré par le ministère de la société chargé des travaux de reboisement, qui en est l'ordonnateur et en dispose dans le cadre d'un programme annuel pour le fonctionnement des stations de reboisement.

Ce fonds est alimenté :
- par la taxe d'aménagement prévue à l'art. 2 ci-dessus
- par les subventions diverses, emprunts ou avances
- par les produits éventuels des activités de l'Office Congolais des Forêts.

Un compte de dépôt B.N.D.C. au Trésor sera ouvert au Trésor, à l'intitulé de ce fonds.

Un décret précisera :
- les modalités de la gestion de ce fonds
- les modalités d'établissement et d'approbation du programme par le Comité de Direction.

Article 28. - Toutes les dispositions réglementaires antérieures à la présente Loi notamment les Lois N° 31/61 du 3 Juin 1961 et N° 37/63 du 4 Juillet 1963 sont abrogées.

Sont abrogées également toutes les dispositions concernant le Fonds Forestier et les taxes de reboisement.

Article 29. - La présente Loi entrera en vigueur à compter du 4 Janvier 1974.

Article 30. - La présente Loi sera exécutée comme Loi de l'Etat.

Fait à Brazzaville, le 4 Janvier 1974

DECRETO-LEY FORESTAL Nº 14/1 981
DE FECHA 29 DE SEPTIEMBRE

GUINEA ECUATORIAL
GUINEE EQUATORIALE

Decreto-Ley Nº 14/1 981

EXPOSICION DE MOTIVOS

La riqueza forestal de Guinea Ecuatorial constituye un don de la naturaleza que el Gobierno debe conservar en beneficio de las generaciones presentes y futuras, sin perjuicio de que un racional aprovechamiento permita obtener los beneficios de que tal riqueza es susceptible sin mengua del propio capital forestal. Tal es el objeto que inspira la necesidad de dictar el presente Decreto-Ley que regula el aprovechamiento de los bosques de Guinea Ecuatorial, estableciendo una serie de principios y normas de actuaciòn que se desarrollan a lo largo de los tres TITULOS que la integran.

El TITULO I regula el régimen dela riqueza y de la propiedad forestal, definiendo en torno a los bienes que nacen de los bosques, al àrea forestal y las industrias forestales, asi como el régimen juridico de las distintas clases de bosques en razòn de su propiedad con alusiòn al régimen de las aguas y de la caza.

El TITULO II aborda las concesiones forestales que constituiràn el régimen normal de aprovechamientos. Para ello, serà necesario la incoaciòn de un expediente administrativo, estableciéndose requisitos que deba contener el Pliego de Bases Administrativas y Técnicas, que sirven de orientaciòn para las actuaciones administrativas, asi como lo diversos tràmites del propio procedimiento.

Finalmente al TITULO III aborda las competencias de la Administraciòn Forestal sobre la riqueza de los bosques, inspiradas en los principios de conservaciòn y ordenaciòn, y la estructura administrativa que pueda irse adaptando a medida que las necesidades lo demanden.

En su virtud, a propuesta del Ministerio de Agricultura, Ganaderia, Pesca y Forestal, y previa deliberaciòn de la Junta Técnica; vengo en sancionar y promulgar el presente Decreto-Ley Forestal :

TITULO I : DEL REGIMEN JURIDICO Y ECONOMICO DE LA RIQUEZA FORESTAL

Capitulo I : Principios generales

Articulo 1. - El presente Decreto-Ley tiene por objeto establecer el régimen juridico y ecònomico que permita el màximo aprovechamiento de los recursos forestales de Guinea Ecuatorial con absoluto respeto a las leyes biològicas y naturales que aseguren su permanencia e incremento en beneficio de las generaciones presentes y futuras.

Capitulo II : De la riqueza forestal

Articulo 2. - En razòn de su aprovechamiento, los bienes de los bosques se clasifican en maderables y no maderables. Estos ùltimos se dividen en los que proceden del àrbol, como leñas (copas y ramas), los jugos (caucho) y las cortezas (taninos medicinales, aromàticas curinales); y las que no proceden del àrbol como los animales (animales libres, aves, peces), los vegetales (pastos hongos, setas, fibras, plantas aromàticas o medicinales); y los minerales (canteras).

En razòn de los servicios que prestan, los bosques se distinguen en servicio al suelo (evitar la erosiòn, laterizaciòn y consecuciòn de la recuperaciòn de los suelos agotados); servicios a las aguas (influir favorablemente en la regulaciòn y calidad); y servicios a la vida silvestre (servir de albergue y hacerla posible).

Articulo 3. - 1° A los efectos de esta Ley se entiende por àrea o especie forestal el terreno cubierto de vegetaciòn que, de manera espontànea o articicial, pueda persistir indefinidamente sin la intervenciòn del hombre.

2° Se entiende por fauna forestal los animales salvajes, asilvestiados o domésticos que vivan la mayor parte de su vida al aire libre y se alimentan con pastizales naturales no cultivados, e igualmente todos los peces fluviales y los cultivados en estanques de forma intensiva o semi intensiva.

Articulo 4. - 1° A los efectos de este Decreto-Ley, se entiende por industrias forestales aquellas que permitan elaborar la materia prima de manera que sea rentable su transporte y su venta a cualquier distancia.

2° Se consideraràn como tales las de madera hasta obtener tabla, tablòn, tablilla, puerta plana, elementos constructivos prefabricados de madera, muebles de madera, pasta celulosa y panel, las de caucho, hasta la elaboraciòn de planchas de goma; las de fibras, hasta la elaboraciòn de cuerdas, cordeles, envases y elementos de calzado y vestido; las de plantas aromàticas o medicinales hasta la obtenciòn del producto de su secado o destilado; las de energia, hasta la obtenciòn del metanol; y las de otros productos anàlogos o similares.

Articulo 5. - 1° Las industrias forestales se clasifican en libres y controladas. Son industrias libres aquellas que se puedan instalar sin ninguna clase de restricciones especificas.

2° Son industrias controladas aquellas que requieran una autorizaciòn especial, ya sea con el fin de regular la capacidad de producciòn comarcal o regional, ya sea para exigir un minimo de mecanizaciòn y técnologia que proporcione puestos de trabajo calificados.

3° Las industrias libres o controladas podràn tener caràcter preferente cuanco asi se declare por producir efectos beneficos para la economia nacional, ya sea por su localizaciòn, ya sea por el sector de la actividad econòmica que realicen. Dicha declaraciòn llevarà aneja los beneficios y preferencias que se determinen.

Capitulo III : De la propiedad forestal

Secciòn 1° - Clases de Bosques

Articulo 6. - En razòn de la titularidad dominical, los bosques pueden ser publicos estatales, comunales y privados.

Articulo 7. - Son bosques pùblicos estatales los que pertenecen al Estado con cualquier tìtolo, los bosques virgenes de primera cosecha que se hayan desarrollado por la sola intervenciòn de la naturaleza, con independencia de su mejora, o no, por tècnicos forestales y aquellos en que la repoblaciones se efectùen por el Estado de cualquiera de las siguientes maneras : a) directamente; b) financiadas por el Estado; c) realizadas por empresas colaboradoras; d) o se realice en virtud de compromisos impuestos a las empresas privadas.

Articulo 8. - Son bosques pùblicos comunales los existentes o que puedan ser objeto de repoblaciòn en las reservas de los poblados, cualquiera que sea su superficie y su régimen de administraciòn por las entidades comunales.

Articulo 9. - Son bosques privados aquellos que pertenezcan a personas fisicas o juridicas y que sean aprovechados a sus expensas, sin perjuicio de las competencias del Estado.

Secciòn 2° - Bosques Pùblicos Estatales

Articulo 10. - 1° Los bosques pùblicos estatales tienen la condiciòn de bienes de domicio pùblico. En consecuencia, no podràn ser enajenados, gravados, ni hipotecados; siendo imprescriptibles e inembargables.

2° Los bosques publicos estatales seràn objeto de aprovechamiento e régimen de gestiòn directa por el Estado o mediante concesiones de tiempo no superior a 20 años y en la condiciones que establece el TITULO II de este Decreto-Ley.

Articulo 11. - 1° Los bosques aprovechados en régimen de gestiòn directo, la administraciòn por simisma, o mediante empresas colaboradoras, deberà ejecutar las repoblaciones, las regeneraciones y toda clase de cuidados necesarios con el fin de obtener el màximo beneficio de los mismos, pero asegurando la permanencia de la riqueza forestal, asi como las mejoras y aplicaciones de su àrea, incluida en caso de necesidad el aprovechamiento maderable del bosque.

2° La enagenaciòn de productos podrà hacerse mediante subasta o adjudicaciòn directa, anual o pluri-anual, ya sea de la madera en pie, en cargadera, a pie de fabrica o sobre parco.

Articulo 12. - En los bosques pùblicos estatales se podràn conceder gratuitamente, o a precios simbòlicos, pequeñas superficies para cultivos de alimentos y àrboles aislados para construcciòn de viviendas y embarcaciones, utilizando para ello las especies apropiadas.

Articulo 13. - En los desbosques para carreteras, vias de saca o para plantaciones agricolas o forestales de caràcter permanente, se señalaràn y cubicaràn los àrboles que hayan de ser objeto de la enajenaciòn al precio de la madera en pie. En tales supuestos, se tomaràn las medidas que garanticen la realidad y operaciòn de tales plantaciones.

Secciòn 3° - Bosques Pùblicos Comunales

Articulo 14. - Los bosques pùblicos comunales tienen la condiciòn de bienes de dominio pùblico. En consecuencia, son inalienables imprescriptibles e inembargables. Tampoco podràn ser objeto de hipòteca, ni avalar ninguna clase de crédito, ni ser objeto de divisiòn, salvo que una ley lo disponga expresamente.

Articulo 15. - 1° El aprovechamiento de los bosques pùblicos comunales podrà realizarse por el Estado, directamente por las entidades comunales o mediante concesiòn.

2° Cuando el aprovechamiento ne se realice por la entidad propietaria del bosque, se procurarà que aquella se efectùe de acuerdo con los deseos y necesidades de ésta, salvo que atentare contra la permanencia misma del Bosque.

Articulo 16. - En todo caso, los bosques pùblicos comunales estaràn sometidos al control y vigilancia del Gobierno, quien velarà por su adecuado aprovechamiento para proporcionar a la entidad propietaria una renta sostenida y creciente, anual o periòdica.

Articulo 17. - Los ingresos que produzcan los bosques pùblicos comunales se destinaràn a financiar necesidades comunitarias, reservàndose el 20% de aquellas para mejora en los propios bosques.

Secciòn 4° - Bosques Privados

Articulo 18. - Corresponde al Estado en los bosques privados la inspecciòn y control precisos para que durante el periodo de desbosque se cumplan las finalidades de permanencia de la riqueza forestal y su adecuado aprovechamiento. Una vez creada la plantaciòn, la competencia del Estado se limitarà a la informaciòn y tutela que asegure la persistencia del bosque. El propietario estarà obligado a comunicar la producciòn del bosque y cuantas incidencias afecten a dicha producciòn, a los meros efectos estadisticos.

Capitulo IV : De las Aguas

Articulo 19. - Las aguas fluviales tendràn el caracter de dominio pùblico y estaràn sometidas al régimen juridico que se establezca. El Gobierno podrà sancionar, e incluso clausurar, aquellas actividades contaminantes de las aguas o que pongan en peligro su riqueza piscicola.

Capitulo V : De la Caza

Articulo 20. - 1° El presente Decreto-Ley regula la protecciòn, conservaciòn y fomento de la riqueza cinegética nacional y su ordenado aprovechamiento, en armonia con los distintos intereses afectados, y se regirà y desarrollarà mediante el Reglamento de caza correspondiente.

2° La caza, como producto primario de la tierra que dependa para su existencia del terreno donde vive, se considera como producto de la misma y, por tanto, pertenece al propietario del terreno.

3° El derecho a cazar corresponde a los propietarios de los terrenos que estan en posesìon de la correspondiente Licencia gubernativa y cumplan los demàs requisitos establecidos en el Reglamento de Caza.

4° Se consideran piezas de caza los animales salvajes que figuren en la relacion que a estos efectos deberà incluirse en el Reglamento de Caza. La condiciòn de piezas de caza no serà aplicable a los animales domésticos salvajes, ni a las especies que se declaren bajo protecciòn especial con el fin de asegurar su conservaciòn, las cuales figuraràn en una relaciòn especial. Las piezas de caza se clasificaràn en dos grupos : caza mayor y caza menor. Este clasificaciòn se harà por el Ministerio de Agricultura, Ganaderia, Pesca y Forestal.

5° A los efectos del presente Decreto-Ley, los terrenos de caza podràn estar sometidos a los siguientes régimenes : parques nacionales, reservas nacionales de caza, cotos privados de caza, refugios de caza, cotos tribales y terrenos sin classificaciòn especial.

6° El ministerio de Agricultura, Ganaderia, Pesca y Forestal, establecerà normas especificas en cada caso para la protecciòn de los cultivos. Dichas normas permitiràn compatibilizar estos cultivos con el ejercicio de la caza y el aprovechamiento ordenado de las especies cinegéticas causantes de daños.

7º En su funciòn de conservador de la riqueza cinegética, el Ministerio de Agricultura, Ganaderia, Pesca y Forestal, señalarà las àreas concretas en donde puede realizarse la caza en todas sus modalidades, para lo cual, los interesados deberàn obtener la autorizaciòn opertiva del mismo, para poder cazar en estas àreas.

TITULO II : DE LAS CONCESIONES FORESTALES

Capitulo I : Disposiciones Generales

Articulo 21. - 1º El aprovechamiento de los bosques pùblicos estatales y comunales podrà llevarse a efecto concesiòn.

2º La concesiòn podrà otorgarse a instancia de los bosques de los particulares o a iniciativa de la Administraciòn, mediante la tramitaciòn del correspondiente expediente de aprovechamiento.

3º Como regla general, la concesiòn de aprovechamiento forestales se harà mediante subasta. Excepcionalmente por razones de urgencia o para fomentar las inversiones, podrà otorgarse la concesiòn por concierto directo.

4º Toda concesiòn de aprovechamientos forestales fijarà con claridad los limites fisicos y la superficie sobre la que recae.

5º Podràn existir las siguientes modalidades de concesiòn : de volumen de madera, de àrbol en pie, en cargaderos, en puertos, a pie de fabrica y anuales o pluri-anuales con limitaciòn establecido en el articulo 10-2 de este Decreto-Ley.

Capitulo II : De los Expedientes para la Adjudicaciòn de Concesiones Forestales mediante Subastas

Articulo 22. - Toda concesiòn de aprovechamientos forestales mediantes subastas requerirà la previa incoaciòn de un expediente administrativo que contendrà un pliego de Bases Administrativas y un Pliego de Bases Técnicas.

Secciòn 1º : Del Pliego de Bases Administrativas

Articulo 23. - El Pliego de Bases Administrativas deberà contener las siguientes circunstancias :

1º Objeto de Concesiòn - Se determinarà con exactitud la ubicaciòn y superficie de la concesiòn con señalamiento concreto de sus linderos y expresiòn del nùmero de hectàreas que abarca, señalàndose en croquis sobre un plano a escala 1:50 000.

2º Canon de la Concesiòn - Se fijarà en funciòn de unidades de moneda por metro cùbico de madera, teniendo el caràcter de minimo y sin que pueda ser inferior al señalado por la Administraciòn forestal. Se revisarà cada dos años segùn las variaciones que experimente el indice de precios al consumo, el valor de la madera en el mercado internacional y la paridad monetaria.

3º Licitaciones - Podràn concurrir a la subasta cualquier persona fisica o juridica, nacional o extranjera con capacidad para obligarse juridicamente y econòmicamente. Tendrà preferencia en la adjudicaciòn los nacionales de Guinea Ecuatorial, en igualdad de condiciones, sean personas fisicas o juridicas. Si se trata de sociedades extranjeras, deberàn acreditar el capital social y los medios de explotaciòn de que disponen.

No podràn concurrir a la subasta :
1. Los que hayan incurrido en actividas calificadas como delictivas en el Còdigo Penal de Guinea Ecuatorial, o estèn sujetos a procesamiento.
2. Los que sean deudores al Estado como consecuencia del incumplimiento de obligaciones econòmicas.
3. Los concesionarios que hayan incumplido las normas reguladoras de otras concesiones, o no hayan explotado los dos tercios de las que sean titulares.
4. Los concesionarios que hayan renunciado a otras concesiones sin haberlas aprovechado, y durante el plazo que restaba para cumplir la concesiòn renunciada, cualquiera que sea la causa alegada.
5. Los expulsados del territorio de Guinea Ecuatorial y los declarados personas no gratas por el Gobierno de la Repùblica.

4° Plazos - Se expresaràn por años, sin que puedan exceder del limite previsto en el articulo 10-2.

5° Fianzas - Se exigirà una fianza provisional para poder concurrir a la subasta equivalente al 10% del canon de superficie que se exija aportar por una vez. Esta fianza podrà prestarse en metàlico o mediante aval bancario extendido por un banco reconocido por el Gobierno Guineano.

El adjudicatario de la concesiòn estarà obligado a presentar una fianza definitiva equivalente al importe del canon de la adjucicaciòn de dos años, en la misma forma prevista en el pàrrafo anterior, y que le serà devuelta o computada al término de la concesiòn segùn casos.

6° Obligaciones del Concesionario - Seràn como minimas las siguientes :
1. El estricto cumplimiento de Bases Técnicas señaladas en el Pliego correspondiente.
2. Pago del canon de la concesiòn.
3. Pago de los gastos del expediente de adjudicaciòn de la concesiòn y de los impuestos que la graven.
4. Iniciar el aprovechamiento concedido en el plazo de cuatro meses mediante la importaciòn del material y maquinaria precisos.
5. Suscribir la concesiòn en el documento oficial correspondiente en el plazo de seis meses.
6. Presentar un proyecto de explotaciòn de licencia oficial en el plazo de seis meses.
7. Proceder a la apertura de trochas en el plazo resultante en dias de multiplicar por tres el perimetro de la concesiòn.
8. Solicitar de la Administraciòn Forestal la mediciòn de las trochas siendo los gastos de esta medicìon de su cuenta.
9. Deslindar en el plazo de seis meses los encalvados existentes en la concesiòn, las fincas autorizadas, las reservas de superficies, para poblados incluyendo todo ello en el plano de la concesiòn.
10. Establecer legalmente en Guinea Ecuatorial en el plazo de un mes, designando un representante legal en la misma.

7° Sanciones al Concesionario - En caso de incumplimiento de sus obligaciones se provearàn sanciones al concesionario, que podràn consistir en :

1. Sanciòn econòmica en funciòn de la gravedad del incumplimiento, en la cuantia que se fije reglamentariamente.
2. Confiscaciòn de la Fianza.
3. Rescisiòn de la concesiòn con confiscaciòn de la fianza, de la maquinaria y material aportado e idemnizaciòn de daños y perjuicios.
4. Prohibiciòn de participar en otras concesiones.

8º Limitaciones - La concesiòn se entenderà otorgada sin perjuicio de las siguientes limitaciones y prohibiciones :

1. Las concesiones no podràn ser objeto de enajenaciòn, ni transmisiòn por actos juridicos "inter-vivos", sin autorizaciòn del Gobierno.
2. No podrà existir ninguna clase de indemnizaciòn por las servidumbres de paso, sobre la superficie de la concesiòn que el Gobierno constituya por considerarla de interés nacional.
3. El concesionario no podrà constituir vias de saca fuera de los limites territoriales de la concesiòn sin la autorizaciòn del Gobierno, en la que constarà el canon de la servidumbre de paso y el precio m^3 de los àrboles maderables que deberà pagar el concesionario.
4. No se podrà constituir servidumbres de paso sobre predios privados sin el consentimiento o acuerdo del propietario o, en su caso, imposiciòn de servidumbre forzosa, previa la correspondiente indemnizaciòn. En caso de conflicto se sometерà al arbitraje del Gobierno, a través de la Administratiòn Forestal.
5. La utilizaciòn de las vias fluviales se sometarà a las normas generales de este tipo tràfico. La madera o las trozas que impidan la navegaciòn deberàn ser retiradas por el concesionario y a su costa.
6. No se podràn utilizar las carreteras nacionales para el transporte de madera sin la autorizaciòn correspondiente y en los términos y condiciones que en la misma se señalen por el Ministerio de Obras Pùblicas, Viviendas, Urbanismo y Transportes.

9º Modificaciones de la Concesiòn y sus prorrogas -

1. Se preveerà la posibilidad de pròrroga de la concesiòn, cuando no haya podido ser explotada en su totalidad por causas imprevistas. La pròrroga serà por un plazo igual e implicarà la revisiòn del canon.
2. Se podrà renunciar a la concesiòn por causas de fuerza mayor que seràn apreciadas por el Gobierno a propuesta de la Administraciòn Forestal. Fuerza de tales casos no se otorgarà nueva concesiòn hasta que no haya transcurrido el periodo por el que fue concedido la concesiòn renunciada.
3. No podrà alegarse escasés de madera para pedir une nueva concesiòn o ampliaciòn de la vigente
4. Todas las instalaciones y maquinaria no retirada de la concesiòn en el plazo de un mes despues de caducada o renunciada, quedarà en propiedad del Gobierno.

10º Causas de extinciòn de las concesiones - Se enumeraràn las causas de la extinciòn de la concesiòn que comprenderàn, al menos, las siguientes :

1. El hecho de no pagar el canon por hectàrea durante un año.
2. La infracciòn repetida de las limitaciones y prohibiciones contenidas en la claùsula 8º del Pliego.

3. La violaciòn reiterada o grave del Pliego de Bases Técnicas.
4. La muerte del concesionario individual, la declaraciòn de quiebra, suspensiòn de pagos o extinciòn de la personalidad juridica de la sociedad concesionaria.
5. El mutuo acuerdo entre el Gobierno y el concesionario.

Secciòn 2° : Del pliego de Bases Técnicas

Articulo 24. - El Pliego de Bases Técnicas contendra, al menos, las siguientes circumstancias :

1. Los concesionarios estàn obligados a cumplir las especificaciones técnicas que a continuaciòn figuran y cualquier otra que se dictara encaminada a mejorar las condiciones selvicolas y economicas de los bosques de Guinea Ecuatorial.
2. Sin perjuicio de su caràcter mòvil de acuerdo con la demanda, se establecerà la siguiente clasificaciòn de las maderas :
   Primera Clase : Okume, Ukola, Palisandro, Oveng, Dussie, Abela, Mbero, Nsangmanguila, Eyo, Etom, Ndongmanguila, Adjap, Abang, Asia.
   Segunda Clase : Akom, Palo Rojo, Abé, Anguekong, Oduma, Ndjong, Afo, Elondo, Ayos, Bocapi.
   Tercera Clase : Calabò, Ceiba, Tom, Akein, Bien, Eyenengui, Esabom, Andjung, Ekun, Alep, Ekuk, Miam, Edum, Elolom, Ebonensok, Aloma, Nsu, Oguomo, Eves, y el resto de especies varias.
3. El concesionario se atendrà en la exploraciòn a los siguientes diametros de cortabilidad :
   A) Calidad primera diametro minimo por encima de las aletas serà de 60 cm.
   B) Calidad segunda y tercera, se establece el diametro minimo de 50 cm.
4. Toda concesiòn obliga al concesionario a presentar un proyecto de explotaciòn que se realizarà de acuerdo con las siguientes normas :
   1° Autor del Proyecto - Serà redactado por el técnico provisto de la correspondiente licencia fiscal.
   Para obtar a esta licencia fiscal los extranjeros solos o agrupados en "consultings" deberàn tener la màxima categoria académica forestal en el apis de origen y la aprobaciòn del Ministerio de Agricultura, Ganaderia, Pesca y Forestal, que revisarà los programas de estudio para juzgar su capacidad. Para los guineanos les bastarà la presentatiòn de sus titulos o diplomas para que el Ministerio de Agricultura, Ganaderia, Pesca y Forestal les autorice si procede.
   2° Contenido del Proyecto -
   - 1. Inventario total consignando especies y diametros por cantones de màs de 100 Has y menos de 500 Has tomando como diametro maderable los 40 cm a estos efectos.
   - 2. Orden de explotaciòn por zonas. No se podrà comenzar la explotaciòn de una zona sin haber agotado la anterior ni volver sobra una zona ya explotada sin autorizaciòn de la Administraciòn Forestal.
   - 3. Se consideraràn las vias de saca existentes y se proyectaràn las que completen la red, incluidos teleféricos o deslizaderos.

- 4. Se consignarà el material necesario para comenzar la explotaciòn, sus periodos de amortizaciòn y costos de horario.
- 5. Se calcularà el coste de extracciòn del metro cùbico y estructura del mismo.
- 6. Se consignarà el minimo de saca para ser la empresa rentable.
- 7. Se procederà a una somera descripciòn botànica y frecuencia con que aparecen las diversas especies.
- 8. Proyecto de vias de saca fuera de la concesiòn.
- 9. Proyectos de la industrias forestales inherentes.

5. Fomento de la industrializaciòn.

1º Con fin de conseguir que el valor añadido en las transformaciones de la madera beneficie a Guinea Ecuatorial, se establece la obligaciòn de instalar industrias de primera transformaciòn con una capacidad de una transformaciòn inicial medida en m^3 que sea el resultado de dividir por dos el nùmero de has. de la concesiòn, para pasar al doble a los dos años, todo ello con caràcter de minimo obligatorio.

2º Podrà sustituirse esta obligaciòn por la de adquirir participaciòn en una sociedad industrial transformadora, cooperativa entre concesionarios e independientes que garantice la transformaciòn citada en el pàrrafo anterior.

6. Cubicaciones y mediciones.

1º La cubicaciòn de àrboles en pie serà efectuada por la empresa y comprobada por la Administraciòn Forestal a cargo de la misma empresa.

2º La cubicaciòn de los àrboles apeados, contada en blanco y su reconocimiento final serà efectuade por el personal cualificado de la Administraciòn Forestal.

7. Duraciòn de la Concesiòn. Las concesiones inferiores a 25 000 Has se adjudicaràn por cinco años; entre 25 000 y 50 000 per cinco años prorrogables automàticamente, previa revisiòn del canon por otros cinco; las superiores a 50 000, por diez años; en el caso de ser mayores de 80 000 Has, estos diez años se prorrogaràn automàticamente a quince o veinte, segùn existencias y demàs circumstancias.

8. Régimen Técnico-Econòmico.

1º Por el ministerio de Hacienda y Comercio se fijarà en cada caso el régimen econòmico aplicables a las explotaciones forestales, y en tanto se proceda a la elaboraciòn de una ley que recoja la imposiciòn interna que estableza los cànones que graven los distintos hechos imponibles, recaerà exclusivamente sobre el tràfico comercial exterior, y cuya cuota tributaria que tendrà el caràcter de màxima serà el resultado de girar el tipo impositivo del 30% fijado en Arancel de Exportaciòn, sobre la base tributaria constituìda por los precios minimos de exportaciòn aprobados conjuntamente por los Ministerios de Agricultura, Ganaderia, Pesca y Forestal y de Hacienda y Comercio, respectivamente.

2º La madera procedente de la ordenaciòn por empresa colaboradora, de acuerdo con el Gobierno, el canon fiscal y forestal igual a las concesiones forestales.

3º En el caso de maderas procedentes de pequeñas fincas para alimento, construcciòn de viviendas y embarcaciones, dado su interés social, podrà concederse gratuitamente.

4º No podrà abandonarse ni destruirse ningùn lote de madera una vez hecha la tala. El pago en concepto de daños y perjuicios serà el doble del valor del lote F.O.B.

5º Toda madera cortada en bosques del Estado sin la correspondiente autorizaciòn, serà decomisada y tendrà una sanciòn economica del tanto al trìplo de la tasaciòn de la madera F.O.B. y en ningùn caso inferior a 6 000 Pipkwele por m³.

Secciòn 3º : Del Procedimiento de Adjudicaciòn

Articulo 25. - 1º La convocatoria de las subastas se harà con la debida publicidad y con el plazo suficiente para que puedan ser objeto de conocimiento por los interesados.

2º El anuncio de la convocatoria deberà indicar el objeto de la concesiòn, el canon, el plazo de la misma y la fianza provisional y se remitirà a los Pliegos de Bases Técnicas y Administrativas para las demàs circunstancias y condiciones de la concesiòn.

Articulo 26. - Los que pretendan licitar, formularàn la correspondiente solicitud de tomar parte en la subasta, dirigida al Presidente del Gobierno y a la que se acompañaràn, en un sobre cerrado y lacrado señalado con la letre A :

1. Documentaciòn que acredite la personalidad del licitador. Tratàndose de Sociedades, tambièn se acompañaràn los documentos que justifiquen su existencia legal e inscripciòn en el Registro Mercantil y los que autoricen al licitador para actuar en su nombre.
2. Justificante de haber constituido fianza.
3. Justificante de poseer, como garantia de solvencia, el capital señalado con tal finalidad en el anuncio de la subasta, y la relaciòn de material ùtil para la explotaciòn forestal, examinado y varado por el Ministerio de Agricultura, Ganaderìa, Pesca y Forestal.
4. Proyecto de explotaciòn y plan de inversiòn.

En otro sobre, cerrado y lacrado y señalado con la letra B : Proposiciones econòmicas.

Articulo 27. - 1º La documentaciòn contenida en el sobre A, serà objeto de examen por la Direcciòn Técnica Forestal e informada por la Administraciòn Forestal o, en su caso, por el Ministerio de Agricultura, Ganaderìa, Pesca y Forestal.

2º Seràn excluidos los licitadores que no cumplan los requisitos contenidos en los pliegos de bases administrativas y técnicas.

Articulo 28. - 1º La apertura del sobre B, se realizarà pùblicamente, en un acto convocado con la suficiente antelaciòn, en los términos que establezca la convocatoria.

2º La adjudicaciòn provisional de la concesiòn se harà al licitador que ofrezca mejores conciciones econòmicas.

3º La adjudicaciòn definitiva corresponde al Presidente de la Repùblica.

Articulo 29. - Toda concesiòn serà objeto de inscripciòn en un Registro pùblico que se llevarà en la Direcciòn Forestal.

Capitulo III : De la Adjudicaciòn Directa de Concesiones Forestales

Articulo 30. - Cuando proceda la concesiòn por concierto directo, los supuestos previstos en el articulo 21.3 de este Decreto-Ley, se procurarà cumplir, en cuanto sea posible, los requisitos establecidos en las Secciones 1º, 2º y 3º del Capitulo anterior, especialmente lo relativo a la personalidad del adjudicatario, garantias, solvencias y proyectos de explotaciòn.

TITULO III : DE LAS COMPETENCIAS DEL GOBIERNO Y ORGANIZACIONES ADMINISTRATIVAS

Capitulo I : De las Competencias del Gobierno

Articulo 31. - Compete al Gobierno de Guinea Ecuatorial en relaciòn con la riqueza forestal.

1. Velar, impulsar y estimular por cuantos medios estén a su alcance, el debido aprovechamiento de recursos forestales para los fines y con las limitaciones establecidas en el articulo 1. del presente Decreto-Ley.
2. Establecer un control permanente sobre las actividas de explotaciòn forestal y sancionar, en su caso, las acciones o omisiones que perjudicen su normal desarrollo.
3. Cuidar el cumplimiento del presente Decreto-Ley y de la normativa legal existente, que se adaptarà permanentemente a las circunstancias del tiempo y a las exigencias del mercado.
4. Procurar la conservaciòn, mejora y expansiòn del àrea forestal, mediante la ordenaciòn de los aprovechemientos, cuidados culturales tendentes a incrementar la cuantia y calidad de la producciòn maderera y poblaciones forestales.
5. Realizar o estimular la investigaciòn sobre los productos prerecolectados, a través de la fisiologia, genetica, selvicultura, ordenaciòn, repoblaciones botànicas o introducciòn de especies.
6. Realizar o estimular la investigaciòn sobre los productos postrecolectados, a través de las caracteristicas fisicas mecanicas de las maderas, su resistencia a los agentes biòticos y abiòticos y nuevos usos de las especies hoy no comerciales.
7. Tramitar, clasificar y, en su caso, autorizar las industrias forestales.
8. Tramitar y adjudicar las concesiones de aprovechamiento o ventas de terrenos con fines especificos forestales y de acuerdo con las disponibilidades de exportaciòn, y necesidades del mercado interior.

9. Informar sobre las necesidades del suelo, altitud y clima del cultivo, semillas, cuidados culturales, producciones y precios.
10. Establecer cultivos pilotos con fines de investigaciòn.
11. Dictar las normas necesarias para evitar el agotamiento de los suelos, asi como para la recuperaciòn de los suelos agotados, acelerando el proceso de la naturaleza.
12. Crear reservas forestales para la protecciòn de la fauna y la flora, con fines cientificos, tecnològicos, de estudio de evoluciòn de las masas, asi como parques para cooperar con Organismos, Gobiernos extranjeros o Fundaciones Internacionales.
13. Fomentar y regular el aprovechamiento de la riqueza piscicola, sea fluvial o lacustre.
14. Regular la caza.
15. Cuantas otras competencias deba asumir por exigirlo asì el interés pùblico y la salvaguarda de la riqueza forestal guineana.

Capitulo II : De la Organizatiòn Administrativa

Articulo 32. - Serà conforme establece en la materia el Reglamento Orgànico y funcional del Ministerio de Agricultura, Ganaderìa, Pesca y Forestal.

Disposicion Adicional

Serà principio inspirado de la actuaciòn administrativa en el aprovechamiento de la riquezà forestal, y en la aplicaciòn del presente Decreto-Ley, que tales aprovechamientos garanticen mediante las correspondientes rotaciones y regeneraciones, la permanencia y transmisiòn a las futuras generaciones, de dicha riqueza. A tales efectos, se establecerà por el Gobierno un Plan Anual que contenga el màximo de la superficie que pueda ser objeto de aprovechamiento.

Disposiciones Finales

1. Se faculta al ministerio de Agricultura, Ganaderìa, Pesca y Forestal para dictar las normas precisas al desarrollo y aplicaciòn del presente Decreto-Ley.
2. El presente Decreto-Ley entrarà en vigor al dia siguiente de su publicaciòn en los medios informativos nacionales.

Asì lo dispongo en la Cuidad de Malabo, a

29 dias de Septiembre de 1981.

POR UNA GUINEA MEJOR,

PRESIDENTE CONSEJO MILITAR SUPREMO

MODELO DE CONCESION

GUINEA ECUATORIAL

GUINEE EQUATORIALE

Modelo de Concesiòn

MODELO DE UN EXPEDIENTE DE CONCESION FORESTAL EN GUINEA ECUATORIAL

Decreto nº_____ de fecha______ por que se otorga una concesiòn Forestal de x Has.

Vista la peticiòn formulada por el subdito ------------ residente en -------------- de que se le otorgue una concesiòn forestal de x Has. en bosque libre del Estado en la Régiòn ---- cuyos linderos son :

Norte
Sur
Este
Oeste

Accidentes Geograficos

Visto asmismo, el expediente instruido al efecto por el Ministerio de ....

Esta Presidencia acuerda otorgarle la concesiòn forestal solicitada de x Has. debiendo sujetarse a las siguientes condiciones :

1. El bosque objeto de la concesiòn serà previamente medido por personal competente de la Direcciòn Técnica Forestal, con cargo a la empresa.

2. La madera procedente de la concesiòn no po rà exportarse en forma de tronco, debiendo toda su procucciòn aserrarse en el Paìs.

3. La empresa concesionaria no podrà apear àrboles fuera de los lìmites que le señale sin la expresa autorizaciòn del Gobierno.

4. Dar cumplimiento a todo lo dispuesto o que disponga la legislaciòn del Paìs en materia de aprovechamiento arboreo, vìas de transformaciòn; contribuciones, cànones Forestales, etc., y sujetarse al Decreto-Ley sobre Inversiòn de Capital Estranjero.

5. Respectar los derechos comunales de los poblados colindantes y enclavados dentro de la concesiòn forestal, dejando una reserva de cuatro Has por cabeza de familia.

6. Comprometerse a ubicar patios con sus consiguientes instalaciones y servicios para las atenciones sociales de sus trabadores.

7. Instalar y poner en funcionamiento un aserradero moderno en el plazo de doce meses a contar desde la fecha de esta concesiòn.

8. Compremeterse a reparar todos los desperfectos que pueda ocacionar el transporte de madera en rollo por carreteras del Estado.

9. Esta concesiòn forestal se otorga por CINCO AÑOS renovables, previa solicitud de la empresa interesada.

10. Esta concesiòn es intransferible.

Asì lo dispongo por el presente, dado en la Cuidad de MALABO, a ______ dìas del mes de _________ del año mil novecientos ochenta y _______

POR UNA GUINEA MEJOR,

LA PRESIDENCIA DE LA REPUBLICA

LOI N° 1/82 DU 22 JUILLET 1982

GABON

Loi d'Orientation en Matière des Eaux & Forêts

Loi N°1/82

L'Assemblée nationale a délibéré et adopté,

Le Président de la République, Chef de l'Etat, promulgue la Loi dont la teneur suit:

TITRE I : DISPOSITIONS GENERALES

Article 1 . - La politique en matière des eaux et forêts mise en oeuvre par la présente loi et ses textes d'application a pour objectif général de promouvoir une gestion rationnelle des ressources des domaines visés à l'art.3, en vue d'accroître la contribution du secteur des eaux et forêts au développement économique, social,culturel et scientifique du pays.

Article 2 . - L'objectif général défini à l'art.1 nécessite la mise en oeuvre :

- d'une politique d'inventaire permanent ayant pour but d'améliorer la connaissance qualitative et quantitative des ressources disponibles;
- d'une politique d'aménagement des ressources visant à assurer un meilleur équilibre entre l'exploitation et le renouvellement de ces ressources;
- d'une politique de reconstitution des ressources en vue d'en garantir la pérennité;
- d'une politique d'exploitation rationnelle ayant pour but l'utilisation optimale des ressources et un meilleur contrôle de leur exploitation;
- d'une politique d'industrialisation visant à promouvoir la transformation d'une plus grande partie de la matière première;
- d'une politique de protection et d'incitation visant à accroître la participation effective des Gabonais dans la mise en valeur des domaines visés à l'art. 3.
- d'une politique de formation et de recherche ayant pour but l'accroissement de la productivité;
- d'une politique de financement soutenu des différents programmes élaborés et des actions entreprises en application des dispositions de la présente loi;
- enfin d'une politique d'information et de vulgarisation en vue de sensibiliser et d'éduquer les usagers et la population.

Article 3 . - Sont soumis au régime juridique et financier établi par la présente loi : le domaine forestier, la faune sauvage, le domaine fluvial, lacustre, lagunaire et maritime en ce qui concerne les ressources halieutiques.

Article 4 . - Les règles de gestion et d'aliénation des domaines visés à l'art. 3 ci-dessus sont définies par la législation foncière et domaniale en vigueur et par les dispositions de la présente loi.

Article 5 . - En vue d'assurer leur subsistance, les collectivités villageoises continuent à exercer leurs droits d'usages coutumiers dans les domaines visés à l'art. 3 selon les modalités fixées par voie réglementaire.

Article 6 . - L'administration des eaux et forêts est chargée de l'application de la politique du gouvernement dans les domaines visés à l'art.3.

L'administration des eaux et forêts est une administration para-militaire; elle assure en matière de forêts, eaux, pêches, faune et chasse les missions de police, de contrôle et de répression. A ce titre, dans l'exercice de leurs fonctions, les agents des eaux et forêts sont astreints au port de l'uniforme, d'insignes distinctifs de grade, d'armes à feu et de munitions selon les modalités fixées par voie réglementaire.

Article 7 . - Sans qu'il puisse être dérogé aux dispositions de l'art. 103 de la Loi N° 14/63 du 8 mai 1963 fixant la composition du domaine de l'Etat et les règles qui en déterminent les modes de gestion et d'aliénation, les agents de l'administration des eaux et forêts sont chargés de l'émission des titres de perception des taxes et redevances prévues par la présente loi. A ce titre, ils ont droit à des ristournes suivant un barême fixé par voie réglementaire.

Article 8 . - Nul ne peut faire des domaines visés à l'art.3, un usage prohibé par les dispositions de la présente loi et par les textes pris pour son application.

TITRE II : DU DOMAINE FORESTIER

Article 9 . - Sont qualifiés de forêts au sens de la présente loi les périmètres comportant ou non une couverture végétale et capables :

- soit de fournir du bois ou des produits végétaux autres qu'agricoles;
- soit d'abriter la faune sauvage;
- soit d'exercer un effet direct ou indirect sur le sol, le climat ou le régime des eaux.

Article 10. - Le domaine forestier comprend :

- les forêts domaniales classées qui constituent le domaine à vocation forestière permanente et déterminée. Ces forêts font partie du domaine public. Aussitôt après leur classement, ces forêts font l'objet d'une délimitation précise.
- les forêts domaniales protégées qui constituent le domaine à vocation forestière non déterminée. Ces forêts font partie du domaine privé.

Article 11. - Font partie des forêts domaniales classées :

- les forêts de production à vocation permanente;
- les périmètres de reboisement;
- les parcs nationaux à vocation forestière;
- les forêts de protection;

- les forêts récréatives;
- les jardins botaniques;
- les arboretums et les sanctuaires de certaines espèces végétales;
- les aires d'exploitation rationnelle de faune.

Article 12. - Le classement des forêts dans l'une des catégories visées à l'art.11 ci-dessus s'effectue par voie réglementaire.

Le texte portant classement d'une forêt dans le domaine public doit préciser à quelle catégorie elle fait partie, le mode de gestion de ses ressources et les restrictions applicables à l'intérieur de cette forêt.

Article 13. - Les forêts domaniales classées doivent couvrir au moins 40% de la superficie totale du territoire national.

Article 14. - Toute forêt domaniale classée doit faire l'objet d'un plan d'aménagement dont les modalités sont fixées par voie réglementaire.

Article 15. - L'administration des eaux et forêts assure la reconstruction des forêts par des programmes de reboisement permanents et soutenus qu'elle détermine périodiquement.

Article 16. - A l'exception des dispositions prévues à l'art. 5 ci-dessus, nul ne peut se livrer à l'exploitation ou à la récolte des produits de la forêt à titre gratuit et sans autorisation préalable délivrée par l'administration des eaux et forêts.

La nature des autorisations et des titres d'exploitation, ainsi que les procédures de leur attribution, sont définies par voie réglementaire.

Article 17. - Au-delà de 15 000 hectares et sous réserve des droits acquis, il ne peut être attribué que des titres d'exploitation visant à créer une industrie locale de transformation de bois.

Article 18. - La politique d'attribution des titres d'exploitation doit permettre le développement d'une industrie locale de transformation de bois.

Le pourcentage de transformation des grumes issues des titres d'exploitation accordés en vue de créer une industrie locale de bois ne peut être inférieur à 75%.

Les bois en grumes issus des permis industriels et non transformés dans les usines du titulaire sont assujettis à une taxe spéciale dont le taux est fixé par la loi de finances sur proposition du ministre chargé des eaux et forêts.

Article 19. - Dans les forêts classées, l'exploitation ne peut se faire qu'en régie ou par vente de coupes en adjudication publique.

Article 20. - Toute personne physique ou morale désirant se livrer à l'exploitation forestière ou à une activité de transformation du bois doit préalablement obtenir son agrément auprès de l'administration des eaux et forêts, selon les modalités fixées par voie réglementaire.

Article 21. - Les permis d'exploitation et les autorisations d'exploiter sont strictement personnels. Ils ne peuvent être cédés, transmis ou transférés que sur autorisation de l'administration des eaux et forêts selon les modalités fixées par voie réglementaire.

Article 22. - Pour des fins de gestion, le domaine forestier du Gabon est divisé en zones dont les limites sont définies par voie réglementaire.

Pour l'exploitation forestière, la première zone est réservée aux nationaux. Fait également partie de la première zone, une bande forestière d'au moins cinq kilomètres de large de part et d'autre des voies férrées, des cours d'eau flottables et des grands axes routiers tels qu'ils seront précisés par un texte réglementaire.

Quelle que soit la zone considérée, l'exploitation des forêts situées aux alentours immédiats des villages est réservée en priorité aux villageois, selon les conditions définies par voie réglementaire.

Article 23. - L'ouverture de toute zone à l'exploitation forestière est subordonnée à un inventaire préalable de celle-ci.

Les titulaires des permis d'exploitation de toute nature sont tenus de fournir à l'administration des eaux et forêts les résultats d'inventaire et les plans d'exploitation de leur permis. De même, toute personne physique ou morale se livrant ou à l'exploitation forestière ou à la transformation du bois est tenue de fournir à l'administration des eaux et forêts les documents statistiques et comptables relatifs à son activité. Les modalités d'application de ces dispositions seront fixées par voie réglementaire.

Les travaux forestiers exécutés par l'administration des eaux et forêts pour le compte des exploitants sont rémunérés au titre d'une redevance dénommée "charges forestières" selon les conditions définies par voie réglementaire.

Le montant de cette redevance est fixé par la loi de finances, sur proposition du ministre chargé des eaux et forêts.

Article 24. - Sous réserve des droits acquis, il ne peut être attribué à un même exploitant forestier plus de 200 000 hectares de permis forestiers quelle qu'en soit la nature.

Les exploitants forestiers dépassant le seuil de 200 000 hectares antérieurement à la date de promulgation de la présente loi, ne peuvent être autorisés à acquérir de nouveaux permis, ni à racheter, ni à transférer leur droit, tant que les superficies totales détenues restent supérieures ou égales à ce seuil.

Article 25. - La participation au capital d'une société d'exploitation forestière et la création d'une société forestière nouvelle par un exploitant forestier installé au Gabon sont soumises à l'autorisation préalable de l'administration des eaux et forêts. Dans tous les cas, ces opérations sont interdites lorsqu'elles concernent les exploitants forestiers dépassant déjà le seuil de 200 000 hectares ou bien lorsqu'elles ont pour effet de porter les superficies forestières détenues par un exploitant au-delà de ce seuil.

Article 26. - Toute exploitation de la forêt est soumise à un cahier des charges comportant des clauses générales et des clauses particulières selon les modalités fixées par voie réglementaire.

Article 27. - Tout exploitant forestier est tenu de respecter les délais d'exploitation.

Si l'exploitation d'un permis n'est pas commencée dans un délai de deux ans après la date de son attribution, celui-ci fait purement et simplement retour aux domaines dans les conditions définies par voie réglementaire. Les redevances et taxes versées à l'Etat lui restent acquises.

Les titulaires des titres d'exploitation en cours de validité disposent du même délai à compter de la date de promulgation de la présente loi pour se conformer aux présentes dispositions.

L'arrêt d'exploitation pendant deux années consécutives entraîne, sauf cas de force majeure dûment constaté, le retrait du permis.

Article 28. - Pour des fins d'aménagement, l'administration des eaux et forêts peut marquer en réserve, assurer la protection de toute espèce végétale, ou édicter des restrictions qu'elle juge utiles à l'intérieur d'une zone forestière concédée ou non.

Article 29. - L'administration des eaux et forêts peut, sans préjudice de la réparation des dommages subis par l'exploitant, soustraire de toute zone forestière concédée, les arbres ou superficies nécessaires à l'exécution des travaux d'intérêt général ou pour des besoins de ses services, selon les modalités fixées par voie réglementaire.

Article 30. - Tout produit forestier brut ou oeuvré est soumis aux règles de normalisation et de classification définies par voie réglementaire.

Article 31. - L'attribution, la possession, le renouvellement, l'échange et le transfert de tout titre d'exploitation, ainsi que toutes les activités relatives à la transformation, la commercialisation et l'exportation des produits de bois, sont soumis au paiement de taxes et de redevances dont l'assiette et le taux sont fixés par la loi de finances sur proposition du ministre chargé des eaux et forêts.

TITRE III : DE LA FAUNE SAUVAGE

Article 32. - La faune sauvage est l'ensemble des animaux que renferme une région donnée; les présentes dispositions s'appliquent à la faune cynégétique.

Article 33. - Pour des fins d'aménagement, de protection et de conservation de la faune, on distingue, d'une part les aires d'exploitation rationnelle de la faune appartenant aux forêts domaniales classées visées à l'art.11 et, d'autre part, la zone protégée d'exploitation de la faune.

Article 34. - Constituent des aires d'exploitation rationnelle de la faune: les réserves naturelles intégrales de la faune, les sanctuaires de faune, les parcs nationaux, les réserves de faune, les jardins zoologiques et les domaines de chasse.

Article 35. - La réserve naturelle intégrale est un périmètre dans lequel le sol, l'eau, la flore et la faune bénéficient d'une protection absolue et dont l'accès n'est autorisé qu'aux chercheurs et aux agents des eaux et forêts.

Article 36. - Le sanctuaire est un périmètre dans lequel une ou plusieurs espèces animales ou végétales nommément désignées bénéficient d'une protection absolue et dont l'accès est réglementé.

Article 37. - Le parc national est une portion du territoire où la flore, la faune, les sites géomorphologiques, historiques et d'autres formes de paysages, jouissent d'une protection spéciale et à l'intérieur de laquelle le tourisme est organisé et réglementé.

Article 38. - La réserve de faune est un périmètre dans lequel la flore et la faune bénéficient d'une protection absolue mais dont l'accès est réglementé.

Article 39. - Le jardin zoologique est une institution publique ou privée caractérisée par l'exhibition d'animaux vivants ou d'expèces rares à des fins récréatives, esthétiques, culturelles ou à des fins de repeuplement.

L'accès au jardin zoologique est réglementé. Les animaux y bénéficient d'une protection absolue.

Article 40. - Le domaine de chasse est une zone où la réglementation de la chasse est plus restrictive en ce qui concerne les limites d'abattage.

Article 41. - La concession des aires d'exploitation rationnelle de la faune est interdite. Toutefois afin d'assurer le développement du tourisme au Gabon, l'exploitation des activités touristiques diverses à l'intérieur des parcs nationaux et des domaines de chasse est autorisée conformément aux dispositions des art. 37 et 40 ci-dessus. A cet effet, les guides de chasse chargés de conduire les touristes exercent l'exclusivité de cette activité à l'intérieur d'un parc national ou d'un domaine de chasse donné, selon les conditions définies par voie réglementaire.

Article 42. - L'exploitation technique des aires d'exploitation rationnelle de faune est de la compétence de l'administration chargée des eaux et forêts. Cette exploitation technique comprend entre autres, l'aménagement de la faune et de la chasse, la réglementation, le contrôle et l'émission des ordres de recette en matière de faune et de chasse.
L'exploitation touristique des aires d'exploitation rationnelle de la faune visées aux art. 37 et 40 précités est de la compétence de l'administration chargée du tourisme. L'exploitation touristique comprend la création et la gestion, à l'intérieur de ces aires, de complexes ainsi que toutes les activités touristiques.

Les recettes dues par l'exploitation touristique ainsi définie sont indépendantes des recettes perçues du fait des taxes et redevances prévues par la présente loi.

Article 43. - La zone protégée d'exploitation de la faune est constituée du reste du territoire national. A l'intérieur de cette zone et à l'exception des forêts domaniales classées, la chasse est autorisée et réglementée.

Article 44. - Conformément à l'art.14 ci-dessus, les aires d'exploitation rationnelle de la faune doivent faire l'objet d'un plan d'aménagement établi par l'administration des eaux et forêts.

Article 45. - Les collectivités locales qui manifestent leur volonté d'appliquer des mesures pratiques susceptibles d'augmenter le capital faunique sur les terres où elles sont usagères, pourront obtenir le classement à leur profit en aires d'exploitation rationnelle de la faune, selon les conditions définies par voie réglementaire.

Article 46. - Sont interdits dans les aires d'exploitation rationnelle de la faune et sous réserve des dispositions des art. 37, 40 et 42 précités :

- la création de villages, de campements, de routes publiques ou privées et de toutes formes d'exploitation susceptibles de modifier l'environnement et ses ressources;
- la pénétration sans autorisation préalable délivrée par l'administration des eaux et forêts;
- la chasse proprement dite sauf dans les domaines de chasse;
- la circulation et le stationnement en dehors des pistes cyclables et balisées;
- les manoeuvres militaires;
- l'empoisonnement des sources, des marigots et des rivières traversant ces aires;
- le pacage des animaux domestiques;
- le séjour de nuit et le camping;
- le survol à moins de 200 mètres d'altitude pour des buts non scientifiques;
- les usages coutumiers.

Article 47. - L'administration des eaux et forêts détermine par voie réglementaire la liste des espèces à protéger intégralement ou partiellement, les périodes de suspension provisoire de la chasse qu'il convient d'imposer pour la protection de certaines espèces, les latitudes d'abattage ainsi que toutes restrictions qu'il est utile d'apporter le cas échéant à la chasse, au commerce ou à la circulation de la viande de chasse, des dépouilles et des trophées de certaines espèces à protéger tout particulièrement.

Article 48. - Toutes les réserves et les aires d'exploitation rationnelle de la faune existant à la date de promulgation de la présente loi seront soumises au nouveau régime juridique établi par cette loi, selon les conditions définies par les décrets d'application qui, en outre, détermineront dans quelle nouvelle catégorie elles sont placées.

Article 49. - Constitue un acte de chasse dans le cadre des dispositions établies par la présente loi le fait de poursuivre, d'approcher, de tirer, de tuer, de capturer, de photographier ou de cinématographier un animal sauvage ou de conduire des expéditions à cet effet.

Article 50. - L'exercice de la chasse doit avoir pour objectif final l'exploitation rationnelle de la faune sauvage et la protection de la nature; certaines espèces devant être protégées soit du fait de leur rareté dans certaines zones où il importe de reconstituer le capital faunique, soit pour leur valeur patrimoniale, touristique, scientifique et l'appoint qu'elles représentent dans l'équilibre de la nature et de l'environnement.

Article 51. - Sans déroger aux dispositions de l'art.5 précité, la chasse coutumière est celle qui est effectuée avec des armes de fabrication locale dont la liste est définie par voie réglementaire; elle se limite à l'abattage des animaux non protégés.

Article 52. - Sont interdits sur toute l'étendue du territoire :
- la poursuite, l'approche ou le tir du gibier en véhicule automobile, bateau à moteur ou aéronef;
- la chasse de nuit avec ou sans engins éclairants;
- les battues au moyen de feu, de filet et de fosse;
- la chasse et la capture au moyen de drogues, d'appâts empoisonnés, de fusils fixes, d'explosifs.

Toutefois, l'administration des eaux et forêts peut, à titre exceptionnel, autorises sous contrôle les procédés de chasse interdits en vue de la protection des personnes et des biens, de la capture des animaux vivants pour le repeuplement de certaines aires ou dans un but scientifique.

Article 53. - En cas d'abus, l'administration des eaux et forêts peut interdire ou réglementer tout procédé autorisé de chasse ou d'approche de la faune susceptible de compromettre la conservation de celle-ci.

L'introduction d'une nouvelle technique de chasse au Gabon doit au préalable être agréée par l'administration des eaux et forêts.

Article 54. - Sous réserve des dispositions de l'art.5 précité, et en dehors des propriétés closes, nul ne peut chasser au Gabon s'il n'est détenteur d'un permis ou d'une licence de chasse délivrée par l'administration des eaux et forêts.

La nature des permis et des licences de chasse, leurs modalités d'attribution, la procédure de retrait ou de suspension et les cas de refus ainsi que les droits et obligations autres que financiers sont fixés par voie réglementaire.

Article 55. - Les permis de chasse ne peuvent être délivrés qu'aux personnes détentrices d'un permis de port d'arme.

La licence de capture est délivrée aux personnes physiques ou morales désireuses de capturer les animaux sauvages pour des fins commerciales, scientifiques ou touristiques.

La licence de chasse d'images est délivrée aux personnes physiques ou morales désireuses de photographier ou cinématographier les animaux sauvages pour des fins commerciales ou professionnelles.

Le permis de chasse, la licence de capture et la licence de chasse d'images sont délivrés, selon le cas, aux personnes ayant contracté pour elles ou pour les personnes commises à cet effet une police d'assurance et ayant pris connaissance de la réglementation en matière de chasse et de protection de la faune.

La nature des taxes et des redevances y afférentes et leur taux sont fixés par la loi de finances, sur proposition du ministre chargé des eaux et forêts.

Article 56. - Les permis et les licences sont strictement personnels et ne peuvent être ni vendus, ni prêtés, ni cédés.

Article 57. - Les permis de chasse n'autorisent, en principe, que l'abattage des mâles adultes des espèces non protégées ou partiellement protégées; dans ce but, les décrets d'application pourront fixer, s'il est nécessaire, les dimensions minimales des dépouilles et trophées et prescrire la protection intégrale des femelles de certaines espèces.

Article 58. - Conformément aux dispositions des art. 34, 35, 36, 37, 38, 39 et 40 précités, la pénétration dans les aires d'exploitation rationnelle de la faune est subordonnée au paiement d'une taxe journalière dont le taux est fixé par la loi de finances, sur proposition du ministre chargé des eaux et forêts.

Dans les mêmes conditions, les abattages d'animaux y sont soumis au paiement d'une taxe d'abattage spécifique selon l'espèce chassée.

En dehors des aires d'exploitation rationnelle de la faune, la taxe d'abattage spécifique définie ci-dessus n'est perçue que pour les animaux partiellement protégés dont la liste est définie par voie réglementaire.

Article 59. - En cas de nécessité en matière de protection de la faune, le ministre chargé des eaux et forêts peut limiter le nombre annuel de permis et de licences de chasse.

Article 60. - Sauf cas de légitime défense et sous réserve de l'art. 51 ci-dessus, la chasse n'est autorisée qu'avec des armes de troisième catégorie conformément aux textes définissant le régime des armes et munitions en vigueur au Gabon.

L'administration des eaux et forêts réglemente le modèle, le calibre des armes de chasse et les munitions. Elle peut également interdire l'emploi de certains modèles d'armes ou de munitions autorisés si les nécessités de protection de la faune le requièrent.

Article 61. - Aucune infraction ne peut être retenue contre quiconque a fait acte de chasse indûment dans la nécessité immédiate de sa défense, de celle d'autrui, de son propre cheptel domestique ou de sa propre récolte.

Toutefois, la preuve de la légitime défense doit être fournie au responsable de l'administration des eaux et forêts le plus proche qui prendra en compte les dépouilles et éventuellement les trophées au profit de l'Etat.

La viande du gibier abattu par suite de la légitime défense sera remise à la collectivité locale la plus proche.

Article 62. - A la suite des dégâts répétés causés aux cultures par certaines espèces, celles-ci pourront être, après enquête de l'administration des eaux et forêts, déclarées nuisibles à l'intérieur d'un périmètre nettement délimité. Les textes réglementaires définiront les conditions dans lesquelles les battues ou tous autres moyens de lutte pourront être organisés.

Article 63. - La détention, la circulation et la commercialisation des produits de la chasse sont réglementés.

Les détenteurs, les transporteurs, les revendeurs et les restaurateurs des produits d'animaux sauvages doivent pouvoir justifier que les produits qu'ils détiennent proviennent d'animaux régulièrement abattus.

Article 64. - La détention de tout animal sauvage vivant et des trophées est soumise à une autorisation dans les conditions définies par voie réglementaire.

Article 65. - L'exportation des animaux sauvages vivants, des dépouilles et des trophées est, d'une part subordonnée à l'obtention d'un certificat d'origine et d'une autorisation à l'exportation délivrés par l'administration des eaux et forêts et d'un certificat sanitaire délivré par le service national d'élevage, selon les conditions définies par voie réglementaire et est soumise, d'autre part, au paiement d'une taxe cynégétique dont le taux est défini par la loi de finances, sur proposition du ministre chargé des eaux et forêts.

Article 66. - Sont interdites : la détention, la circulation et la commercialisation des animaux sauvages vivants ainsi que des dépouilles et trophées dont les caractéristiques et les dimensions ne sont pas conformes aux textes d'application de la présente loi.

Les pointes d'ivoire trouvées doivent être remises au responsable du poste des eaux et forêts le plus proche. Le Trésor est tenu de verser au déposant une prime correspondant à la moitié de la valeur mercuriale de l'ivoire en vigueur à la date de dépôt, au vu d'un titre de paiement émis par le receveur des domaines.

Les titulaires de permis de grande chasse sont autorisés à conserver par devers eux une paire de pointes d'ivoire par an. Tous autres trophées doivent être remis à l'administration des eaux et forêts.

La détention des trophées est soumise à déclaration dont les modalités sont fixées par voie réglementaire.

Article 67. - Pour les besoins de protection de la faune sauvage, l'administration des eaux et forêts peut, en cas de nécessité, faire appel à des personnes physiques ou morales reconnues pour leur compétence, leur moralité et l'intérêt qu'elles portent aux problèmes de protection de la nature, dans les conditions définies par voie réglementaire.

TITRE IV : DES RESSOURCES HALIEUTIQUES

Article 68. - La pêche s'exerce dans les domaines maritimes, lagunaire, fluvial et lacustre. Elle consiste dans la capture de tout animal à respiration branchiale.

Article 69. - En vue d'une exploitation rationnelle des ressources halieutiques, la délivrance de toute autorisation de pêche est subordonnée à l'évaluation préalable des stocks existants.

Article 70. - Sous réserve des dispositions de l'art.5 précité, nul ne peut se livrer à la pêche au Gabon à titre gratuit et sans autorisation préalable délivrée par l'administration des eaux et forêts.

La nature des autorisations de pêche ainsi que la procédure de leur attribution et de leur retrait sont définies par voie réglementaire.

L'attribution et le renouvellement des autorisations de pêche sont soumis au paiement d'une redevance dont l'assiette et le taux sont fixés par la loi de finances, sur proposition du ministre chargé des eaux et forêts.

Article 71. - Pour des fins de conservation et de protection des ressources halieutiques, l'administration des eaux et forêts réglemente les techniques, les moyens, les périodes et les lieux de pêche.

Article 72. - Les armateurs de navires de pêche sont tenus de fournir régulièrement à l'administration des eaux et forêts les statistiques de pêche ainsi que les documents comptables relatifs à leurs activités selon les modalités définies par voie réglementaire.

Article 73. - Conformément aux dispositions de l'art. 1 de la présente loi, l'exploitation des ressources halieutiques doit s'accompagner d'investissements au sol et contribuer au développement du secteur industriel. Dans les eaux sous juridiction gabonaise, l'exploitation des ressources halieutiques par des non-nationaux s'effectue dans le cadre des entreprises conjointes de pêche industrielle ou artisanale, selon les modalités fixées par voie réglementaire et sous réserve des accords internationaux.

Article 74. - Sous réserve des dispositions de l'art.71 ci-dessus la pêche dans les rivières, fleuves, lagunes, estuaires et à l'intérieur d'une bande de trois milles à partir de la côte est réservée à la pêche artisanale dans les conditions définies par voie réglementaire.

Les ressortissants étrangers qui se livrent à la pêche dans ces zones sont assujettis au paiement d'une redevance spéciale dont le taux et l'assiette sont fixés par la loi de finances sur proposition du ministre chargé des eaux et forêts.

Article 75. - L'introduction d'une nouvelle technique de pêche au Gabon doit être préalablement agréée par l'administration des eaux et forêts.

Article 76. - L'aquaculture représente l'ensemble des activités relatives à l'élevage d'animaux à respiration branchiale, de mammifères ou de reptiles au mode de vie aquatique. Elle constitue, lorsqu'elle est soutenue et exploitée de façon rationnelle, une source de revenus et de protéines indispensables au développement et au mieux-être des populations.

Article 77. - L'administration des eaux et forêts établit des programmes de recherche, d'expérimentation, de formation et de vulgarisation en vue de favoriser le développement de la pêche et de l'aquaculture.

Elle crée des fermes ou des stations d'aquaculture dont l'accès est réglementé. La pollution des eaux y est interdite.

Article 78. - Aucun produit de pêche et de l'aquaculture ne peut être mis en vente sans qu'il soit délivré par l'administration des eaux et forêts un certificat d'origine selon les modalités définies par les textes réglementaires.

TITRE V : DISPOSITIONS ECONOMIQUES ET FINANCIERES

Article 79. - Pour tenir compte des données de la conjoncture, de la politique menée dans ces secteurs et des besoins de l'Etat, la loi de finances détermine annuellement et sur proposition du ministre chargé des eaux et forêts, le taux et l'assiette des redevances, droits et taxes applicables sur les activités relatives aux bois, aux forêts, à la faune, à la chasse et aux ressources halieutiques.

Le recouvrement des redevances, droits et taxes ci-dessus est confié à l'administration des domaines.

Article 80. - En vue de préparer l'exploitation prochaine des plantations forestières artificielles par l'administration des eaux et forêts, il est crée une régie dont les attributions et l'organisation seront définies par voie réglementaire.

Article 81. - Afin de permettre aux Gabonais de participer plus efficacement à la mise en valeur des domaines visés à l'art. 3 précité, le gouvernement prendra des mesures facilitant la mise en oeuvre d'une exploitation de groupe de ces domaines, selon les formes qui seront définies par les textes réglementaires.

TITRE VI : DISPOSITIONS DIVERSES

Article 82. - Les grumes abandonnées le long des cours d'eau, plages, routes, parcs, deviennent propriété de l'Etat dans les conditions définies par voie réglementaire.

Article 83. - L'introduction sur le territoire national de tout végétal ou animal, des dépouilles et trophées, est soumise à l'autorisation préalable de l'administration des eaux et forêts, sur présentation d'un certificat phytosanitaire ou zoosanitaire délivré par un organisme compétent selon les modalités fixées par voie réglementaire.

TITRE VII : DISPOSITIONS REPRESSIVES (art.84 à 116 omis)

TITRE VIII : DISPOSITIONS FINALES

Article 117. - Des textes réglementaires détermineront les mesures de toute nature nécessaires à l'application de cette loi.

Article 118. - La présente loi, qui abroge toutes dispositions antérieures, sera enregistrée, publiée selon la procédure d'urgence et exécutée comme loi de l'Etat.

Fait à Libreville, le 22 juillet 1982

Le Président de la République
Chef de l'Etat

LOI N° 10/75 DU 18 DECEMBRE 1975

GABON

Portant Création de l'Ecole Nationale des Eaux et Forêts

LOI N° 10/75

L'Assemblée Nationale a délibéré et adopté :

Le Président de la République, Chef du Gouvernement Promulgue la Loi dont la teneur suit :

Article 1. - Il est créé une Ecole Nationale des Eaux et Forêts en abrégé ENEF en remplacement de l'Institut National des Etudes Forestières (INEF).

L'Ecole Nationale des Eaux et Forêts est un établissement Public doté de la personnalité morale et jouissant de l'autonomie financière. Elle est placée sous la tutelle administrative du Ministre des Eaux et Forêts et sous le contrôle universitaire du Ministre de l'Education Nationale.

Article 2. - L'Ecole Nationale des Eaux et Forêts dispense un enseignement moyen et supérieur spécialisé, destiné à former les cadres des Eaux et Forêts des secteurs Public, Parapublic, et privé.

Article 3. - L'Ecole Nationale des Eaux et Forêts est habilité à recevoir des élèves de nationalité étrangère présentés par le Gouvernement de leur pays d'origine sous réserve de satisfaire aux conditions prévues à l'art.15.

Article 4. - Le budget de l'ENEF est approuvé chaque année par arrêté conjoint du Ministre des Eaux et Forêts et du Ministre de l'Economie et des Finances après avis du Conseil d'Etablissement prévu à l'art.10.

Article 5. - Les ressources du budget de l'ENEF sont constituées par une subvention annuelle inscrite au budget de l'Etat, par les contributions payées par les pays qui envoient les élèves en formation dans l'établissement, ainsi que par des dons et legs de toute nature.

Article 6. - L'ENEF est soumise au contrôle de l'Inspection des Etablissements publics et sociétés d'Etat conformément à la Loi N° 22/66 du 27 Juin 1966.

Chapitre I : DU DIRECTEUR

Article 7. - L'Ecole Nationale des Eaux et Forêts est placée sous l'autorité d'un Directeur nommé par décret pris en conseil des Ministres.
Le Directeur est nécessairement un Ingénieur des Eaux et Forêts.

Article 8. - Le Directeur est investi des pouvoirs les plus étendus pour assurer le fonctionnement de l'ENEF, notamment :
- Pour proposer à l'approbation du conseil d'Etablissement un règlement intérieur et assurer la discipline;

- pour établir les horaires et les programmes d'enseignement ainsi que les modalités des examens;
- Pour signer toutes correspondances, viser toutes pièces de dépense, passer les marchés nécessaires au fonctionnement de l'Etablissement conformément à la réglementation en vigueur;
- Pour arrêter les livres comptables, contrôler la régularité des opérations inscrites et vérifier la caisse;
- Pour recruter le personnel dans la limite des postes budgétaires disponibles, et proposer leur rémunération au conseil d'Etablissement par analogie avec les traitements en vigueur dans la Fonction Publique Gabonaise;
- pour préparer des réunions du Conseil d'Etablissement, l'ordre du jour, et en dresser les procès-verbaux.

Article 9. - L'organisation de l'Ecole Nationale des Eaux et Forêts fera l'objet d'un arrêté conjoint du Ministre des Eaux et Forêts, du Ministre de l'Education Nationale et du Ministre de l'Economie et des Finances.

Chapitre II : DU CONSEIL D'ETABLISSEMENT

Article 10. - Un Conseil d'Etablissement est constitué auprès de l'Ecole Nationale des Eaux et Forêts, il se réunit au moins deux fois par an sur convocation de son président. Il est chargé notamment :
- de donner son avis sur les moyens d'adapter les programmes d'enseignement au programme de développement du pays;
- de coordonner les activités de l'Ecole Nationale des Eaux et Forêts avec celles des services publics et parapublics et des entreprises privées intéressées;
- d'examiner le projet de budget présenté par le Directeur;
- d'établir en fin d'année scolaire la liste des élèves ayant mérité un diplôme de fin d'études.

Article 11. - La composition du Conseil d'Etablissement est fixée comme suit :

- Le Ministre des Eaux et Forêts	Président
- Le Ministre de l'Education Nationale	V. Président
- Le Ministre de l'Economie et des Finances ou son Représentant	Membre
- Le Ministre du Commerce et de l'Industrie ou son Représentant	"
- Le Ministre du Plan ou son Représentant	"
- Le Ministre de la Fonction Publique et de la Réforme Administrative ou son Représentant	"
- Le Conseiller à la Présidence de la République chargé des questions forestières	"
- Le Secrétaire Général du Ministère des Eaux & Forêts	"
- Le Recteur de l'Université Nationale	"
- Le Directeur des Forêts	"
- Le Directeur des Chasses et Pêches	"
- Le Directeur du Reboisement	"
- Le Directeur des Etudes et de la Recherche au Ministère des Eaux & Forêts	"
- Le Directeur de l' Enseignement des Eaux & Forêts	"

- Un Représentant des Organismes de Recherche Scientifique et Technique — Membre
- Le Président du Syndicat des Exploitants Forestiers — "
- Le Directeur Général de l'Office National des Bois du Gabon — "
- Le Directeur de l'Ecole Nationale des Eaux & Forêts — "
- Deux membres du personnel enseignant de l'ENEF — Membres

Le Conseil peut appeler en consultation toute personne dont la présence serait jugée souhaibable.

Article 12. - Un Conseil intérieur est institué au sein de l'ENEF. Sa composition et ses attributions sont fixées par le règlement intérieur.

Chapitre III : DES ETUDES

Article 13. - L'Ecole nationale des Eaux et Forêts comprend deux sections :
- Une section moyenne destinée à former les conducteurs des travaux des Eaux et Forêts;
- Une section supérieure destinée à former les Ingénieurs des Techniques des Eaux et Forêts.

Article 14. - La sanction des études est constituée par les diplômes suivants :
- Diplôme de Conducteur des travaux des Eaux et Forêts pour la section moyenne;
- Diplôme d'Ingénieur des Techniques des Eaux et Forêts pour la section supérieure.

Ces diplômes sont délivrés conjointement par le Ministre des Eaux et Forêts et le Ministre de l'Education Nationale à l'issue de la scolarité.

Les élèves qui n'ont pas obtenu le diplôme reçoivent un certificat de scolarité délivré par le Directeur de l'établissement.

Article 15. - Les modalités d'entrée dans chacune des sections , les conditions de délivrance de diplôme et la durée des études sont fixées par décret.

Article 16. - Toutes dispositions contraires sont abrogées.

Article 17. - La présente Loi sera exécutée comme Loi de l'Etat.

Libreville le 18 Décembre 1975

Par le Président de la République
Chef du Gouvernement

STRUCTURE AND LIST OF PROVISIONS OF
THE FORESTS ORDINANCE
(Chapter 157; Amended Version)

AN ORDINANCE FOR THE PROTECTION OF FORESTS AND FOR THE CONSTITUTION AND PROTECTION OF FOREST RESERVES

1. - Short Title and Application

2. - Interpretation :
- Cattle
- Competent Forest Authority
- Drift Timber
- Forest Officer
- Forest Offence
- Forest Produce
- Forest Reserve
- Native Court
- Native Appeal Court
- Private Lands
- Property Mark
- River
- Timber
- Tree
- Tribal or Stool Lands

3. - Appointment of Forest Officers

4. - Power to constitute Forest Reserves

5. - Notification of intention to create a Forest Reserve; and appointment of Reserve Settlement Commissioner

6. - Restriction on Acquisition of Rights during interim period

7. - Notice of Enquiry

8. - Power to make Bye-Laws before beginning of Enquiry and consequent Stay of Proceedings

9. - Procedures for Enquiry on and Settlement of Existing Rights

9A.- Survey of Land in Dispute

10. - Reserve Settlement Commissioner to be a Court

11. - Special power to exclude Land from proposed Forest Reserve

12. - Commutation of Rights

13. - Restriction on Payment of Compensation

14. - Extinction of Rights

15. - Judgment on Completion of Enquiry

16. - Appeal from Judgment of Reserve Settlement Commissioner

17. - Order constituting Forest Reserve

18. - 1) Owership of Land within Forest Reserve
2) Method of Management of Forest Reserve
3) Option as to Method of Management
4) Fetish Trees and Groves

19. - Transfer of rights in a Forest Reserve

20. - Power to declare Forest no longer reserved

21. - Power to District Commissioner to stop ways and Water-Courses in a Forest Reserve

22. to 33. - "REPEALED by FOREST PROTECTION, Decree NRCD 243/74 of 1974"

34. - Special Provisions Relating to Forest Reserves constituted by Bye-Laws

35. - Regulations :
- Fees
- Forest Produce
- Protection of Forest Produce
- Leases and Permits
- Management, etc. of Forest Reserves
- Marks
- Officers
- Rewards
- Survey

FORESTS ORDINANCE OF 30 MARCH 1927
CHAPTER 157 - LAWS OF GHANA;
1951 EDITION AND AMENDMENTS

GHANA

FORESTS ORDINANCE

Long Title	AN ORDINANCE FOR THE PROTECTION OF FORESTS AND FOR THE CONSTITUTION AND PROTECTION OF FOREST RESERVES
Date of Commencement	30th March, 1927
Short Title and application	1. This Ordinance may be cited as the "Forests Ordinance" and shall apply to Ghana
Interpretation	2. In this Ordinance, unless the context otherwise requires - "Bye-laws" includes rules and regulations; (Added by 21 of 1949, section 2)
Cattle	"Cattle" includes sheep, goats, horses, mules, donkeys, camels, and pigs;
Competent Forest Authority	"Competent Forest Authority" includes the Chief Conservator of Forests and any Forest Officer acting within the scope of the functions assigned to him by the Chief Conservator of Forests;
Drift timber	"Drift timber" means timber, either afloat or stranded, bearing no marks of ownership, or timber obviously out of control;
Forest Officer	"Forest Officer" means any officer of the Forestry Department;
Forest Offence	"Forest Offence" means any offence punishable under this Ordinance;
Forest Produce"	"Forest Produce" includes the following, if found in or brought from a Forest Reserve (i) Timber, charcoal, rubber, wood, oil, resin, and natural varnish; (ii) Trees and leaves, flowers and fruit, and all other parts and produce not herinbefore mentioned of trees; (iii) Plants not being trees (including grass, creepers, reeds and moss) and all parts and produce of such plants; (iv) Wild animals and skins, tusks, horns, bones, silk, honey, and wax, and all parts and produce of wild animals; and (v) Peat, surface soil, and minerals, other than minerals within the meaning of any Ordinance regulating the working of minerals; (Amended by 21 of 1949, section 2)

Forest Reserve — "Forest Reserve" means a Forest Reserve constituted under section 17;

Native Court — "Native court" means a court constituted as a Native Court under the provisions of any Ordinance, but, notwithstanding the provisions of any Ordinance to the contrary, shall not include such Native Court, when sitting as a Native Appeal Court;

(Substituted by N°45 of 1954, section 2a)

Native Appeal court — "Native Appeal Court" means a court constituted as Native Appeal Court under the provisions of any Ordinance, and sitting as such;

(Inserted by N°45 of 1954, Section 2b)

Private lands — "Private lands" means any land alienated form tribal or stool lands and owned or held by any individual or group of individuals;

Property mark — "Property mark" means a mark placed on trees or timber to denote that, after all purchase money due on such trees or timber has been paid, the registered holder of the property mark has or will have a right in such trees or timber;

(Substituted by 21 of 1949, Section 2)

River — "River" includes streams, canals, creeks, and other, Channels, natural and artificial;

Timber — "Timber" includes trees when they have fallen or have been felled, and all wood, whether it has been cut up, or fashioned or hollowed out for any purpose or not;

Tree — "Tree" includes palms, bamboos, stumps, brushwood, and canes;

Tribal or Stool lands — "Tribal or Stool lands" means lands at the disposal of any native community or of any chief acting in accordance with native customary law.

Appointment of Forest Officers

3. The President, or any officer empowered by the President in this behalf, may from time to time appoint Forest Officers by name or as holding an office, to carry out all or any of the purposes of this Ordinance, or to do anything required or allowed by this ordinance to be done by a Forest Officer.

Power to constitute Forest Reserves

4. Subject as is provided in section 34, it shall be lawful for the President to constitute as a Forest Reserve any of the following lands, namely -
(1) Lands the property of the Government;
(2) Tribal or Stool lands, at the request of the Native Authority;
(3) Private lands, at the request of the owner;
(4) Lands in respect of which the President is, on the advice of the Chief Conservator of Forests, satisfied that the forests thereon ought in the public interest to be protected from injury

or destruction, as the case may be, or that forest growth should be established thereon, in order -

(a) to safeguard the water supply of the district; or
(b) to assist the well-being of the forest and agricultural crops grown on the said lands or in the vicinity thereof; or
(c) to secure the supply of forest produce to the inhabitants of villages situated on the said lands or in the vicinity thereof.

Notification of intention to create a Forest Reserve and appointment of Reserve Settlement Commissioner

5. 1) When it is proposed to constitute any area of land of Forest Reserve, a notice shall be published in the Gazette -
(a) Specifying as nearly as may be the situation and limits of the land;
(b) Stating the reasons for constituting the Forest Reserve; and
(c) Appointing a Reserve Settlement Commissioner.

2) If, owing to absence, illness, or any other cause, the Reserve Settlement Commissioner is unable to perform or complete his duties, the President may, by notice published in the Gazette, appoint any person to act on his behalf or in succession to him.

Where the Reserve Settlement Commissioner has, prior to such appointment, begun to hold, but has not completed, an enquiry under the provisions of section 9 of this Ordinance, the person appointed to act on his behalf or in succession to him shall not be obliged to begin the enquiry *de novo* but may, if he so thinks fit, and with the consent of all the parties thereto, continue and complete the enquiry so begun but not completed.

(Amended by 21 of 1949, section 4)

Restriction on acquisition of rights during interim period

6. During the interval between the publication of the notice referred to in section 5. (1) and the date of an Order made under section 17 -
(a) in the case of any Tribal or Stool land comprised in such notice no alienation of the same and no acquisition of rights in or over the same or any produce thereof shall take place;
(b) in the case of any other land comprised in such notice, no right shall be acquired in or over the same or any produce thereof except by succession or with the consent in writing of the Reserve Settlement Commissioner (who may give or with-hold such consent in his discretion) and under an instrument in writing approved by him and made by some person in whom the right to make such instrument was vested at the date of publication of the notice or by his successor in title;

(c) no clearing, cutting or burning shall be made on any land comprised in such notice except with the permission of a Forest Officer not below the rank of an Assistant Conservator of Forests.

Notice of Inquiry

7. Upon the publication of the said notice the Reserve settlement Commissioner shall forthwith cause the particulars contained therein to be made known in the district or districts in which the land is situated by causing the same to be read and interpreted in the local native language or languages at a convenient session of the Magistrate's Court in the said district or districts, as the case may be, and by causing the notice to be posted outside such Court, and also by informing the Native Authority or Native Authorities concerned. The Reserve Settlement Commission shall also and in the manner aforesaid fix a period being not less than six months within which and a place or places to or at which, any person or native community claiming any right affecting the land or rights over the land which it is proposed to constitute a Forest Reserve shall either send in a written statement of his or their claim to the Reserve Settlement Commissioner or appear before him and state orally the nature and extent of his or their alleged rights.

(Amended by 19 of 1936, s.8, & by 21 of 1949, s.6)

Power to make bye-laws before beginning of enquiry and consequent stay of proceedings

8. Notwithstanding anything contained in any preceding section, the Native Authorities concerned may, at any time between the publication of a notice under section 5 and the beginning of an enquiry by the Reserve Settlement Commissioner under section 9, make bye-laws constituting the area of land specified in the notice aforesaid a Forest Reserve as if such notice had not been published, and from the date of publication of such bye-laws all further proceedings for constituting the said area of land a Forest Reserve under this Ordinance shall be discontinued, subject always to the operation of subsection (3) of section 34;

(Added by 38 of 1929, s.3 and amended by 21 of 1949 s.7)

Procedures for Enquiry on and Settlement of Existing Rights

9. (1) Save as is otherwise provided in this section, the Reserve Settlement Commissioner shall enquire into and determine the existence, nature and extent of the rights in respect of which he has received any claim under section 7 of this Ordinance and for the purpose of such enquiry he may divide the proposed Forest Reserve into as many portions of land as he may deem expedient and make a separate enquiry and determination in respect of each such portion.

If in the course of any enquiry any other rights affecting the proposed Forest Reserve are alleged to exist or are brought to his notice, the Reserve Settlement Commissioner shall enquire into and determine their existence, nature and extent.

(2) If in the course of any enquiry made under subsection (1) of this section a dispute arises as to the ownership of any land which lies within the proposed Forest Reserve, the Reserve Settlement Commissioner shall try and determine such dispute, either in the course of such enquiry or at a separate enquiry, and shall incorporate his finding in his judgment given under the provisions of section 14 of this Ordinance.

(3) Notwithstanding the provisions of subsection (2) of this section, where such a dispute is determinable by a Native Court, the Reserve Settlement Commissioner may, in his discretion, refer the dispute to the appropriate Native Court, and upon such reference it shall be deemed for all purposes to be a civil suit brought before such Native Court, and, subject to the provisions of subsection (4) of this section, shall be tried and determined accordingly.

(4) Where a dispute has been referred to a Native Court for determination in accordance with the provisions of subsection (3) of this section, the Reserve Settlement Commissioner shall accept and adopt the judgment of that Native Court or, in a case where there has been an appeal to a Native Appeal Court in accordance with the provisions of subsection (5) of this section, the judgment of that Native Appeal Court, and he shall incorporate the same in his judgment given under the provisions of section 14 of this ordinance :

Provided that where the Native Court does not give judgment within three months after the date of the reference thereto or in the case of an appeal from the Native Court the Native Appeal Court does not give judgment within three months after the date of the determination by the Native Court of such dispute, the Reserve Settlement Commissioner, after giving the Native Court or the Native Appeal Court, as the case may be, notice in writing of his intention so to do, may where the Native Court has not given judgment, proceed in accordance with the provisions of subsection (2) of this section as though there had been no such reference thereto, or where the Native Appeal Court has not given judgment, accept and adopt the judgment of the native Court in accordance with the provisions of this subsection as though there had been no such appeal and the proceedings in the Native Court or the Native Appeal Court, as the case may be, shall cease and determine from the date of such notice.

(5) Notwithstanding the provisions of any Ordinance under which a Native Court or Native Appeal Court is constituted, there shall be no appeal from the judgment of a Native Court given under the provisions of subsection (3) of this section, other than -
(a) to a Native Appeal Court, where the Ordinance constituting the Native Court provides in such a case for an appeal to a Native Appeal Court and an appropriate Native Appeal Court has been duly constituted under such Ordinance; or
(b) in accordance with the provisions of section 16 of this Ordinance.

(6) Where there has been an appeal to a Native Appeal Court in accordance with the provisions of subsection (5) there shall be no appeal from the judgment of that Court, save in accordance with the provisions of section 16 of this Ordinance.

(Subsections 2-6 substituted by N°45 of 1954, section 3)

Survey of land in dispute

9 A. Where there is any dispute as to the ownership of any land which lies within any proposed Forest Reserve -
(a) the expenses of any survey which is necessary for the determination of the dispute shall be borne by the Government of Ghana;
(b) the Reserve Settlement Commissioner may direct the competent Forest Authority to survey the boundaries of any such land and to do such other acts as may be reasonably necessary for the survey and demarcation of such land.

(Inserted by N°45 of 1954, section 4)

Reserve Settlement Commissioner to be a Court

10. (1) For the purpose of exercising jurisdiction conferred upon him by section 9, the Reserve Settlement Commissioner shall be and form a Court. The Court so formed shall be called the Court of the Reserve Settlement Commissioner, and shall be a Court within the meaning of that term in the Courts Decree, and so far as is not inconsistent with the provisions of this Ordinance the proceedings therein shall be governed and regulated accordingly.

(2) Every enquiry in the Court of the Reserve Settlement Commissioner shall be entitled "In the Court of the Reserve Settlement Commissioner of Ghana"

(Substituted by 3I of 1935, section 20)

Special power to exclude land from proposed Forest Reserve

11. The Reserve Settlement Commissioner may, after consultation with a Forest Officer not below the rank of Assistant Conservator of Forests, exclude any land from the proposed Forest Reserve either by altering the external boundary thereof or by demarcating the land within such external boundary:

Provided that no such alteration or demarcation shall have the effect of increasing the area of the proposed Forest Reserve.

(Substituted by 21 of 1949, s.9)

Commutation of rights

12. (1) If, after consultation with a Forest Officer not below the rank of Assistant Conservator of Forests, the Reserve Settlement Commissioner finds it impossible having due regard to the establishment and maintenance of the proposed Forest Reserve, to permit wholly or in part the exercise of any established right within the Forest Reserve, he shall assess a lump sum amount by the payment of which to the holder of the right such right may be commuted.

(2) If any such assessment is increased by the Court on appeal under section 16 of this Ordinance, the President, if he deems it inexpedient that such increased assessment should not be paid, may direct that the right in question shall not be commuted and that the exercise thereof shall be permitted.
(Substituted by 21 of 1949, s.10)

Restrictions on payment of compensation

13. Notwithstanding anything in this Ordinance contained, no compensation shall be payable in respect of -
(a) any restriction upon the exercise of rights in Tribal or Stool lands which lie within a proposed Forest Reserve;

(b) any restriction, whether in whole or in part, upon the exercise of the rights of any person to collect forest produce, hunt, fish, set traps, obstruct the channel of any river, pasture cattle or light fires, where such restriction is for the protection and orderly management of the Forest Reserve.
(Added by 21 of 1949, s.11)

Extinction of rights

14. Every right in or over any land in respect of which no claim has been made under section 7, or of which no knowledge has been acquired at any enquiry, shall be extinguished unless the claimant shall have satisfied the Reserve Settlement Commissioner before the delivery of his judgment that he had good reason for not preferring his claim within the period fixed under section 7; in which event the Reserve Settlement Commissioner may defer his judgment until he has decided such claim.
(Amended by 38 of 1929, S.7)

Judgment on completion of enquiry

15. (1) Whenever the Reserve Settlement Commissioner holds an enquiry he shall deliver his judgment upon the completion thereof.

(2) Such judgment shall -
(a) describe the limits of the land in respect of which the enquiry has been held;
(b) describe the limits of the land which the Reserve Settlement Commissioner recommends for reservation;
(c) specify, with all necessary particulars, the rights in respect of which the Reserve Settlement Commissioner has received claims under sections 7, 9 and 14 of this Ordinance and any other rights alleged to exist, or brought to his notice at, or after, the enquiry, in favour of any person or native community;
(d) specify those claims which the Reserve Settlement Commissioner considers not to have been established;
(e) admit or prohibit, in whole or in part, the exercise of all rights which he considers to have been established;
(f) specify any lump sum amount or amounts assessed under subsection (1) or section 12 of this Ordinance.

Appeal from judgment of Reserve Settlement Commissioner

16. Any person who has made a claim under section 7, 9 or 14 of this Ordinance, and the Chief Conservator of Forests, may within six months from the date of judgment delivered by the Reserve Settlement Commissioner under the provisions of section 15 of this Ordinance appeal from such judgment to the Court of Appeal and such appeal may relate to such part of the judgment as was incorporated therein under the provisions of subsections (2) or (4) of section 9 of this Ordinance.
(Amended by 45 of 1954, section 4)

Order constituting Forest Reserve

17. (1) As soon as the Reserve Settlement Commissioner shall have delivered his judgment under section 15 the President may make an order constituting the land in respect of which the Reserve Settlement Commissioner has in his judment described the limits recommended for reservation a Forest Reserve. Such order shall set forth the limits and situation of the land which constitutes the Forest Reserve and all rights affecting the same as set forth in the judgment of the Reserve Settlement Commissioner; and such order shall be published in the Gazette, and shall thereupon come into operation.
(Amended by 21 of 1949, s.14)

Ownership of land within Forest Reserve

18. (1) The ownership of land within a proposed Forest Reserve shall not be altered by its constitution as a Forest Reserve.

Method of management of Forest Reserve

(2) Every Forest Reserve shall be managed in one of the following ways -
(a) By the owner or owners under the direction of the Forestry Division; or
(b) By the Government for the benefit of the owner or owners.
In the latter case there shall be paid to the owner or owners in such proportion as the President shall decide the whole of the gross yearly revenue of the Forest Reserve accruing under this Ordinance, subject to the deduction of such sum not exceeding one-third of such gross revenue as may at the direction of the Conservator of Forests be reserved for expenditure on the improvement of the forest in the interest of the owner or owners. If such deduction be made the Chief Conservator of Forests shall render an account of its expenditure to the owner or owners concerned.
(Amended by 21 of 1949, s.15)

Option as to method of management

(3) The method of management of a Forest Reserve shall be at the option of the President who shall have power at any time to vary the method of management if and whenever a variation appears to him necessary or desirable, Prodided that the owners shall be at liberty to refuse to undertake the management under the method specified in subhead (a) of subsection (2).

Fetish trees and groves

(4) Where any recognised fetish grove or fetish tree is included in a Forest Reserve, the Forest Officers concerned shall not wilfully interfere therewith, and shall do their utmost to ensure that the fetish character thereof is respected.
(Section amended by 38 of 1929, s.10; and 10 of 1932, s.2)

Transfer of rights in a Forest Reserve

19. Rights in a Forest Reserve may not be alienated by sale, lease, mortgage, charge, or transfer, unless and until the right-holder shall have given a written motification of his intention to the Chief Conservator of Forests.

Power to declare forest no longer reserved

20. When in the opinion of the President there is no longer any need to maintain any particular land as a Forest Reserve, the President may, by order published in the Gazette, direct that from a date to be fixed by such order any land or any portion thereof reserved under this Ordinance shall cease to be a Forest Reserve.

Power to District Commissioner to stop ways and water-courses in a Forest Reserve

21. A District Commissioner may, on the request of the Chief Conservator of Forests, stop any public or private way or water-course in a Forest Reserve, provided that a substitute for the way or water-course so stopped, which the District Commissioner deems to be reasonably convenient, already exists or has been provided or constructed by the Forestry Department in lieu thereof.

22. Offences and penalties.

23. Section 22 not to prohibit exercise of admitted rights.

24. Penalty for counterfeiting or defacing marks on trees or timber, and for altering boundary.

25. Penalties for forest offences not specifically penalised.

26. Prevention of offences.

27. Power to arrest without warrant.

28. Power to seize forest produce.

29. Persons bound to assist Forest Officers and District Commissioners.

30. Forfeiture of instruments connected with forest offences.

31. Burden of proof.

32. Indemnity for acts done in good faith.

33. Operation of other laws not barred.

(Sections 22 to 33 omitted; replaced by the Provisions of the Forest Protection Decree N.R.C.D. 243 of 1974)

34. Special provisions relating to Forest Reserves constituted by bye-laws.

(Omitted, no more applicable)

Regulations

35. (1) It shall be lawful for the President to make regulations for the further, better, or more convenient effectuation of any of the provisions or purposes of this Ordinance, and in particular (but without derogating from the generality of the provision last aforesaid) with respect to any or all of the following matters -

Fees

(1) The prescription of fees for any purpose under this Ordinance;

Forest produce

(2) Prescribing the conditions of sale of any forest produce taken from a Forest Reserve, and the manner of its collection;

Protection of forest produce

(3) Providing for the protection of forest produce in a Forest Reserve, and prescribing the time at which and the manner and place in which rights which under this Ordinance for the time being are, or are treated as admitted rights, may be exercised;

Leases and permits

(4) Prescribing the form of leases or permits dealing with forest produce, the conditions under which they may be granted, and providing for their issue, production, revocation, and return;

Management, etc, of reserves

(5) Providing for the management, utilisation, and the protection, of Forest Reserves;

Marks

(6) Prescribing the marks which may be used by Forest Officers for the purpose of carrying out the provisions of this Ordinance;

Officers

(7) Prescribing powers and duties of Forest Officers, and providing for the maintenance of discipline;

Rewards

(8) Prescribing the rewards to be paid to Government servants and informers out of the proceeds of fines and compensations;

Survey

(9) Prescribing for the survey and demarcation of Forest Reserves or of any land the survey or demarcation of which is required for the purposes of this Ordinance. (Subsection amended by 21 of 1949, s.17)

(2) All regulations made under subsection (1) shall be published in the Gazette, and shall thereupon have the like force and effect as if enacted herein, either immediately or on and from such later date as may therein or in that regard be provided.

(3) Any person who contravenes any regulation made under subsection (1) or the conditions of any permit issued thereunder for which no fine or term of imprisonment is expressly provided in the regulations, shall be liable to a fine not exceeding fifty pounds or to imprisonment with or without hard labour for any term not exceeding six months.

ACT Nº 12 OF 29th JUNE 1960

GHANA

Act 12 / 1960

THE FORESTS IMPROVEMENT FUND ACT, 1960

An Act to establish an improvement fund for the Forest reserves and to provide for the control of the revenue derived from forest reserves.

Short title and commencement

1.-This Act may be cited as the Forests Improvement Fund Act, 1960 and shall come into operation on the first day of July, 1960.

Interpretation

2.-In this Act unless the context otherwise requires -

"Administrator of Stool Lands" includes the Receiver of Stool revenue;

"forest reserve" means a forest reserve constituted under section 17 of the Forests Ordinance (Cap.157) or under bye-laws made by a Local Authority;

"Fund" means the Fund established under section 3 of this Act;

"Minister" means the Minister responsible for forests in consultation with the Minister responsible for Finance.

"revenue" includes rents, dues, royalties, fees for silvicultural work and fines.

Establishment of Fund

3.-1) There is hereby established a fund to be known as the Forests Improvement Fund which shall be under the control of the Minister responsible for forests.

2) All moneys in the Forest Reserves General Deposit Account and any other deposit account maintained for any forest reserve shall, on the coming into operation of this Act and without further authority than this section, be transferred to the Accountant-General who shall open an Account for the Fund.

Payments into Fund

4.-All revenues, and other payments, including moneys from the Government by way of grants-in-aid or endowment or otherwise due in respect of any forest reserve shall be collected by the Chief Conservator of Forests and paid into the Fund.

Disbursement from Fund

5.-1) It shall be lawful for the Chief Conservator of Forests with the prior approval of the Minister to make payments from time to time from the Fund -

a) for costs incurred in connection with exploitation and silvicultural work; and

b) to landowners of forest reserves, through the appropriate authority, the royalties which by agreement the landowners are entitled to be paid.

Accounts and audit

6.-1) The Chief Conservator of Forests shall keep proper records in relation to the accounts and shall for each financial year not later that the 30th day of September following the end of the financial year prepare a statement of accounts in such form as the Auditor-General may direct.

2) The accounts shall be audited by the Auditor-General and published in such manner as the Minister may determine.

3) The Minister shall lay a copy of the statement of accounts prepared by the Chief Conservator of Forests with a copy of any report made by the auditor on the statement of accounts before Parliament.

Schemes of improvement

7.-The Chief Conservator of Forests shall on the coming into operation of this Act prepare a scheme covering a period of not less than five and not more than ten years for the improvement of forest reserves for the approval of the Minister.

Stool land revenue

8.-In the application of this Act to forest reserves subject to the provisions of the Akin Abuakwa (Stool Revenue) Act, 1958 (N°8), the Ashanti Stool lands Act, 1958 (N°28) and the Stool lands Control Act, 1959 (N°79) the Chief Conservator of Forests shall be deemed to be acting for and on behalf of the Stool lands Administrator for the purposes of the collection of revenue and accordingly any moneys in excess of the moneys spent by the Chief Conservator of Forests in the improvement of such forest reserves including moneys paid to landowners shall be paid by the Chief Conservator of Forests to the Administrator of Stool lands under the said Acts.

Payment of Forest fines

9.-Any moneys paid by way of fines to a forest officer or any moneys derived from the sale of forest produce or instruments seized under the Forest Offences (Compounding of Fines) Act, 1959, N°83) shall be paid into the Fund and accordingly section 4 of that Act is hereby amended by substituting for the words"Consolidated Fund" the words "Forests Improvement Fund".

Regulations

10.-The Minister may by legislative instrument make regulations prescribing the mode of collection of moneys due under this Act and the manner in which payments shall be made from the Fund.

THE CONCESSIONS ACT, 1962

An Act to provide that the provisions of the Concessions Ordinance shall cease to apply in respect of stool lands, to continue in force certain existing concessions subject to their terms and to provide for purposes connected therewith or incidental thereto.

Date of Assent : 14th June, 1962

Be it enacted by the President and the National Assembly in this present Parliament assembled as follows :

The Ordinance except certain provisions, to cease to apply to stool lands

1.- Subject as hereinafter provided in this Act the Ordinance, other than section 30, Part 4 and section 40, shall cease to apply to stool lands with effect from the date of the commencement of this Act.

Certain existing concessions to continue in force subject to their terms

2.- Every concession in respect of any stool land which is in force on the day, immediately preceding the date of the commencement of this Act shall continue in force on and after that date according to the terms of the concession :

Provided that -

a) such terms may be varied by agreement between the parties to the concession, subject to the written consent of the Minister; and

b) all persons authorised by the Minister may use for any purpose any road made on any land specified in such concession and may enter and inspect such land for any purpose which the Minister may think appropriate.

3.- to 13.- OMITTED

Regulations

14.- 1) The Minister may by legislative instrument make regulations -

a) in respect of the records to be maintained for the purposes of this Act and for the imposition of any fees for such purposes;

b) in respect of any matter relating to the Tribunal including the payment of fees and allowances to the members thereof; and

c) for the purposes of giving effect to the principles and provisions of this Act.

2) Any statutory instrument made under the provisions of the Ordinance and in force on the day immediately proceding the date of the commencement of this Act shall continue in force on and after that date as if the instrument was made under this Act.

15.- OMITTED

Forest reserves and timber concessions

16.- 1) All lands referred to in subsection (2) or subsection (4) of section 4 of the Forests Ordinance (Cap.157) and which have been constituted or proposed to be constituted as forest reserves under that Ordinance and all lands deemed to be constituted as forest reserves under subsection (7) of this section are hereby vested in the President in trust for the stools concerned:
Provided that all rights, customary or otherwise, in such lands validly existing immediately before the commencement of this Act shall continue on and after such commencement subject to this Act and any other enactment for the time being in force.

2) All lands which in the future shall be proposed to be constituted as forest reserves under the Forests Ordinance (Cap.157) shall become vested in the President in trust for the stools concerned with effect from the date of the publication of the notice relating to such land and precribed under section 5 (1) of that Ordinance.

3) Any land, other than land referred to in the preceding subsections, subject to the Administration of Lands Act, 1962 and in respect of which rights have been granted with respect to timber or trees under any concession and in force immediately before the commencement of this Act are vested in the President in trust for the stools concerned, subject to the terms of the concession, this Act and any other enactment for the time being in force.

4) All rights with respect to timber or trees on any land other than land specified in the preceding subsections of this section are vested in the President in trust for stools concerned.

5) It shall be lawful for the President to execute any deed or do any act as a trustee in respect of lands or rights referred to in this section.

6) Any revenue from lands or rights vested in the President under this section or derived under subsection (11) shall be collected, paid in and disbursed as provided by the Administration of Lands Act, 1962.

7) Forest reserves established under any laws relating to local government shall be deemed to be forest reserves constituted under the Forest Ordinance (Cap.157) and paragraph 27 of Part II of the First Schedule to the Local Government Act, 1961 (Act 54) is hereby rescinded.

8) The Forest Ordinance (Cap.157) shall apply _mutatis mutandis_ to any land outside a forest reserve in respect of which rights relating to timber or trees have been or shall be granted.

9) Section 30 of the Concessions Ordinance and the Concessions (Timber Restriction) Order, Nº 55 of 1939, shall apply mutatis mutandis to all land in Ghana in respect of which rights relating to timber or trees have been or shall be granted.

10) The Minister may terminate the concession of any holder, if he is found guilty of an offence under section 30(4) of the Concessions Ordinance (Cap.136)

11) REPEALED BY FOREST PROTECTION DECREE Nº 243/1974

Interpretation

17.- In this Act, unless the context otherwise requires - "Concession" and "certificate of validity" shall have the same meanings as in the Ordinance (Cap.136) and, for the resolution of doubts, shall include concessions excluded in whole or in part from the provisions of the Ordinance;

"Minister" means the Minister to whom functions under this Act have been assigned by the Presdient;

"Ordinance" means the Concessions Ordinance (Cap.136);

"stool lands" means stool lands to which the Administration of Lands Act, 1962, applies;

"terms of concession" include, where applicable, the terms of a certificate of validity; and

"Tribunal" means the Tribunal established under section 8 of this Act.

FOREST PROTECTION DECREE, 1974

In persuance of the National Redemption Council (Establishment) Proclamation, 1972 this Decree is hereby made :

Forest offences

1. (1) Any person who in a Forest Reserve without the written authority of the competent forest authority -
(a) fells, uproots, lops, girdles, taps, injures by fire or otherwise damages any tree or timber;
(b) makes or cultivates any farm or erects any building;
(c) sets fire to any grass or herbage, or kindles a fire without taking due precaution to prevent its spread;
(d) makes or lights a fire contrary to any order of the Chief Conservator of Forests;
(e) causes any damage by negligence in felling any tree or cutting or removing any timber;
(f) in any way obstructs the channel of any river, stream, canal or creek;
(g) hunts, shoots, fishes, poisons water or sets traps or snares;
(h) subjects to any manufacturing process, collects, conveys or removes any forest produce;
(i) pastures cattle or permits any cattle to trespass,

shall be guilty of an offence and liable on summary conviction to a fine not exceeding ₵1,000.00 or to imprisonment not exceeding five years or to both: PROVIDED that for a second or subsequent offence under this section an offender shall be liable on summary conviction to a fine not exceeding ₵5,000.00 or to imprisonment not exceeding ten years or to both.

(2) A person convicted of an offence under subsection (1) (a) or (1) (h) of this section shall, in addition to any other punishment imposed under this section, be liable to pay to the Commissioner responsible for Lands twice the commercial value of each tree or of the timber or forest produce which is the subject-matter of the offence.

(3) A person convicted of an offence under this section shall, in addition to any other punishment imposed under this section, be liable to pay to the person whose rights have been infringed such compensation as the court may direct.

(4) Nothing contained in this section shall prohibit the exercise in a Forest Reserve by any person of any right which under the Forest Ordinance (Cap. 157) for the time being is, or is treated as, an admitted right.

Offences relating to marks

2. Any person who -

(a) knowingly counterfeits or fraudulently uses upon any timber or standing tree a mark used by Forest Officers or any registered property mark to indicate that such timber or tree is the property of some person; or

(b) without the written permission of a Forest Officer alters, defaces or obliterates any mark placed on a tree or on timber; or

(c) alters, moves, destroys or defaces any boundary mark of any Forest Reserve,

shall be guilty of an offence and liable on summary conviction to a fine not exceeding ₵1,000.00 or to imprisonment not exceeding five years or to both: PROVIDED that for a second or subsequent offence under this section an offender shall be liable on summary conviction to a fine not exceeding ₵5,000.00 or to imprisonment not exceeding ten years or to both.

Persistent offenders to be banned

3. (1) Any person who has been convicted three times of an offence under this Decree shall be deemed to be prohibited from owning, operating or participating in any timber business or timber concession, and all permits and property marks held by him under any enactment relating to forests, trees, or timber shall be deemed to be forfeited.

(2) Any person who contravenes a prohibition imposed on him by subsection (1) of this section shall be guilty of an offence and liable on conviction to imprisonment not exceeding ten years without the option of a fine.

Duties of Forest Officers

4. (1) Every Forest Officer shall take all necessary steps to prevent the commisison of an offence under this Decree.

(2) Where the Commissioner responsible for Lands is satisfied that any Forest Officer has aided, condoned or connived at the commission of any offence under this Decree, he may order that such Forest Officer be summarily dismissed.

Arrest and seizure

5. (1) Any Forest Officer may arrest without warrant any person whom he reasonably suspects to have committed or to have been concerned in any offence under this Decree, if such person fails to give his name and address or gives a name or address which is believed to be false, or if there is reason to believe that he may abscond.

(2) A person arrested under this section shall within forty-eight hours be brought before a Magistrate, if not sooner released.

(3) Where there is reason to believe that an offence has been committed under this Decree, any Forest Officer may seize all forest produce to which the offence relates together with all instruments, vehicles and other articles suspected to have been used in committing the offence.

(4) A Forest Officer who seizes anything under this section shall place thereon a mark indicating that it has been seized and shall report the seizure to the Commissioner responsible for Lands.

Forfeiture and disposal of articles

6. (1) A court which convicts any person of an offence under this Decree shall order that all forest produce, instruments, vehicles and other articles in respect of which or by means of which the offence was committed (including anything seized under section 5) shall be forfeited to the Republic.

(2) Any vehicle or other article which is seized under section 5 and whose owner cannot be ascertained shall, after the expiration of fourteen days from the date of seizure, be deemed to be forfeited to the Republic.

(3) A vehicle or other article which is seized under section 5 and whose owner has been ascertained shall, if no prosecution is brought under this Decree, be restored to its owner.

(4) Anything which is forfeited to the Republic under this section may be sold or otherwise disposed of by the Commissioner responsible for Lands, and the proceeds applied for forest rehabilitation:

PROVIDED that where a vehicle is forfeited and the Commissioner is satisfied that the owner was in no way implicated in the offence, the Commissioner may restore the vehicle to such owner.

Burden of proof

7. The burden of proof that any forest produce has not been taken in contravention of this Decree shall lie upon the person in whose possession it is found.

Persons bound to assist Forest Officers

8. (1) Every person who exercises any right in or is permitted to take any forest produce from a Forest Reserve, and every person who is employed in a Forest Reserve, shall be bound to give to any Forest Officer without delay any information he may have regarding the commission or intended commission of any offence under this Decree, and shall assist any such officer -

(a) to extinguish any fire in the Reserve;
(b) to prevent any fire occurring nearby from spreading to the Reserve;
(c) to prevent the commission of any offence under this Decree, and to assist in discovering any offender.

(2) Any person who contravenes subsection (1) shall be guilty of an offence and liable on conviction to a fine not exceeding ₵50.00.

Indemnity for acts done in good faith

9. No action shall lie against any person in respect of any act done by him in good faith in the execution or intended execution of his powers or duties under this Decree.

Interpretation

10. In this Decree, unless the context otherwise requires:
"cattle" includes cows, sheep, goats, pigs and horses;
"forest produce" includes the following, if found in or brought from a Forest Reserve: -
(a) timber, charcoal, rubber, wood, oil, resin and natural varnish;
(b) trees, plants, leaves, flowers and fruit, and all other parts and produce of trees and plants;
(c) wild animals and birds and their skins, and all other parts and produce of wild animals and birds;
(d) eggs, snails, crabs, fish, silk, honey and wax;
(e) peat, surface soil, and minerals other than minerals within the meaning of any enactment regulating the working of minerals;

"Forest Reserve" means a Forest Reserve constituted under section 17 of the Forests Ordinance (Cap.157)

"timber" includes trees when they have fallen or have been felled, and all wood, whether it has been cut up or fashioned or hollowed out for any purpose or not;

"tree" includes palms, bamboos, stumps, brushwood and canes.

Repeals

11. The following enactments are hereby repealed: -

Section 22 to 33 of the Forests Ordinance (Cap.157)
Forest Offences (Compounding of Fines) Act, 1959 (N°83)
Forest Offences (Compounding of Fines) (Amendment) Act, 1962 (Act 99)
Section 16(11) of the Concessions Act, 1962 (Act 124).

Made this 12th day of February, 1974

Date of Gazette notification : 22nd February, 1974

N.R.C.D. 273 of
23 AUGUST 1974

GHANA

Decree N° 273 / 1974

TREES AND TIMBER DECREE, 1974

In pursuance of the National Redemption Council (Establishment) Proclamation, 1972 this Decree is hereby made :

PART I : PROPERTY MARKS

Locality marks

1.- 1) The Chief Conservator of Forests shall divide Ghana into such areas as he thinks fit, and shall allot to each area a distinctive mark to be known as a locality mark.

2) The Chief Conservator of Forests shall keep at his office a record of each area and the locality mark allotted , which shall be open to public inspection without charge during office hours.

Felling of trees for export

2.- No person shall cut or fell any growing tree for export in log form or for conversion in a mill unless he has first registered a property mark at the office of the Chief Conservator of Forests, endorsed for the locality in which he proposes to cut or fell.

Export of logs

3.- No person shall export any log unless it is marked with the cutter's property mark registered at the office of the Chief Conservator of Forests.

Registration of property marks

4.- 1) An application to register a property mark shall be made in writing to the Chief Conservator of Forests stating the name and address of the applicant, the proposed property mark, the area within which the applicant proposes to cut trees for export or for conversion in a mill and the locality mark.

2) If the Chief Conservator of Forests approves a proposed property mark, he shall on payment of the applicant of a registration fee of Ȼ100.00, register it in a Register of Property Marks to be kept for the purpose.

3) A registered property mark shall expire on the 31st day of December after the date of registration, but may be renewed from year to year on application therefor made within one month after the expiration of the registration or, in case of a renewal of registration, within one month after the expiration of the renewal, and on payment of a renewal fee of Ȼ50.00.

4) A renewal of registration shall be valid until the 31st day of December next following the date of renewal.

5) On registering a property mark the Chief Conservator of Forests shall give the applicant a certificate of registration; and on renewal of registration he shall issue to the applicant a new certificate endorsed with the renewal.

Refusal and cancellation of marks

5.- 1) The Chief Conservator of Forests may in his discretion refuse to register a property mark, or cancel the registration of a property mark.

2) Any person aggrieved by a refusal to register or cancellation of the registration of a property mark may appeal in writing to the Commissioner within thirty days after such refusal or cancellation, and the Commissioner may, if he thinks fit, direct the property mark to be registered or restored to the register, as the case may be.

Marking of stump and logs

6.- 1) Every person who cuts or fells a growing tree for export in log form or for conversion in a mill shall as soon as possible mark clearly with white waterproof paint or a deep-cutting scribe :

a) the stump thereof with his registered property mark and with a number (to be known as the stump number);

b) each of the logs therefrom at both ends with the locality mark of the area in which the tree is situated, his registered property mark, the number of the tree, and a log number (hereinafter referred to as "the log number").

2) Stump numbers shall run consecutively from number 1, number 1 being applicable to the first tree felled for export or for conversion in a mill by any person after he has registered his property mark.

3) Log numbers shall run consecutively from number 1 upwards, the butt log being numbered 1.

4) Letters and figures comprising marks made in pursuance of this section shall in all cases be not less than 2½ inches in height and, except in the case of scribed marks, not less than ½ inch in width.

Production of certificate

7.- Any person who fells or cuts a growing tree for export in log form or for conversion in a mill shall, on demand, produce the certificate of registration of his property mark to any police officer or Forest Officer.

Logs not duly marked

8.- No person shall buy, sell, export, or be in possession of any log which is not duly marked in accordance with the provisions of this part.

Transfer etc. of property mark

9.- No person shall, without the written permission of the Chief Conservator of Forests, loan, borrow or otherwise transfer or obtain any registered property mark to the use of which he is not entitled.

Stump to be shown 10.- Every person having a registered property mark shall on demand show to any police officer or Forest Officer the stump of any tree felled by him for export in log form or for conversion in a mill or give such information as will enable him to find such stump without difficulty.

Offences 11.- 1) Any person who contravenes or fails to comply with any of the foregoing provisions of this Decree shall be guilty of an offence and liable on summary conviction to a fine not exceeding ₵5 000.00 or to imprisonment not exceeding five years or to both.

2) When, a stump or log has been marked with a registered property mark, the onus of proof that it has been marked in accordance with this Part shall be on the registered holder of the property mark.

3) Where any person is convicted of an offence under sub-section (1) of this section, the court may in addition to the punishment imposed order that the whole or any part of the trees or timber in respect of which the offence was committed shall be forfeited and disposed of as the court may direct, and may order that any licence or permit held under this Decree or the regulations by the person convicted shall be forfeited.

4) Where any person is convicted of an offence under sub-section (1) of this section the Chief Conservator of Forests may cancel the registration of any property mark allotted to him.

PART II : PROTECTED AREAS

Protected areas 12.- 1) To prevent the waste of trees or timber in any area outside a Forest Reserve, the Commissioner may by executive instrument declare any area (other than a Forest Reserve) which consists wholly or mainly of standing trees or timber to be protected area with effect from a date four weeks after the publication of the instrument or such later date as may be specified in the instrument.

2) The Commissioner shall keep each protected area under review and if it appears to him that the control exercisable under this Part can conveniently be withdrawn from any part of a protected area the Commissioner shall revoke the instrument as respects that part of the protected area and that part shall accordingly cease to be a protected area.

Farming in protected area 13.- On the making of an instrument under section 12 any person engaged in farming in the protected area shall give written notice of that fact to the Commissioner, who if satisfied that the notice is correct shall grant a licence authorising him to continue farming within the area specified in the notice subject to any conditions imposed by the Commissioner in the interest of the protected area.

Offences in protected area

14.- Any person who is not exercising rights under a concession and who in any protected area without the written consent of the Commissioner -
a) fells, uproots, lops, girdles, taps, injures by fire or otherwise damages any tree or timber; or
b) makes or cultivates any farm or erects any building; or
c) sets fire to any grass or herbage, or kindles a fire without taking due precautions to prevent its spread;
shall be guilty of an offence and liable on summary conviction to a fine not exceeding one thousand cedis or to imprisonment not exceeding five years or to both.

Control of protected area

15.- The Commissioner may be legislative instrument make regulations :
a) imposing duties on persons who hold concessions in a protected area;
b) for permitting farming in protected areas;
c) for the appointment of forest guards; and
d) for the payment of fees by holders of concessions, and for applying any part of such fees towards the expense of guarding protected areas.

PART III : GENERAL

Arrest of offenders

16.- 1) Any police officer or Forest Officer may arrest without warrant any person whom he reasonably suspects to have committed or to have been concerned in any offence under this Decree, if such person fails to give his name and address or gives a name and address which is believed to be false, or if there is reason to believe that he may abscond.

2) A person arrested under this section shall within forty-eight hours be brought before a Magistrate, if not sooner released.

Regulations

17.- 1) The Commissioner may by legislative instrument make regulations -
a) Controlling or prohibiting the cutting or felling of any trees of smaller girth than prescribed in the regulations;
b) for the marking of trees that may be cut or felled;
c) for the control of the transit or export of timber, and for the salving and disposal of drift timber;
d) for the prescription for any purposes of a standard method for use in calculating the volume of any tree or timber;
e) for the control or prohibition of the purchase, sale, export or possession of timber cut, felled, collected or moved in contravention of the regulations;
f) for the protection of trees or timber;
g) for the imposition of fees for anything done for the the purposes of this Decree and the regulations;
h) otherwise for carrying out the principles and purposes of this Decree.

2) Any person who contravenes or fails to comply with any regulation made under this section or the conditions of any licence or permit issued or granted thereunder for which no penalty is expressly provided in the regulations shall be guilty of an offence and liable on summary conviction to a fine not exceeding one thousand cedis or to imprisonment not exceeding two years or to both.

3) Where any person is convicted of an offence against any regulation made under this section, the court may in addition to the punishment imposed order that the whole or any part of the trees or timber in respect of which the offence was committed shall be forfeited and disposed of as the court may direct, and may order that any licence or permit held under the regulations by the person convicted shall be forfeited.

Interpretation

18.- In this Decree, unless the context otherwise requires :
"Commissioner" means the Commissioner responsible for lands;
"Forest Reserve" means a Forest Reserve constituted under section 17 of the Forest Ordinance (Cap.157)
"mill" means a factory or conversion plant used to process logs or parts of trees into products of wood;
"property mark" means a mark placed on trees or timber to denote that after all purchase money due thereon has been paid the registered holder of the property mark has or will have a right in such trees or timber;
"protected area" means an area declared for the time being under this Decree to be a protected area;
"timber" includes trees when they have fallen or have been felled, and all wood, whether it has been cut up or fashioned or hollowed out for any purpose or not.

Repeals and saving

19.- The following enactments are hereby repealed : -
- Trees and Timber Ordinance (Cap. 158);
- Trees and Timber (Amendment) Act, 1957 N° 40);
- Protected Timber Lands Act, 1959 (N° 34).

Notwithstanding the above repeals, the following instruments as subsequently amended shall continue in force as if made under the corresponding provisions of this Decree until modified or revoked : -

- Trees and Timber (control of Cutting) Regulations,	1958,	(LN 368)
- Trees and Timber (Measurement) "	"	(LN 388)
- Timber Lands (Protected Areas) "	1959	(LN 311)
- Trees and Timber (Control of Measurement) "	1960	(LI 23)
- Trees and Timber (Control of Export logs) "	1961	(LI 130)

Notwithstanding the repeal of the Protected Timber Lands Act, 1959 (N°34), all instruments made under that Act to declare a protected area and in force immediately before the commencement of this Decree shall continue in force as if made under section 12 of this Decree, until modified or revoked.

Made this 23rd day of August, 1974

Date of Gazette notification : 30th August, 1974

L.I. N° 1089 OF 31 AUGUST 1976 GHANA GHA/28

L.I. 1089 /1976

FOREST FEES REGULATIONS, 1976

In exercise of the powers conferred on the Commissioner responsible for Lands by section 35 of the Forests Ordinance (Cap.157) and in pursuance of section 16(8) of the Commissions Act, 1962 (Act 124) these Regulations are made this 31st day of August, 1976.

1. - Any person who after the commencement of these Regulations cuts or takes away or extracts timber from land to which the Ordinance applies shall pay the royalty specified in the First Schedule to these Regulations in relation to the timber cut or taken away or extracted by him. *Royalty to be paid*

2. - After the commencement of these Regulations the rents payable for a timber lease or licence shall subject to paragraph 3 of these Regulations, be as specified in the Second Schedule to these Regulations. *Rent to be paid*

3. - 1) Any person who has cut or taken away or extracted timber before the Ist of July, 1976 shall pay the royalty or rent outstanding, if any, by 30th Semptember 1976. *Outstanding royalty or rent*

2) Any person who holds a timber lease or licence dated before the Ist July, 1976 and whose royalty or rent, as the case may be, in that regard, is in arrears by 30th September, 1976, shall after that date, pay, in respect of the arrears, the royalty or rent, as the case may be, specified in the First or Second Schedule to these Regulations.

4. - 1) Any person who cuts or takes away or extracts timber in any year in any forest reserve constituted under section 17 of the Forest Ordinance (Cap.157) or under bye-laws made by a Local Authority shall for that year pay a fee of seven cedis fifty pesewas for every hectare in such reserve over which he is permitted in that year to exercise rights of cutting or taking away or extracting timber. *Payment of fees by persons exploiting forest*

2) Any person who cuts or takes away or extracts timber in any year on any land outside a forest reserve shall for that year pay a fee of two cedis fifty pesewas for every hectare over which he exercises rights of cutting or taking away or extracting timber.

5. - 1) The fees charged under paragraph 4 of these Regulations shall be paid to the Forestry Department and shall be credited by that Department to the Forests Improvement Fund and shall be used to meet the cost of silvicultural work. — Forests Improvement Fund

2) The Forestry Department shall in each year publish a statement showing the amount credited to the said account and the amount withdrawn from the account to meet the costs of silvicultural work.

6. - 1) Where in the opinion of the Chief Conservator of Forests the value of timber in any area in a forest reserve or outside a forest reserve does not justify the levying of the fees specified in paragraph 4, he may reduce the amount payable in either case or waive the payment of the fees. — Waiver of Payment

7. - Nothing in these Regulations shall affect anything in any other enactment or any lawful claim or remedy against any person who cuts or takes away or extracts timber in the absence of a title and for the avoidance of doubts, payment of the royalties and rents prescribed in these Regulations shall not be deemed to relieve any person from any requirements of holding a title. — Other enactments not affected

8. - In these Regulations unless the context otherwise requires : — Interpretation

"timber" includes trees when they have fallen or have been felled and all wood whether it has been cut up or fashioned or hollowed out for any purpose or not;

"title" means valid and effective rights obtained in compliance with the law, to cut or take away or extract timber from any parcel of land referred to in these regulations;

"fees" includes dues, rents and royalties in respect of any interest in land.

9. - The Forests (Fees for Silvicultural Works) Regulations, 1959 (L.N.87) is hereby revoked — Revocation

10. - These Regulations shall be deemed to have come into force on the Ist day of July, 1976. — Commencement

GHA/30

FIRST SCHEDULE

Royalties

Local Name	Trade Name	Scientific Name	Royalty ₵
Odum	Iroko	Maclura excelsa	35.00
Edinam	Gedu Nohor	Entandrophragma angolense	30.00
Penkwa	Sapele	" cylindricum	45.00
Efuobrodedwo	Utile	" utile	54.00
Mahogany	African Mahogany	Khaya spp.	35.00
Baku	Makore	Tieghemella heckelii	48.00
Kusia	Opepe	Nauclea diderrichii	28.00
Kokrodua	Afrormosia	Pericopsis elata	52.00
Dubinibiri	African Walnut	Lovoa trichilioides	24.00
Emeri	Idigbo	Terminalia ivorensis	21.00
Wawa	Obeche	Triplochiton scleroxylon	15.00
Nyankom	Niangon	Heritiera utilis	14.00
Candollei	Omu	Entandrophragma candollei	35.00
Kwabohoro	Guarea	Guarea spp.	15.00
Kaku	Ekki	Lophira alata	28.00
Dahoma	Dahoma	Piptadeniastrum africanum	10.00
Kyenkyen	Antiaris	Antiaris africana	15.00
Aprono	Mansonia	Mansonia altissima	18.00
Subata	Abura	Mitragyna spp.	14.00
Danta	Danta	Nesogordonia papaverifera	14.00
Wansenwa	Avodire	Turraeanthus africanus	14.00
Ofram	Afara	Terminalia superba	18.00
Bonsamdua	Ayan	Distemonanthus benthamianus	10.00
Kokoti	Kokoti	Anopyxis klaineana	15.00
Papao	Afzelia	Afzelia africana	18.00
Wonton	Wonton	Morus mesozygia	12.00
Onyina	Ceiba	Ceiba pentandra	12.00
Denya	Okan	Cylicodiscus gabunensis	16.00
Otie	Illomba	Pycnanthus angolensis	6.00
Hyedua	Ogea	Daniellia ogea	22.00
Hyeduanini	Ovangkol	Guibourtia ehie	22.00
Entedua	Bubinga	Copaifera salikouna	22.00
Kyere	Pterygota	Pterygota macrocarpa	12.00
Asamfona	Aningeria	Aningeria robusta	12.00
Esa	Celtis	Celtis spp.	8.00
Teak	Teak	Tectona grandis	10.00
Bediwunua	Canarium	Canarium schweinfurthii	9.00
Cedrela	Cigar box Cedar	Cedrela odorata	15.00
Okuro	Albizia	Albizia spp.	15.00
		Any other species	6.00

SECOND SCHEDULE

Rents

25 pesewas per hectare per annum.

COMMISSIONER RESPONSIBLE FOR LANDS

Date of Gazette notification : 17th Sept. 1976

L.I. N° 23 OF 31st MARCH 1960 and
L.I. N° 1090 OF 31st AUGUST 1976

GHANA

L.I. 23 / 1960 and
L.I. 1090 / 1976

THE TREES AND TIMBER (CONTROL OF MEASUREMENT) REGULATIONS 1960

as amended by

THE TREES AND TIMBER (CONTROL OF MEASUREMENT) - AMENDMENT - REGULATIONS, 1976

Cap.158 n°40 — In exercise of the powers conferred upon the Government-General by section three of the Trees and Timber Ordinance as amended by the Trees and Timber (Amendment) Act, 1957 the following regulations are hereby made :

Commencement — 1. These Regulations shall come into operation on the first day of April, 1960.

Certificate of measurement L.N.388/58 — 2.1) Any transaction whatsoever affected by the provisions of the Trees and Timber (Measurement) Regulations, 1958, shall not be deemed to be valid unless a Certificate of measurement in the form set out in the First Schedule to these Regulations has been issued by the Commissioner for Lands or any person authorised by him.

2) The fees for the issue of the Log Measurement Certificate shall be as specified in the Second Schedule to these Regulations.

Payment of Freight.. — 3. The payment of freights on logs transported shall be computed by reference to a certificate of measurement.

Penalty — 4. Any person who transports a log by rail or exports any log without a certificate of measurement first had and obtained commits an offence and shall be liable on summary conviction to a fine not exceeding hundred New Cedis or to a term of imprisonment not exceeding six months.

FIRST SCHEDULE : Form for Log Measurement / Grading Certificate

OMITTED

SECOND SCHEDULE:

Initial issue of Log Measurement Certificate	₵ 1.00
Remeasurement Fee per Log	₵ 5.00
Replacement of Certificate	₵ 10.00

Dated at Accra this 31st day of March, 1960

SUPREME MILITARY COUNCIL
DECREE N° 128 OF
9th SEPTEMBER 1977

GHANA

Decree N°128/1977

TIMBER INDUSTRY AND GHANA TIMBER MARKETING BOARD (AMENDMENT) DECREE, 1977

Be it enacted by the Supreme Military Council as follows :

Only Ghana Timber Marketing Board to Export timber and timber products by land

1. - No person other than the Ghana Timber Marketing Board in this Decree referred to as the"Board " shall export overland timber, or any of the timber products specified in the Schedule to this Decree.

Prohibition of unauthorised persons from exporting timber and timber products by sea and air

2. - No person except :
a) the Board;
b) a person authorised in writing by the Board, shall export by sea or air timber or any of the timber products specified in the Schedule to this Decree.

Offences for contravening section 1 or 2

3. - 1) Any person who contravenes section 1 or 2 of this Decree shall be guilty of an offence and liable on summary conviction to a term of imprisonment not less than fifteen years and not more than thirty years.
- Provided that the court convicting the accused may, if satisfied on grounds stated that the offence was trivial or that there are special circumstances relating to the offender which would render unjust the application of the minimum penalty prescribed by this section, impose on him a term of imprisonment less than the said minimum or a fine not exceeding ₵ 20 000.00 or both.

2) unless the court otherwise decides on grounds stated, all goods involved in the commission of an offence under this section shall be forfeited to the State

Board to fix export and local prices of timber products

4. - 1) Notwithstanding any enactment to the contrary the Board shall, to the exclusion of all other persons have the right to fix the prices at which the timber products specified in the Schedule to this Decree and any other timber products prescribed by the Commissioner by regulations made under section 12 of this Decree, shall be exported.

2) The Board shall advise the Prices and Incomes Board and the Commissioner as to the price at which such timber products shall be sold within Ghana.

Power to demand returns

5. - The Board may by writing request any manufacturer of any timber products to submit to the Board returns of such products manufactured by him and of sales made of such products.

Power to prescribe percentage of timber products to be exported, etc.

6. - The Board may also give directions to any manufacturer of timber products relating to the percentage of such products manufactured by him which should be exported and the percentage to be sold in Ghana.

Registration of dealers in timber products

7. - No person shall, after three months from the commencement of this Decree operate as a retailer, wholesaler or dealer of or in plywood, timber, flush doors and any other timber products prescribed by the Commissioner by regulations made under section 12 of this Decree unless he is registered with the Board and has a valid licence from the Board so to do.

General penalties

8. - Any person, who :
a) fails without reasonable excuse proof of which shall be on him to submit any return requested by the Board under section 5 of this Decree or submits in response to any such request any return which he knows to be false or which he has no reason to believe to be true;
b) contravenes any direction given to him by the Board under section 6 of this Decree;
c) contravenes the provision of section 7 of this Decree;
shall be guilty of an offence and shall be liable upon summary conviction to a fine not exceeding ₵5 000.00 or one year's imprisonment or both; and in the case of a continuing offence to a further fine not exceeding ₵50.00 in respect of each day on which the offence continues.

Offences by bodies of persons

9. - 1) Where a body of persons commits an offence under this Decree or under regulations made thereunder then :
a) in the case of a body corporate other than a partnership the directors and the Secretary of that body shall also be deemed to have committed the offence;
b) in the case of a partnership the partners shall also be deemed to have committed the offence.

2) No person shall be deemed to have committed an offence under subsection (1) of this section if he proves to the satisfaction of the Court that the offence was committed without his knowledge or connivance and that he took all necessary steps which he was required to take to prevent the commission of the offence having regard to all the circumstances.

Power of Board to suspend manufacturers

10. - 1) The Board may by writing suspend for a time specified by the Board any manufacturer who has been convicted of an offence under this Decree or under any regulations made thereunder from the manufacture of timber products.

2) Any person aggrieved by a decision of the Board under this section may appeal to the Commissioner against the decision.

3) The Commissioner may upon such appeal, confirm, reverse or vary the decision of the Board and the Commissioner's decision on the matter shall be final.

Amendment of Schedule

11. - The Commissioner may, after consultation with the Board and with the prior approval of the Supreme Military Council, by legislative instrument amend the Schedule to this Decree.

Regulations

12. - 1) The Commissioner may on the advice of the Board by legislative instrument make regulations for carrying the provisions of this Decree into full effect.

2) Without prejudice to the generality of subsection 1) of this section regulations may be made thereunder :

a) for regulating the marketing of timber and timber products;
b) for registration of exporters of timber and timber products;
c) for regulation of the conveyance and haulage of timber and timber products;
d) for the registration of carpenters and other persons who depend on wood work for their livelihood for the purpose of enabling them to obtain timber products for the pursuit of their occupations;
e) to regulate the supply of timber products to persons requiring them for buildings purposes in the case of towns or places near the land borders of Ghana;
f) for providing that timber and timber products shall be exported only through particular designated exit points;
g) for prescribing fees in respect of anything to be done under this Decree;
h) for precribing any thing required and authorised to be precribed by this Decree;
i) for prescribing in relation to a contravention of any of the regulations, a penalty not exceeding a fine of ₵20 000.00 or ten years imprisonment or both and prescribing where the Commissioner thinks fit a minimum penalty and requiring the forfeiture of any goods involved in the commission of the offence.

Interpretation 13. - In this Decree,
"Board means the Ghana Timber Marketing Board; and "Commissioner" means the Commissioner responsible for Trade.

SCHEDULE :
Logs (in the round, squared or boulles), Plywood, Veneer, Lumber, Flush doors, Furniture, Furniture parts.

Made this 31st day of August, 1977

CHAIRMAN OF THE SUPREME MILITARY COUNCIL

Date of Gazette notification: 9th Sept. 1977

A.F.R.C.D. 47 OF 21 SEPT. 1979 GHANA

Decree N° 47 / 1979

ECONOMIC PLANTS PROTECTION DECREE, 1979

BE it enacted by the Armed Forces Revolutionary Council as follows :

Destruction of specified plants prohibited

1. - Any person who,
a) without the written authority of the Commissioner, or
b) for purposes other than for horticultural husbandry,
intentionally destroys or causes the destruction of any specified plant shall be guilty of an offence.

Payment of compensation

2. - Omitted

Commissioner to prescribe values and rates of compensation

3. - Omitted

Grant of felling rights prohibited

4. - 1) No felling rights in respect of timber trees shall be granted where such timber trees stand in farms where specified plants are cultivated.
2) Where immediately before the publication of this Decree in the Gazette there existed any felling rights in respect of timber trees standing in farms where specified plants are cultivated, such rights shall be deemed to be void and shall forthwith cease to have any effect.

3) Any person who grants or purports to grant or acts or purports to act in pursuance of any felling rights in contravention of the provisions of this section shall be guilty of an offence.

Penalties

5. - Omitted

Amendment of Schedule

6. - Omitted

Interpretation

7. - In this Decree, unless the context otherwise requires: "Commissioner" means the Commissioner responsible for matters relating to cocoa; "specified plant" means any plant specified in this Schedule * to this Decree.

Made this 21st day of September, 1979,

CHAIRMAN OF THE ARMED FORCES REVOLUTIONARY COUNCIL

Date of Gazette notification : 22nd Sept. 1979

* The Schedule refers to Cocoa.

Law 42/ 1982

PROVISIONAL NATIONAL DEFENCE COUNCIL (ESTABLISHMENT) PROCLAMATION (SUPPLEMENTARY AND CONSEQUENTIAL PROVISIONS) LAW 1982

Section 34. - Forestry Commission

1) There shall be established a Forestry Commission which shall be responsible for :

a) a review of the national practices relating to forests and forestry resources and the formulation of recommendations of national policy on forests and the exploitation of forestry resources including game and wildlife with special reference to establishing national priorities having due regard to the national economy;

b) monitoring the operation of such policy as the Council may adopt relating to forests and forestry resources and reporting to the Council;

c) ensuring that forests are maintained and protected as an economic resource and their natural and artificial regeneration are developed;

d) ensuring that needless waste and destruction of the forest and associated natural resources, avoided;

e) monitoring the operations of all bodies or establishements concerning forests, forestry resources, timber and the marketing thereof, and reporting to the Council;

f) ensuring for national decision-making a firm basis of comprehensive data collection on forests, forestry resources and timber and the marketing thereof, and the technologies for exploiting such resources;

g) co-operating and liaising with national and international organisations and bodies all over the world on matters of forestry conservation and utilization;

h) advising the Lands Commission on any grants of public land that affect forests and forestry resources;

i) advising the Secretary responsible for forestry and wildlife; and

j) such other functions as the Council may assign to it.

2) The composition of the Forestry Commission shall be determined by the Council.

3) The Forest Products Research Institute which became a Division of the Ghana Forestry Commission established by the Ghana Forestry Commission Act, 1980 (Act 405) and in existence before the coming into force of this Law shall continue to be a Division of the Forestry Commission established under this section.

4) The Forestry Department and the Department of Game and Wildlife shall continue in existence as government departments subject to the control of the Secretary responsible for forestry and wildlife and shall not be Divisions within the Forestry Commission established under this section.

5) The Ghana Timber Marketing Board which became a Division of the Ghana Forestry Commission established under the Ghana Forestry Commission Act, 1980 (Act 405) shall be reconstituted as statutory board.

Section 64. - Dissolvement of Agencies
Except as otherwise provided in this Law the following bodies which were in existence immediately before the coming into force of the Proclamation are hereby dissolved -
- Ghana Fisheries Commission
- Ghana Forestry Commission
- Lands Commission
- Local Government Grants Commission
- National Council for Higher Education
- National Development Commission
- National Security Council
- Parliamentary Service
- Press Commission.

Section 65. - Repeal of Enactments
The following enactments are hereby repealed -
....
Ghana Forestry Commission Act, 1980 (Act 405)

Section 67. - Date of Commencement
The provisions of this Law shall be deemed to have come into force on the 31st day of December, 1981; unless the Council otherwise decides in a specific instance or except where the context otherwise provides or specific provision is made for the commencement of any provision of the Proclamation in which case such provision shall, unless otherwise provided, come into force on the day this law is made.

Made this 30th day of December, 1982

LOI N° 65/425 DU 20 DEC. 1965
portant Code Forestier

IVORY COAST

COTE D'IVOIRE

Loi N° 65/425

L'Assemblée Nationale a adopté

Le Président de la République promulgue la Loi dont la teneur suit

TITRE I : DEFINITIONS

Article 1. - Sont considérées comme forêts les formations végétales dont les fruits exclusifs ou principaux sont les bois d'ébénisterie, d'industrie et de service, les bois de chauffage et à charbon et qui accessoirement peuvent produire d'autres matières telles que bambous, écorces, latex, résines, gommes, graines et fruits.

Article 2. - Sont considérés comme périmètres de protection :

- Les versants montagneux protégés de l'érosion par leur couverture végétale
- Les terrains où pourraient se produire des ravinements et éboulements dangereux
- les bassins versants des sources.

Article 3. - Sont considérés comme reboisement, les terrains plantés de main d'homme en espèces ne donnant pas de produits agricoles, ainsi que les forêts naturelles enrichies artificiellement en essence de bois d'oeuvre par des travaux de plantation ou de sylviculture.

Article 4. - Les formations végétales définies aux articles ler, 2 et 3 constituent le domaine forestier.

Le domaine forestier comprend :
- Le domaine forestier de l'Etat
- Le domaine forestier des particuliers et des collectivités.

TITRE II : DU DOMAINE FORESTIER DE L'ETAT

Chapitre Premier : Généralités

Article 5. - Le domaine forestier de l'Etat comprend les catégories suivantes :
- forêts classées
- forêts protégées
- périmètres de protection
- reboisements.

Article 6. - Les forêts classées avant la date de promulgation de la présente loi le demeurent.

Pourront, en outre être classées les forêts indispensables :

- à la stabilisation du régime hydrographique et du climat
- à la conservation des sols
- à la satisfaction des besoins du pays en bois à usages industriels et traditionnels
- à la préservation des sites et à la conservation de la nature
- à la salubrité publique
- à la défense nationale.

Chapitre II : Des droits d'usage

Article 7 . - Les droits d'usage comprennent :
- ceux portant sur le sol forestier
- ceux portant sur les fruits et les produits de la forêt naturelle
- ceux à caractère commercial portant sur certains fruits et produits de la forêt naturelle.

Section I : Les droits d'usage portant sur le sol forestier
Domaine classé, périmètres de protection, reboisements

Article 8 : - Le domaine classé, les périmètres de protection et les reboisements sont affranchis de tous droits d'usage portant sur le sol forestier.

Les défrichements, qu'il s'agisse d'abattage ou de débrousaillement de la végétation ligneuse, suivis ou non d'incinération sont interdits dans le domaine classé, les périmètres de protection et les reboisements

Ils ne peuvent être autorisés temporairement en vue de l'établissement de cultures que sur les terrains destinés à être enrichis en essences forestières de valeur.

Domaine protégé

Article 9 . - Les droits d'usage portant sur le sol forestier ne peuvent s'exercer que dans le domaine forestier protégé.

Article 10: - Tout citoyen ivoirien quelles que soient son ethnie et sa région d'origine peut exercer ce droit sur l'ensemble du domaine forestier protégé à condition de se conformer aux dispositions domaniales et après avoir obtenu l'autorisation de l'autorité chargée de la gestion du domaine rural.

Article 11. - L'emprise des forêts classées sera choisie de telle sorte que des surfaces suffisantes de forêts protégées soient laissées à la disposition des populations pour assurer leurs besoins usagers en produits forestiers et l'extension de leurs cultures en relation avec l'accroissement démographique et la substitution progressive d'une agriculture sédentaire intensive aux cultures itinérantes traditionnelles.

Article 12. - Les droits d'usage portant sur le sol forestier peuvent être réglementés pour la mise en oeuvre de plans d'aménagement ruraux et de modernisation de l'agriculture.

Article 13. - Les droits d'usage portant sur le sol forestier peuvent être suspendus temporairement quand l'Etat donne aux boisements une destination qui en exclut l'exercice :

1) Délivrance de permis temporaires d'exploitation de bois d'oeuvre ou de vente de coupes dans des régions encore peu habitées et dépourvues de cultures
2) Constitution de réserves de bois d'oeuvre où l'exploitation forestière précèdera obligatoirement les défrichements et les cultures.

Section 2 : Les droits d'usage portant sur les fruits et les produits de la forêt naturelle

Domaine protégé

Article 14. - Les droits d'usage portant sur les fruits et les produits de la forêt naturelle s'exercent librement dans le domaine protégé.

Domaine classé

Article 15. - Dans le domaine classé, les droits d'usage portant sur les fruits et produits forestiers sont limités :
1) au ramassage du bois mort
2) à la cueillette des fruits et des plantes alimentaires ou médicinales
3) à l'exploitation des bois d'industrie et de service destinés à la construction des habitations traditionnelles et des bois d'oeuvre pour le façonnage des pirogues
4) au parcours de certains animaux, qui peut être interdit dans la mesure où il présente un danger pour les peuplements.

Article 16. - Ces droits sont exercés exclusivement par les populations riveraines et restent toujours subordonnés à l'état des boisements

Périmètres de protection et reboisements

Article 17. - Les périmètres de protection et les reboisements sont affranchis de tous droits d'usage.

Section 3 : Les droits d'usage à caractère commercial

Domaine protégé

Article 18. - L'exploitation commerciale par les usagers des produits issus des palmiers, karités, kolatiers, kapokiers, rotins et autres plantes ayant crû naturellement peut se faire librement dans les forêts protégées sous réserve que les récolteurs ne détruisent pas les végétaux producteurs.

Domaine classé

Article 19. - Dans les forêts classées, l'exploitation commerciale est subordonnée à la délivrance d'un permis d'exploitation spécial indiquant les lieux et les modalités de la cueillette.

Article 20. - Les citoyens ivoiriens riverains de la forêt qui en font la demande sont prioritaires pour l'attribution du permis. S'ils ne font pas valoir ce droit tout autre citoyen, quelles que soient son ethnie et sa région d'origine peut en bénéficier.

Article 21. - Dans tous les cas prévus aux art. 16 & 18, les usagers pourront être tenus de contribuer au prorata des droits dont ils jouissent, à l'entretien des forêts et à la protection des végétaux producteurs.

Périmètres de protection et reboisements

Article 22. - Les périmètres de protection et les reboisements sont affranchis de tous droits d'usage à caractère commercial.

Section 4 : Espèces protégées

Article 23. - Sont interdits dans le domaine forestier de l'Etat, sauf autorisation spéciale, l'abattage, l'arrachage et la mutilation des essences forestières dites protégées.

Chapitre III : De l'exploitation du domaine Forestier de l'Etat

Article 24. - L'exploitation des forêts du domaine par les services publics ou les particuliers peut être faite :
- soit en régie
- soit par vente de coupes
- soit par permis temporaire d'exploitation
- soit par permis de coupe d'un nombre limité d'arbres, de pièces, de mètres cubes ou de stères.

Article 25. - Pour aider à l'exécution des plans de développement économique et social du pays, l'autorité administrative pourra fixer ou réglementer :
1) Les volumes annuels des coupes de bois d'oeuvre en fonction de la possibilité des peuplements.
2) Les contingents de la production de bois en grumes destinés aux besoins internes du pays, à ses industries de transformation et à l'exportation.
3) La transformation du bois en produits semi-finis ou finis.
4) Le transport, la commercialisation, le conditionnement des bois et des produits dérivés.

TITRE III : DU DOMAINE FORESTIER DES PARTICULIERS ET DES COLLECTIVITES

Article 26. - Les particuliers et les collectivités propriétaires de forêts immatriculées à leur nom, y exerceront les droits résultant de leur titre de propriété. Ils ne pourront toutefois en pratiquer le défrichement qu'en vertu d'une autorisation administrative. Cette autorisation ne peut être refusée que si le défrichement est susceptible de compromettre :
1) le maintien des terres sur les pentes
2) la défense du sol contre les érosions et les envahissements des cours d'eau
3) la protection des sources et de leurs bassins de réception
4) la protection des côtes et la constitution d'écrans contre la violence des vents
5) la conservation des sites classés
6) la salubrité publique
7) la défense nationale.

Article 27. - En cas d'infraction à l'art. précédent les propriétaires pourront être mis en demeure de rétablir en nature de bois les lieux défrichés dans un délai n'exédant pas cinq années.

Article 28. - Si les délais fixés pour la remise en état des lieux ne sont pas respectés dans les conditions prévues à l'art. précédent, il pourra y être procédé par autorisation administrative aux frais du propriétaire.

Article 29. - Le respect du domaine forestier, le reboisement et la reforestation sont un devoir pour tout citoyen. Il doit être rempli par les collectivités et les particuliers indépendamment des opérations que se réserve l'Etat.

Des terrains domaniaux seront mis, à cet effet, à leur disposition. Des plants et des graines d'essences forestières leur seront fournis ainsi que l'encadrement nécessaire à la bonne exécution des travaux. Ils devront par la suite assurer l'entretien des boisements ainsi constitués et leur protection contre les incendies et autres dégradations dans le cadre des directives qui leur seront données.

Article 30. - Ces boisements seront soumis au même régime que les reboisements.

Article 31. - Sous réserve des obligations prévues à l'alinéa 2 de l'art. 29, les collectivités au bénéfice desquelles est entrepris le reboisement en ont l'usufruit de plein droit.

Toutefois, l'exploitation devra être exécutée conformément aux réglements établis par l'autorité administrative.

Les produits de cette exploitation pourront, soit être consacrés à la satisfaction des besoins de la collectivité, soit être livrés au commerce.

TITRE IV : REGLEMENTATION DES FEUX

Article 32. - Il est interdit d'abandonner un feu susceptible de se communiquer à la végétation.

Article 33. - Il est interdit de porter ou d'allumer du feu en dehors des habitations et des bâtiments d'exploitation à l'intérieur et à la distance de 500 mètres de forêts domaniales situées en bordure ou dans la zone des savanes. Cependant, des fours à charbon peuvent être établis dans ces régions dans les conditions fixées par l'autorité administrative.

Article 34. - Il est interdit d'allumer des feux de brousse.

Toutefois, à titre transitoire, l'autorité administrative pourra fixer des périodes pendant lesquelles, suivant les régions, les mises à feu seront autorisées.

Ces dernières ne pourront être pratiquées que par la méthode dite des "feux précoces".

Article 35. - Les infractions aux dispositions du présent titre sont passibles des peines prévues à l'art. 50 ci-après.

TITRE V : REPRESSION DES INFRACTIONS (Art. 36 - 49 omis)

TITRE VI : INFRACTIONS ET PENALITES (Art. 50 - 60 omis)

TITRE VII: MODALITES D'APPLICATION

Article 61. - Les modalités d'application du présent Code seront fixées par voie réglementaire notamment en ce qui concerne :
- la procédure de classement et de déclassement des forêts domaniales
- les conditions d'exploitation des forêts domaniales, la procédure d'attribution, de renouvellement ou d'annulation des autorisations d'exploiter
- les modalités de gestion et de constitution des forêts des particuliers et des collectivités et de l'aide qui peut éventuellement leur être apportée par la puissance publique
- les conditions dans lesquelles s'effectuera la remise en état des forêts particulières ou de collectivités indûment défrichées
- les modalités de la représentation de l'Administration devant les juridictions répressives et la procédure applicable en matière de transaction; les modalités de mises à feu autorisées; les possibilités de transaction sous forme de travaux d'intérêt forestier.

Article 62. - La présente Loi sera exécutée comme loi de l'Etat et publiée au Journal Officiel de la Rébublique de Côte d'Ivoire.

Fait à Abidjan, le 20 décembre 1965

Décret N° 66-422 DU 15 SEPTEMBRE 1966

portant création d'une société d'Etat, dénommée "Société pour le Développement des plantations forestières" SODEFOR

COTE D'IVOIRE

IVORY COAST

Décret N° 66-422

LE PRESIDENT DE LA REPUBLIQUE,

Sur le rapport du ministre délégué à l'Agriculture;
Vu la loi 62-82 du 22 Mars 1962, autorisant la création par décret de sociétés d'Etat, modifiée par la loi des Finances N° 63-22 du 5 Février 1963 (art.12);
Vu le décret 66-47 du 8 Mars 1966, portant attribution du ministre de l'Agriculture;
Vu le décret 63-277 du 12 Juin 1963, réglementant le contrôle des sociétés d'Etat;

Le Conseil des ministres entendu;
Décrète :

Article premier. - Il est institué en Côte d'ivoire, dans les conditions prévues par les lois susvisées des 22 Mars 1962 et 5 Février 1963, une société d'Etat, dotée de la personnalité civile et de l'autonomie financière appelée "Société pour le Développement des Plantations Forestières" (SODEFOR).

Cette société a la qualité de commerçant et sera inscrite au registre du commerce.

L'administration et la disposition de son patrimoine sont soustraites aux règles domaniales.

Article 2. - La SODEFOR a pour objet d'étudier et de proposer au Gouvernement de la Côte d'Ivoire, toutes les mesures tendant à assurer l'exécution des plans de développement de la production forestière et des industries connexes, soit par intervention directe, soit en coordonnant, en dirigeant et en controlant l'action des différents organismes publics ou privés intéressés.

Article 3. - Il est fait obligation à la société d'utiliser, mais seulement dans la mesure compatible avec la réalisation de son objet, le personnel fourni par le Service civique dans les conditions prévues par les statuts.

Article 4. - Le capital social fixé à 50 millions de francs CFA est constitué au moyen d'une dotation inscrite au B.S.I.E. et pourra faire l'objet d'augmentations dans les mêmes conditions.

Article 5. - Les dépenses effectuées par la SODEFOR pour la réalisation de son objet seront couvertes au moyen de :
- son capital social;
- Du produit de toute taxe à caractère fiscal ou para-fiscal instituée pour financer les opérations de reboisement;
- Des dotations ouvertes au Budget spécial d'investissement et d'équipement;
- Des emprunts.

En outre, elle pourra recourir aux moyens usuels de crédit à moyen terme et à court terme.

Article 6. - La société est administrée par un conseil d'administration de dix membres, composé comme suit :

Représentants :		
	- du ministre de l'Intérieur	1
	- du ministre délégué à l'Agriculture	2
	- du ministre délégué aux Affaires Economiques et Financières	1
	- du ministre délégué au Plan	1
	- de l'Assemblée Nationale	2
	- du Conseil Economique et Social	1
	- du secteur privé et des Instituts de Recherche	2

Les représentants du secteur privé et des Instituts de Recherche sont nommés par le ministre délégué à l'Agriculture.

Le conseil d'administration élit un président pris dans son sein.

Article 7. - Auprès de la société est désigné un commissaire aux comptes nommé par décret sur proposition du ministre délégué aux Affaires Economiques et Financières.

Ce commissaire exécute sa mission dans les conditions prévues par la réglementation en vigueur concernant les sociétés anonymes.

Il adresse son rapport sur les comptes de la société au président du conseil d'administration, au ministre délégué aux Affaires Economiques et Financières et au ministre délégué à l'Agriculture.

Les comptes de la société ne deviendront définitifs qu'après avoir été approuvés par le ministre délégué aux Affaires Economiques et Financières. Ils seront soumis à l'Assemblée nationale dans les conditions prévues par la loi.

La société est soumise au contrôle de l'Etat dans les conditions prévues au décret susvisé du 12 Juin 1963.

Article 8. - La société est soumise au contrôle du commissaire du Gouvernement désigné par arrêté du ministre délégué au Plan. Ce contrôle s'effectue dans les conditions prévues par la loi N°62-255 du 31 Juillet 1962, en particulier :

- Le commissaire du Gouvernement a entrée aux séances du conseil d'administration et du comité technique;
- Le commissaire du Gouvernement a tous pouvoirs d'investigation sur pièces et sur place;
- Le commissaire du Gouvernement a le pouvoir de faire suspendre l'application d'une décision du conseil, à charge d'en rendre compte sans délai au ministre délégué au Plan.

Article 9. - Sauf dissolution anticipée, la durée de la société est fixée à quatre-vingt-dix-neuf ans. Elle pourra être prolongée par décret.

La réalisation de l'actif et le réglement du passif sont poursuivis conformément au droit des sociétés anonymes. L'actif ne fait retour au fonds spécial prévu par la loi qu'après remboursement aux organismes d'aide extérieure des reliquats de leurs avances respectives.

Article 10. - Sont approuvés les statuts de la société joints au présent décret.

Article 11. - Le présent décret sera publié au Journal Officiel de la République.

Fait à Abidjan, le 15 Septembre 1966

DECRET N° 72/114 DU 9 FEVRIER 1972

portant création de périmètres d'approvisionnement en matière ligneuse des industries du bois

COTE D'IVOIRE

IVORY COAST

Décret N° 72/114

LE PRESIDENT DE LA REPUBLIQUE

VU le rapport du Secrétaire d'Etat chargé de la reforestation
VU la Loi N° 65/425 du 20 Décembre 1965, portant Code Forestier et notamment ses art. 25 et 61
VU le décret N° 66/421 du 15 Septembre 1966, réglementant l'exploitation des bois d'oeuvre et d'ébénisterie, de service, de feu et à charbon;
VU le décret N° 71/476 du 23 Septembre 1971, portant attributions du ministre de l'Agriculture;
VU le décret N° 71/479 du 23 Septembre 1971, complété par le décret N° 71/621 du 23 Novembre 1971, fixant les attributions du Secrétaire d'Etat chargé de la Reforestation et portant organisation du secrétariat d'Etat

Le Conseil des ministres entendu
Décrète :

Article premier. - Pour aider à l'exécution des plans de développement industriel, les permis temporaires d'exploitation de bois d'oeuvre et d'ébénisterie sont regroupés en périmètres d'approvisionnement des industries du bois.

Article 2. - Ces périmètres qui sont au nombre de vingt-six sont définis comme suit : ....
(suit la description des limites topographiques de chaque périmètre)

Article 3. - La production de bois d'oeuvre et d'ébénisterie de chaque périmètre est destinée en priorité à l'approvisionnement des industries du bois qui y sont installées.

A l'intérieur de chaque périmètre, toute nouvelle attribution de permis temporaires d'exploitation de bois d'oeuvre et d'ébénisterie sera faite soit aux industriels qui y sont (ou seront) installés, soit à des groupements d'exploitants forestiers liés à ces industriels par des contrats d'approvisionnement qui seront soumis au visa du Secrétaire d'Etat chargé de la Reforestation.

Article 4. - Chaque périmètre fera l'objet d'un réglement d'exploitation dont les modalités seront définies par arrêté du Secrétaire d'Etat chargé de la Reforestation.

Article 5. - Le Ministre de l'Agriculture et le Secrétaire d'Etat chargé de la Reforestation sont chargés, chacun en ce qui le concerne, de l'exécution du présent décret qui sera publié au Journal Officiel de la République de Côte d'Ivoire.

Fait à Abidjan, le 9 Février 1972

DECRET N° 72/125 DU 9 FEVRIER 1972

portant création d'un contrat de fermage pour certains permis temporaires d'exploitation

COTE D'IVOIRE

IVORY COAST

Décret N° 72/125

LE PRESIDENT DE LA REPUBLIQUE

Sur le rapport du Ministre de l'Agriculture et du Secrétaire d'Etat chargé de la Reforestation

VU la Loi N° 65/425 du 20 Décembre 1965, portant Code Forestier et ses textes d'application

VU le Décret N° 71/476 du 23 Septembre 1971, portant attributions du Ministre de l'Agriculture

VU le Décret N° 71/479 du 23 Septembre 1971, complété par le décret N° 71/621 du 23 Novembre 1971, fixant les attributions du Secrétaire d'Etat chargé de la Reforestation et portant organisation du Secrétariat d'Etat

VU le Décret N° 66/50 du 8 Mars 1966, réglementant la profession d'exploitant forestier

Le Conseil des ministres entendu;
Décrète :

Article Premier. - Les attributaires de permis temporaires d'exploitation définis au Titre IV du décret N° 66/421 du 15 Septembre 1966, dont les moyens en personnel et en matériel sont reconnus notoirement insuffisants pour entreprendre l'exploitation, pourront affermer leurs permis à un autre exploitant appelé fermier, sous réserve que le contrat de fermage soit au préalable approuvé par le Secrétaire d'Etat chargé de la Reforestation.

Article 2. - Le fermier se substitue entièrement à l'attributaire pour tous les droits et obligations qui découlent, d'après les textes, de la possession d'un permis temporaire d'exploitation.

Article 3. - Le fermier est chargé du règlement de tous les droits et taxes applicables à la concession d'un permis temporaire d'exploitation, à l'exception de la taxe d'attribution du permis qui reste à la charge de l'attributaire.

Article 4. - Avant toute exploitation, le fermier devra présenter au Secrétariat d'Etat chargé de la Reforestation un dossier complet concernant chaque permis et comprenant :

a) Un plan 1/5 000 où les arbres exploitables seront repérés par rapport aux layons d'inventaire, avec leur diamètre mesuré à hauteur d'homme.

b) Un tableau récapitulatif des arbres classés par essence et par catégorie de diamètre.

c) Un programme d'exploitation du permis.

Article 5. - Après examen et approbation du dossier, le Secrétaire d'Etat chargé de la Reforestation délivrera au fermier une autorisation d'exploitation assortie d'un cahier des charges particulier destiné à sauvegarder l'avenir du peuplement et à empêcher le gaspillage. Ces dispositions concerneront notamment la qualité des essences abattues, le rythme d'exploitation et la fourniture aux industries locales.

L'autorisation d'exploitation est donnée pour un an. Elle est renouvelable sur décision de l'Administration.

Article 6. - Les infractions au présent décret, notamment le non-respect des prescriptions du cahier des charges particulier, entraîneront l'annulation de cette autorisation.

Dans ce cas, le contrat de fermage devient caduc.

Article 7. - En cas de rupture de contrat de fermage, le fermier sera déchu de ses droits sans indemnite.

En tant que créancier privilégié, l'Administration se réserve de récupérer ses créances par saisie des bois abattus et du matériel du fermier.

Article 8. - Si, par application du Code Forestier et des textes réglementaires qui s'y rapportent, le permis temporaire d'exploitation est annulé, le fermier ne pourra faire valoir aucun droit de préemption pour le reprendre à son compte personnel.

Article 9. - Si, après un délai de trois mois à compter de la date du présent décret, il est constaté qu'un attributaire de permis fait exploiter ses chantiers par un tiers, sans qu'un contrat de fermage n'ait été approuvé par décision du Secrétaire d'Etat chargé de la Reforestation, le permis temporaire d'exploitation sera automatiquement annulé et les taxes et redevances versées ne seront pas remboursées.

Article 10. - Le Ministre de l'Economie et des Finances, le Ministre de l'Agriculture et le Secrétaire d'Etat chargé de la Reforestation sont chargés, chacun en ce qui le concerne, de l'exécution du présent décret qui sera publié au Journal Officiel de la République de Côte d' voire.

Fait à Abidjan, le 9 Février 1972

DECRET N° 72/606 DU 18 SEPTEMBRE 1972

portant création des sociétés civiles de groupement d'exploitants forestiers

LE PRESIDENT DE LA REPUBLIQUE

VU le décret N° 71/476 du 23 Septembre 1971 portant attribution du Ministre de l'Agriculture

VU le décret N° 71/477 du 23 Septembre 1971 portant organisation du ministère de l'Agriculture

VU le décret N° 71/479 du 23 Septembre 1971 fixant les attributions du Secrétariat d'Etat chargé de la Reforestation et portant organisation du Secrétariat d'Etat

VU la Loi N° 65/426 du 20 Décembre 1965 portant Code Forestier et ses textes d'application

Le Conseil des ministres entendu
Décrète :

Article Premier. - Le Groupement d'Exploitants Forestiers est une société civile à durée de vie limitée formée entre personnes physiques et morales régie par les art. 1832 à 1834 et 1841 à 1872 du Code Civil, à l'exclusion des troisième, quatrième et cinquième alinéas de l'art. 1865.

Les exploitants forestiers peuvent y apporter les permis temporaires d'exploitation qui leur ont été attribués.

Article 2. - Le Groupement d'Exploitants Forestiers a pour objet la gestion des permis temporaires d'exploitation par les associés.

Article 3. - La constitution d'un Groupement d'Exploitants Forestiers et l'apport de permis à cette société, font l'objet d'une autorisation préalable donnée par arrêté du Secrétaire d'Etat chargé de la Reforestation, après examen d'un dossier comprenant notamment la liste des permis apportés, le projet de statut conforme au statut type annexé au présent décret et la définition du mode d'exploitation envisagé.

Article 4. - L'arrêté autorisant la constitution d'un Groupement d'Exploitants Forestiers définit la ou les zones forestières dans lesquelles est installé le Groupement.

Article 5. - Les attributaires de permis temporaires d'exploitation qui veulent constituer un Groupement d'Exploitants Forestiers sont tenus d'apporter la totalité des permis dont ils sont titulaires et qui sont situés dans la ou les zones forestières du Groupement définies par l'Administration.

Article 6. - Les permis temporaires d'exploitation apportés par les associés d'un Groupement d'Exploitants Forestiers sont transférés au nom du Groupement.

La durée de validité de ces permis est fixée à 5 ans à compter de la date de constitution du Groupement.

La durée de validité des permis dont un Groupement d'Exploitants Forestiers est titulaire peut être exceptionnellement augmentée lorsque le Groupement participe à l'implantation ou la modernisation d'une industrie de transformation dont il assure l'approvisionnement.

Article 7. - Les Groupements d'Exploitants Forestiers sont tenus de constituer un fonds de réserve d'investissement au moyen d'un prélèvement obligatoire sur les bénéfices dont le taux sera fixé par les statuts.

Article 8. - Le Ministre de l'Agriculture et le Secrétaire d'Etat chargé de la Reforestation sont chargés, chacun en ce qui le concerne, de l'application du présent décret qui sera publié au Journal Officiel de la République de Côte d'Ivoire.

Fait à Abidjan, le 18 Septembre 1972

DECRET N° 78-231 DU 15 MARS 1978

fixant les modalités de gestion du Domaine forestier de l'Etat

COTE D'IVOIRE

IVORY COAST

Décret N°78-231

LE PRESIDENT DE LA REPUBLIQUE

Sur le rapport du ministre des Eaux et Forêts
Vu la Constitution de la République de Côte d'Ivoire
Vu les décrets N° 77-482 du 20 Juillet 1977 et N° 78-125 du 16 Février 1978, portant nomination des membres du Gouvernement;
Vu la Loi N° 65-425 du 20 Décembre 1965, portant Code Forestier;

Le Conseil des ministres entendu,
Décrète :

TITRE PREMIER : GENERALITES

Article premier. - Le Domaine forestier de l'Etat, tel qu'il est défini aux art. 5 et suivants de la Loi N° 65-425 du 20 Décembre 1965, portant Code Forestier, est subdivisé en Domaine forestier permanent de l'Etat et en Domaine forestier rural de l'Etat.

Article 2. - Le Domaine forestier permanent de l'Etat produit du bois et garantit l'équilibre écologique. Le Domaine forestier rural de l'Etat constitue une réserve de terres pour les opérations agricoles et, en attendant son aménagement, est exploité pour son bois.

TITRE II : DU DOMAINE FORESTIER PERMANENT DE L'ETAT

Article 3. - Le Domaine forestier permanent de l'Etat s'étend sur la zone dite forestière et sur la zone dite savane. Des dispositions seront prises pour qu'il couvre une surface de forêt naturelle non dégradée de 3 millions d'hectares en zone forestière et 1,7 million d'hectares en zone de savane.

Article 4. - Le Domaine forestier permanent de l'Etat comprend :
- Les forêts qui ont été classées avant la publication du présent décret, à l'exclusion de celles qui sont visées aux art. 9 et 10 ci-dessous et les périmètres de protection; la liste de ces forêts et périmètres figure à l'annexe I du présent décret;
- Les forêts qui présentent encore le caractère de massif forestier et qui seront incorporées dans le Domaine permanent par arrêté du ministre des Eaux et Forêts, postérieurement à la date de publication du présent décret; elles s'ajouteront à la précédente liste.

Article 5. - Le Domaine forestier permanent de l'Etat est affranchi de tous droits d'usage, autres que ceux prévus aux art. 15 et 16 de la Loi N° 65-425 du 20 Décembre 1965, portant Code forestier; les défrichements y sont interdits conformément à l'art. 8 de ce Code, et réprimés selon les dispositions de l'art. 50.

Article 6. - L'exploitation forestière dans le Domaine forestier permanent de l'Etat se poursuit conformément aux dispositions de la Loi N° 65-425 du 20 Décembre 1965 et ses textes d'application, cependant des mesures sont prises pour :

- Définir, délimiter et surveiller efficacement la totalité du Domaine forestier permanent de l'Etat afin de garantir l'intégrité de sa surface et sa vocation forestière;
- Organiser rationnellement l'exploitation afin d'assurer la pérennité de l'approvisionnement en bois d'oeuvre du pays;
- Assurer le renouvellement des peuplements par des opérations de reboisement correspondant aux besoins en bois, à long terme du pays.

Pour la mise en application de ces mesures, on se réferera aux directives de la Loi-Plan 1976-1980.

Article 7. - Le Domaine forestier permanent de l'Etat, aussi bien dans ses surfaces que dans ses limites, ne pourra être réduit que par décret pris en Conseil des ministres.

Pour assurer le maintien de l'équilibre écologique, des terrains non forestiers pourront être inclus dans le Domaine permanent par arrêté conjoint du ministre des Eaux et Forêts et du ministre de l'Agriculture, en vue de leur reboisement.

Article 8. - Un arrêté du ministre des Eaux et Forêts précisera la liste des forêts du Domaine permanent qui seront consacrées en priorité aux opérations de reboisement.

TITRE III : DU DOMAINE FORESTIER RURAL DE L'ETAT

Article 9. - Le Domaine forestier rural de l'Etat comprend :
- Les forêts classées avant la publication du présent décret et inscrites sur une liste qui figure à l'annexe II du présent décret;
- Les forêts non classées du Domaine forestier de l'Etat qui ne font pas l'objet d'un statut particulier, tel que Parc national ou Réserve.

Article 10. - Les forêts classées du Domaine forestier rural de l'Etat feront l'objet de plans d'aménagement agricole. Elles seront déclassées progressivement au moment de la mise en oeuvre de ces plans par arrêté conjoint du ministre des Eaux et Forêts, du ministre de l'Agriculture et du ministre de l'Economie, des Finances et du Plan.

Un calendrier de mise en valeur sera établi : il permettra de programmer la récupération de tous les bois d'oeuvre de ces forêts avant les défrichements.

Article 11. - En application de l'art.12 de la Loi N° 65-425 du 20 Décembre 1965, les forêts non classées du Domaine forestier rural de l'Etat feront aussi l'objet d'un calendrier de mise en valeur, dans la mesure où leur superficie et la rentabilité de la récupération du bois le justifieront.

Article 12. - Des dispositions seront prises pour que l'exploitation du bois d'oeuvre et si possible des autres produits ligneux soit aussi complète que possible.

Pour ce faire, les titulaires de chantiers situés dans ce domaine auront l'obligation de les exploiter en priorité.

Lorsqu'une zone aura été délimitée en vue de son défrichement par tranches annuelles successives, les titulaires des chantiers situés dans cette zone auront l'obligation de vider la totalité du bois d'oeuvre commercialisable inclus dans ces chantiers avant le début des opérations de défrichement.

Lorsque le défrichement n'aura pas été programmé ou quand il l'aura été et que les exploitants concernés n'auront pas vidé la totalité du bois d'oeuvre existant dans la tranche annuelle en cours de défrichement, la récupération de ce reliquat incombera au concessionnaire qui fait exécuter le défrichement.

TITRE IV : DISPOSITIONS TRANSITOIRES

Article 13. - Pendant une durée de trois ans à dater de la publication du présent décret, la liste des forêts ou les limites de certaines forêts du Domaine forestier permanent de l'Etat pourront être modifiées en fonction des résultats d'études éventuelles de vocation des sols qui ne pourront être entreprises qu'avec l'autorisation conjointe du ministre des Eaux et Forêts, du ministre de l'Agriculture et du ministre de l'Economie, des Finances et du Plan.

La modification ne pourra en aucun cas réduire la surface totale du Domaine forestier permanent de l'Etat : toute soustraction de surface devra donc être compensée par l'apport d'une surface forestière équivalente. Par ailleurs, la modification ne pourra :
- Concerner les forêts visées à l'art. 8 ci-dessus;
- Entraîner une dislocation de la forêt concernée;
- Porter sur une surface inférieure à 1 000 hectares;
- Réduire une forêt classée à une surface inférieure à 5 000 hectares.

Un arrêté conjoint du ministre des Eaux et Forêts, du ministre de l'Agriculture et du ministre de l'Economie, des Finances et du Plan sanctionnera les modifications éventuelles.

Article 14. - Tant que les limites du Domaine forestier permanent ne seront pas entièrement matérialisées sur le terrain, l'autorité administrative évitera d'attribuer ou de laisser occuper des terres situées à proximité des forêts dont les limites ne sont pas encore matérialisées sur le terrain et dont les structures de surveillance ne sont pas encore mises en place.

Article 15. - En attendant l'établissement des premiers calendriers de défrichement, l'autorité administrative veillera :
- A ce que les attributions de terres soient effectuées dans les îlots forestiers disséminés dans les zones de cultures ou dans les blocs forestiers dont la superficie est inférieure à 1 000 Hectares;
- A ce que le représentant local du ministère des Eaux et Forêts ait connaissance de ces implantations, afin qu'il puisse prendre les mesures nécessaires à la récupération des bois d'oeuvre.

Article 16. - La mise en place du dispositif de planification de l'exploitation forestière dans le Domaine forestier permanent sera progressive, mais devra être achevée avant le ler Juillet 1980. En attendant que ce dispositif soit opérationnel, des mesures techniques ponctuelles seront prises pour régulariser la production du bois et favoriser la création d'industries fabriquant des produits semi-finis ou finis.

TITRE V : DISPOSITIONS DIVERSES

Article 17. - Une carte constamment à jour représentant les forêts classées des Domaines forestiers permanent et rural de l'Etat, sera mise à la disposition de toutes les préfectures et sous-préfectures, afin que les autorités administratives puissent participer efficacement à la protection ou à la mise en valeur de ces forêts classées.

Article 18. - Le ministre des Eaux et Forêts, le garde des Sceaux, ministre de la Justice, le ministre de l'Intérieur, le ministre de l'Economie, des Finances et du Plan et le ministre de l'Agriculture sont chargés, chacun en ce qui le concerne, de l'application du présent décret qui sera publié au Journal Officiel de la République de Côte d'Ivoire.

Fait à Abidjan, le 15 Mars 1978,

A N N E X E S
au décret N°78-231 du 15 Mars 1978 fixant les modalités de gestion du Domaine forestier de l'Etat.

ANNEXE I : Liste des forêts classées antérieurement à la date de publication du présent décret, incluses dans le Domaine forestier permanent.

A. - ZONE FORESTIERE : 2 404 270 Hectares

= Région forestière d'Abidjan .... 718 670 hectares
suit dénomination et superficie des forêts

= Région forestière de San-Pédro .... 661 200 hectares
suit dénomination et superficie des forêts

= Région forestière de Man .... 643 750 hectares
suit dénomination et superficie des forêts

= Région forestière de Daloa .... 205 340 hectares
suit dénomination et superficie des forêts

= Région forestière de Bouaké .... 163 110 hectares
suit dénomination et superficie des forêts

= Région forestière de Bondoukou .... 12 200 hectares
Baya-Kokoré 12 200 hectares

B. - ZONE DE SAVANE : 1 222 190 Hectares

= Région forestière de Man .... 20 000 hectares
suit dénomination et superficie des forêts

= Région forestière de Daloa .... 289 400 hectares
suit dénomination et superficie des forêts

= Région forestière de Bouaké .... 521 980 hectares
suit dénomination et superficie des forêts

= Région forestière de Bondoukou .... 22 800 hectares
Bélé-Fima 22 800 hectares

= Région forestière de Korhogo .... 368 010 hectares
suit dénomination et superficie des forêts

TOTAL GENERAL .... 3 626 460 hectares

ANNEXE II : Liste des forêts classées antérieurement à la date de publication du présent décret qui sont déclassées selon les dispositions de l'art. 10 et incluses dans le Domaine forestier rural.

ZONE FORESTIERE : 713 750 hectares

= Région forestière d'Abidjan .... 242 050 hectares
suit dénomination et superficie des forêts

= Région forestière de San-Pedro .... 57 140 hectares
suit dénomination et superficie des forêts

= Région forestière de Man .... 39 700 hectares
suit dénomination et superficie des forêts

= Région forestière de Daloa .... 145 260 hectares
suit dénomination et superficies des forêts

= Région forestière de Bouaké .... 218 200 hectares
suit dénomination et superficies des forêts

= Région forestière de Bondoukou .... 11 400 hectares
Tankessé 11 400 hectares

Publié au journal Officiel N° 23 du
25 Mai 1978

DECRET N° 81-735 DU 2 SEPTEMBRE 1981

fixant les attributions du ministre des Eaux et Forêts et portant organisation du ministère

LE PRESIDENT DE LA REPUBLIQUE

Vu la Constitution de la République de Côte d'Ivoire;
Vu le décret N° 81-56 du 2 Février 1981, portant nomination des membres du Gouvernement;

Le Conseil des ministres entendu;
Décrete :

Article premier. - Le ministre des Eaux et Forêts exerce, conformément aux dispositions législatives et réglementaires en vigueur, les attributions dévolues au Gouvernement en matière de politique forestière, d'économie des produits forestiers, de pisciculture, et pêche dans les eaux continentales et de protection de la faune, des sols et des eaux.

A ce titre, avec le concours des organismes publics et privés compétents, il a pour mission d'assurer :

- la constitution, la délimitation, la conservation, le renouvellement, l'aménagement et la gestion du patrimoine forestier national;
- le maintien de l'intégrité du domaine forestier de l'Etat et le contrôle de l'exploitation et de l'administration des forêts, des collectivités publiques et privées soumises au régime forestier;
- l'application des règles de gestion des forêts domaniales en vue de les aménager pour accroître leur potentiel de production de bois et leur efficacité pour la conservation du milieu naturel et la protection des sols et des eaux;
- toutes les opérations se rapportant à l'inventaire du domaine forestier national et des autres formations boisées tant publiques que privées;
- la programmation et le développement des plantations forestières par la définition et la mise en oeuvre du plan national de reboisement, la coordination et le contrôle de l'exécution des travaux;
- l'utilisation et la valorisation de la production forestière en vue d'une meilleure rentabilité et d'une plus grande économie de la matière première;
- le contrôle de l'exportation des produits ligneux et leur conformité avec les normes en vigueur;
- l'étude, l'organisation et le développement des industries du bois;
- le contrôle et le recouvrement des taxes;
- l'établissement des statistiques forestières;
- la constitution, le classement, la conservation, l'aménagement et la gestion des Parcs nationaux, des réserves analogues ainsi que leur promotion scientifique, éducative, récréative et touristique;

- la gestion des ressources cynégétiques, l'application de la réglementation de la chasse et de ses produits, la protection de la faune sauvage;
- la protection des sols, des eaux et de la végétation par l'utilisation de méthodes rationnelles en vue de favoriser la conservation des climats, lutter contre l'érosion et les feux sauvages, restaurer le couvert végétal naturel et plus généralement, assurer la pérennité des ressources naturelles;
- le développement de la pisciculture et de la pêche en eaux continentales;
- l'instruction et le suivi des affaires contentieuses de son ressort;
- l'organisation pédagogique, la gestion, l'équipement et le contrôle des établissements de formation spécialisée en matière forestière, de faune, de pisciculture et de pêche en eaux continentales;
- la tutelle technique de tout établissement public ou de tout organisme d'Etat dont les objectifs entrent dans le cadre des attributions fixées par le présent décret.

Article 2. - Dans l'exercice de ses attributions le ministre des Eaux et Forêts dispose de :

- la direction du Cabinet qui comporte :
  le bureau des Etudes de Programmation
  le service autonome des Statistiques
- l'Inspection générale des Eaux et Forêts;
- la direction de l'Enseignement et de la Formation forestière :
  Sous-direction de la Documentation
- la direction de la Conservation du Domaine forestier :
  Sous-direction de la Délimitation, du Cadastre et des Inventaires.
- la direction du Reboisement et de la Conservation des Sols et des Eaux.
- la direction de la Production forestière :
  Sous-direction des Attributions des Permis temporaires d'Exploitation forestière et des Autorisations annuelles.
- la direction des Industries forestières :
  Sous-direction du Contrôle des Industries
- la direction des Parcs nationaux et des Réserves :
  Sous-direction des Aménagements et des Inventaires
- la direction de la Chasse :
  Sous-direction des Ressources cynégétiques
- la direction de la Pisciculture et de la Pêche en Eaux continentales
  Sous-direction de la Vulgarisation et de l'Animation des Pêcheurs nationaux
- la direction du Contrôle forestier et du Contentieux :
  Sous-direction du Contentieux et des Taxes
- la direction des Affaires administratives et financières :
  Sous-direction du Personnel.

Article 3. - Le territoire national est divisé en régions forestières. Chaque région a à sa tête, un directeur.
Les directeurs de régions forestières exercent dans leurs circonscriptions les attributions dévolues au ministre des Eaux et Forêts.

Article 4. - Les régions forestières sont divisées en inspections forestières.

Les chefs d'inspections forestières remplissent auprès des préfets de leurs circonscriptions les fonctions de chef du service départemental des Eaux et Forets.

Article 5. - Les inspections forestières sont sub-divisées en cantonnements. Les cantonnements peuvent être spécialisés, leur compétence s'exerçant soit en matière forestière, soit en matière de pisciculture et de pêche en eaux continentales, soit dans plusieurs de ces domaines.

Article 6. - Des arrêtés du ministre des Eaux et Forêts détermineront, en tant que de besoin, l'organisation du Cabinet de l'Inspection générale, des directions centrales et des services extérieurs de son départment.

Article 7. - Par décret ultérieur, il sera créé auprès du ministère des Eaux et Forêts, un fonds d'intervention et d'action destiné au financement des opérations reconnues nécessaires à la mise en oeuvre de la politique forestière définie aux termes du présent décret.

ARticle 8. - Le ministre des Eaux et Forêts est chargé de l'exécution du présent décret qui abroge toutes dispositions antérieures contraires, et notamment le décret N° 78-689 du 18 Août 1978, et qui sera publié au Journal Officiel de la République de Côte d'Ivoire.

Fait à Abidjan le 2 Septembre 1981

LOI N° 81- 127 DU 31 DECEMBRE 1981
du Budget Général de Fonctionnement
pour l'Exercice 1982

COTE D'IVOIRE

IVORY COAST

Loi N° 81- 127

ARTICLE 8 MODIFIANT DROIT UNIQUE DE SORTIE SUR LES BOIS TRANSFORMES ET MODIFIANT LA TAXE D'ABATTAGE

Article 8. - Droit unique de sortie sur les bois - Taxe d'abattage.

a) les droits de sortie sur les bois transformés désignés ci-après sont modifiés comme suit :

Tarif N°	Désignation des Produits	Droit
44-05	Bois simplement sciés longitudinalement tranchés ou déroulés, d'une épaisseur supérieure à 5 mm	
" -51	- Bois sciés des espèces Aboudikrou, Assaméla, Acajou, Sipo, Makoré, Dibétou, Niangon, Bété, présentés en lots homogènes de pièces de dimension identiques	2%
" -59	- Autres bois sciés présentés en lots homogènes de dimensions identiques	2%
" -61	- Bois sciés des espèces Aboudikrou, Assaméla, Acajou, Sipo, Makoré, Dibétou, Niangon, Bété autrement présentés	4%
" -69	- Autres bois sciés autrement présentés	4%
" -79	- Autres bois feuillus tropicaux sciés	4%
" -90	- Autres bois non dénommés simplement sciés	4%
44-14	Bois simplement sciés longitudinalement, tranchés ou déroulés, d'une épaisseur égale ou inférieure à 5 mm : feuilles de placage et bois pour contre-plaqués, de même épaisseur	
" -39	- Autres bois tranchés	1%
" -69	- Autres bois déroulés	1%
44-15	Bois plaqués ou contre-plaqués, même avec adjonction d'autres matières : bois marquetés ou incrustés	
" -20	- Bois plaqués	1%
" -20	- Bois plaqués constitués exclusivement de feuilles de placage	1%
" -39	- Autres bois contre-plaqués à âme épaisse, panneautée, lattée ou lamellée	1%

b) Code Général des Impôts, appendice VII
Taxe forestière (Taxe d'abattage)

1° Le montant de la taxe d'abattage est modifié comme suit :
Article premier. - ....

3) Le montant de la taxe d'abattage est fixé par mètre cube de bois utilisable et commercialisable selon les tarifs ci-après :

Désignation	Catégorie 1	Catégorie 2	Catégorie 3
Bois en grumes exportés	600	400	200
Bois en grumes vendus aux usines locales	300	200	100

Les bois provenant des permis de coupe sont imposés au double des taux précédents par mètre cube utilisable.

2° La répartition des essences par catégorie est modifiée comme suit :
Article 2. - Les essences actuelles exploitées se répartissent dans les catégories suivantes :

Catégorie 1	Catégorie 2	Catégorie 3
Aboudikrou	Amazakoué	Autres essences
Acajou	Bahia	
Aninguéri	Fraké	
Assaméla	Framiré	
Avodiré	Kotibé	
Bété	Koto	
Bossé	Samba	
Dibétou		
Iroko		
Lengué		
Kossipo		
Makoré		
Niangon		
Sipo		
Tiama		

A titre exceptionnel, il ne sera retenu sur le Fromager et l'Ilomba exportés en grumes que la moitié de la taxe d'abattage prévue pour les bois de la catégorie 3.

Publié au Journal Officiel 7 Numéro Spécial
8 Février 1982

DECRET N° 82/70 DU 13 JANVIER 1982

fixant les conditions d'approvisionnement en bois des industries locales et d'exportation de bois et de produits ligneux

COTE D'IVOIRE

IVORY COAST

Décret N°82/70

LE PRESIDENT DE LA REPUBLIQUE

Sur le rapport conjoint du Ministre du Plan et de l'Industrie, du Ministre des Eaux et Forêts, du Ministre du Commerce et du Ministre de l'Economie et des Finances,

Vu la loi 64-292 du 1er Août 1964, relative aux obligations des commerçants et à la modification des art. 147 et 150 du Code Pénal,

Vu la Loi N°65-425 du 20 Décembre 1965, portant Code Forestier,

Vu le décret N°72-543 du 28 Août 1972, portant obligation aux exportateurs de bois agréés d'assurer l'approvisionnement des usines;

Vu le décret N°76-281 du 20 Avril 1976 déterminant les conditions d'entrée en Côte d'Ivoire des marchandises étrangères de toutes origine et de toute provenance, ainsi que les conditions d'exportation et de réexpédition des marchandises à destination de l'étranger,

Vu le décret N°81-56 du 2 Février 1981, portant nomination des membres du Gouvernement,

Vu le décret N°81-525 du 1er Juillet 1981, déterminant les attributions du Ministre du Commerce,

Vu le décret N°81-523 du 1er Juillet 1981, définissant les attributions du Ministre du Plan et de l'Industrie et portant organisation de son ministère,

Vu le décret N°81-735 du 2 Septembre 1981, fixant les attributions du Ministère des Eaux et Forêts et portant organisation du Ministère,

Vu l'arrêté N°001/MINEFOR/COM du 2 Février 1979 portant application des dispositions du décret N°78-234 du 20 Mars 1978, réglementant la profession d'exportateur en bois ou de produits ligneux,

Vu le décret N°81-465 du 24 Juin 1981 fixant les attributions du Ministre de l'Economie et des Finances et portant organisation du Ministère,

Le Conseil des Ministres entendu,

Décrete :

Article 1. - L'exportation de Côte d'Ivoire de bois en grumes est effectuée par des entreprises ou coopératives agréées, dans des conditions fixées par les dispositions du décret N°76-281 du 20 Avril 1976 et par celles du présent décret.

Les entreprises ou coopératives agréées sont désignées dans le présent décret par l'expression "Exportateur de bois agréé".

Article 2. - L'annexe B du décret N°76-281 du 20 Avril 1976 est modifiée comme indiqué à l'annexe 1 du présent décret.

Article 3. - Toute exportation de bois en grumes quelle qu'en soit l'essence est soumise à l'application d'un quota.

Ce quota est fonction du volume de produits finis ou semi-finis élaborés dans les usines ivoiriennes.

Toutes ces activités de transformation du bois sont prises en compte pour le calcul du quota.

Le quota annuel de chaque société industrielle agréée est fonction du rapport existant entre sa propre production industrielle et celle de la production industrielle nationale de produits ligneux.

L'application de ce pourcentage au volume de bois en grumes à l'exportation, arrêtée conjointement par les Ministres du Plan et de l'Industrie, des Eaux et Forêts, du Commerce, de l'Economie et des Finances, déterminera le quota en volume attribué à chaque société industrielle.

Les quotas attribués à chaque industriel agréé sont librement transférables en totalité ou en partie entre industriels agréés, industriels agréés et exportateurs de bois agréés, ainsi qu 'entre exportateurs de bois agréés entre eux.

Les quotas non utilisés au cours de l'année civile de leur attribution ne pourront être reportés sur une période ultérieure.

Article 4. - L'exportation du bois en grumes peut être pratiquée par toute société commerciale, agréée en qualité d'exportateur de bois, qu'elle soit ou non propriétaire d'un établissement de transformation du bois.

L'agrément d'exportateur de bois est attribué par le Ministre du Commerce sur proposition du Ministre des Eaux et Forêts.

Les conditions de l'obtention de l'agrément d'exportateur de bois seront définies par l'arrêté d'application du présent décret.

Article 5. - Trimestriellement les entreprises industrielles doivent remettre au Ministère des Eaux et Forêts un relevé de :
- leur consommation de matière première ligneuse,
- leur volume de production,
- transfert de quota.

Trimestriellement les exportateurs agréés doivent remettre au Ministre des Eaux et Forêts un relevé :
- des quotas qui leur auront été transférés,
- un justificatif de leurs exportations de grumes.

Le Ministère des Eaux et Forêts transmettra à la commission visée à l'art.6 le résultat de l'analyse de ces documents.

Article 6. - Il est créé une commission consultative interministérielle du bois de 8 membres nommés par arrêté conjoint du Ministre du Plan et de l'Industrie, du Ministre des Eaux et Forêts, du Ministre du Commerce, du Ministre de l'Economie et des Finances et sur proposition des autorités dont ils relèvent.

Cette commission présidée par un des représentants du Ministère des Eaux et Forêts, est composée comme suit :
- 2 représentants du Ministre du Plan et de l'Industrie,
- 2 " du Ministre des Eaux et Forêts,
- 1 représentant du Ministre de l'Economie et des Finances,
- 1 " du Ministre du Commerce,
- 1 " du Ministre de l'Agriculture,
- 1 " du Ministre de la Marine.

Cette commission pourra valablement délibérer si cinq de ses membres sont présents.

Elle pourra se faire assister lors de ses travaux par des représentants des organismes concernés.

Article 7. - La commission aura pour attribution, outre l'établissement et le contrôle des quotas d'exportation visés aux art. 3, 4, et 5 ci-dessus, de faire toutes recommandations susceptibles de promouvoir les activités de transformation industrielle du bois.

La commission se réunira une fois par trimestre et aussi souvent que nécessaire et sur convocation de son président.

Article 8. - Les infractions aux art. 2, 3, et 4 ci-dessus constituent des exportations sans déclaration de marchandises prohibées et sont constatées et réprimées conformément aux dispositions des art. 31, 296 et 287 du Code Général des Douanes.

Le Directeur Général des Douanes en avise le Président de la Commission interministérielle du bois, visée à l'art. 6 du présent décret.

Indépendamment des peines prévues par les dispositions légales et réglementaires, notamment le Code Général des Douanes, l'auteur de l'une des infractions définies aux art. 2, 3 et 4 ci-dessus, est passible du retrait d'agrément d'exportateur de bois agréé, pris par arrêté conjoint du Ministre du Plan et de l'Industrie, du Ministre des Eaux et Forêts, du Ministre du Commerce, sur proposition du Président de la commission interministérielle du bois.

Article 9. - Le présent décret entre en vigueur à la date de la signature de l'arrêté interministériel pris pour son application, il abroge à compter de cette date, toutes dispositions antérieures contraires, notamment le décret N° 72-543 du 28 Août 1972, portant obligation aux exportateurs de bois agréés d'assurer l'approvisionnement des usines, et le décret N° 78-234 du 20 Mars 1978 réglementant la profession d'exportateur en bois et en produits ligneux.

Article 10. - Le Ministre du Plan et de l'Industrie, le Ministre des Eaux et Forêts, le Ministre du Commerce, le Ministre de l'Economie et des Finances, sont chargés, chacun en ce qui le concerne, de l'exécution du présent décret qui sera publié au Journal Officiel de la République de Côte d'Ivoire.

ANNEXE : Omis

Fait à Abidjan le 13 Janvier 1982

Publié au Journal Officiel N°8 du 11 Février 1982 page 140.

IVC/28

ORDONNANCE N° 82/71 DU 13 JANVIER 1982
portant modification du tarif des
droits de sortie des bois en grumes

COTE D'IVOIRE

IVORY COAST

Ordonnance N°82/71

LE PRESIDENT DE LA REPUBLIQUE,

Sur le rapport du ministre du Plan et de l'industrie, du ministre des Eaux et Forêts, du ministre du Commerce et du ministre de l'Economie et des Finances,

Vu la Constitution de la République, notamment son article 45;

Vu le tarif des droits d'entrée et de sortie promulgué par Ordonnance N°73-315 du 3 Juillet 1973, ratifié par la Loi N°73-577 du 22 Septembre 1973, ensemble les textes qui l'ont modifié et complété;

Vu la Loi N° 64-291 du ler Août 1964, portant Code des Douanes et notamment ses articles 11, 12, et 196;

Vu le traité instituant la Communauté Economique de l'Afrique de l'Ouest signé à Abidjan le 17 Avril 1973 et notamment son article 16;

Vu la décision N°1-74 CM de la Communauté Economique de l'Afrique de l'Ouest en date du 8 Mars 1974, portant mise en vigueur d'une nomenclature douanière et statistique unifiée, modifiée et complétée par la décision N° 15-78 CM du 21 Août 1978;

Vu la loi N° 63-524 du 26 Décembre 1963, portant aménagement et codification des textes fiscaux et les textes modificatifs subséquents;

Vu l'Ordonnance N° 79-10 du 5 Janvier 1979, portant modification du tarif des droits d'entrée et de sortie;

Vu l'urgence;

Le Conseil des ministres entendu,

Ordonne :

Article premier. - Le tableau des droits de sortie des bois en grumes, équarris et en plots (position tarifaire 44-03, 44-04, 44-05) est modifié et complété conformément à l'annexe I à la présente Ordonnance.

Article 2. - Le ministre de l'Economie et des Finances est chargé de l'exécution de la présente Ordonnance qui entrera en vigueur à compter du ler Février 1982.

Article 3. - La présente Ordonnance sera publiée selon la procédure d'urgence et exécutée comme loi de l'Etat.

Fait à Abidjan, le 13 Janvier 1982

TABLEAU DES DROITS UNIQUES DE SORTIE - Chapitre 44 -

Tarif n°	Désignation des produits		D.U.S
44-03		Bois bruts écorcés ou simplement dégrossis	
" -01	Aboudikrou	en grumes	44%
" -02	Acajou	"	44%
" -03	Avodiré	"	30%
" -04	Bossé	"	30%
" -05	Sipo	"	44%
" -06	Dibétou	"	44%
" -07	Iroko	"	44%
" -08	Makoré	"	44%
" -09	Tiama	"	44%
" -10	Niangon	"	44%
" -11	Samba	"	36%
" -12	Bété	"	44%
" -13	Framiré	"	36%
" -14	Lengué	"	30%
" -15	Ilomba	"	36%
" -16	Fraké	"	36%
" -17	Assamela	"	36%
" -18	Essessang	"	24%
" -19	Fromager	"	30%
" -20	Aninguéri	"	44%
" -21	Kossipo	"	44%
" -22	Amazakoué	"	44%
" -23	Ako	"	24%
" -24	Koto	"	36%
" -25	Azobé	"	24%
" -26	Badi	"	30%
" -27	Kotibé	"	36%
" -28	Aiélé	"	24%
" -29	Akossika	"	24%
" -30	Ba	"	24%
" -31	Bahia	"	36%
" -32	Bi	"	24%
" -33	Dabéma	"	24%
" -34	Difou	"	24%
" -35	Emien	"	24%
" -36	Faro	"	36%
" -37	Iatandza	"	24%
" -38	Kékélé	"	24%
" -39	Kondroti	"	36%
" -40	Lohonfé	"	24%
" -41	Lotofa	"	30%
" -42	Mélegba	"	24%
" -43	Movingui	"	30%
" -44	Pocouli	"	24%
" -45	Poco	"	24%
" -46	Tali	"	24%
" -47	Vaa	"	24%

44-03-59	Autres bois feuillus tropicaux en grumes à l'exclusion des bois figurés du n° 44-03-60	24%
" -60	Bois feuillus tropicaux figurés en grumes des espèces Acajou, Sipo, Makoré et Tiama	44%
" -71	Bois de trituration	11%
" -72	Bois de mine	11%
" -73	Poteaux de conifères	11%
" -74	Poteaux autres	11%
" -90	Autres bois bruts non dénommés ailleurs	24%

44-04 Bois simplement équarris

(suit la même liste des Essences et D.U.S. applicable aux tarifs 44-03-01 au 44-03-47)

" -59	Autres bois feuillus tropicaux équarris à l'exclusion des bois figurés du n°44-04-60	24%
" -60	Bois feuillus tropicaux équarris figurés des espèces Acajou, Sipo, Makoré et Tiama	44%
" -90	Autres bois simplement équarris	24%

44-05 Bois simplement sciés longitudinalement, tranchés ou déroulés d'une épaisseur supérieure à 5 millimètres

(suit la même liste des Essences et D.U.S. applicable aux tarifs 44-03-01 et 44-03-47)

" -59	Autres bois tropicaux en plots	24%

FORESTS ACT

Approved April 17, 1953

LIBERIA

Forests Act 1953

STRUCTURE AND LIST OF PROVISIONS OF THE FOREST ACT

AN ACT FOR THE CONSERVATION OF THE FORESTS OF THE REPUBLIC OF LIBERIA

I - TITLE

II - INTERPRETATION :
- Forests
- Forest Products
- Commercial Use
- Fish Resources
- Wildlife Resources

III - ESTABLISHMENT OF THE BUREAU OF FOREST CONSERVATION

IV - POLICIES AND OBJECTIVES OF CONSERVATION

V - FUNCTIONS OF THE BUREAU OF FOREST CONSERVATION

VI - ESTABLISHMENT OF GOVERNMENT FOREST RESERVES

VII - ESTABLISHMENT OF NATIVE AUTHORITY FOREST RESERVES

VIII- ESTABLISHMENT OF COMMUNAL FORESTS

IX - APPOINTMENT OF FOREST OFFICERS

X - ESTABLISHMENT OF NATIONAL PARKS

XI - POWER TO SET REGULATIONS AND RULES
- Power to prescribe the form of licences, permits, agreements, and other instruments dealing with forest, recreational, fish and wildlife resources
- Power to control transport and export of forest products and levy forest fees

XII - PENALTIES AND PROSECUTION

XIII- EFFECTIVE DATE AND PUBLICATION

FORESTS ACT
Approved April 17, 1953

LIBERIA

Forests Act 1953

AN ACT FOR THE CONSERVATION OF THE FORESTS OF THE REPUBLIC OF LIBERIA

WHEREAS, our forests are among our greatest natural ressources and may best contribute to our economic and social welfare by being devoted to their most productive use for the permanent good of the whole people, and

WHEREAS, no program now exists for the protection, development, and utilization of these resources, and they therefore remain almost wholly unproductive, yet suffer from gradual depletion, and

WHEREAS, the conservation and utilization of these resources should be brought about promptly, efficiently, and wisely, under such restrictions as will insure perpetual benefits from this heritage, Therefore :

It is enacted by the Senate and House of Representatives of the Republic of Liberia in Legislature assembled :

Section I. That this Act be cited as "An Act For the Conservation of the Forests of the Republic of Liberia".

Section II. That in this Act, the following words have the meaning indicated unless the context otherwise requires:

Forests - All areas supporting woody vegetation other than planted or cultivated crops, regardless of the composition, age or density of the vegetative cover.

Forest Products - The materials yielded by forests, as follows :

(a) Trees, which include seedlings, saplings, brushwood, palms and canes;
(b) Timber, including trees fallen or cut down, stumps and wood in any shape or form;
(c) Charcoal, wild rubber, wood oil, resin and gums;
(d) Leaves, flowers, fruits, seeds and all other parts of trees not hereinbefore mentioned;
(e) Plants, other than trees (including grass, vines, reeds and moss) growing in the forest which are not cultivated for agricultural purposes, and all parts and products of such plants.

Commercial use - Any use other than the direct use for private purposes, whether such other use involves barter, sale, or any other disposition of forest products.

Fish Resources - All non-mamal aquatic forms of animal life found living in waters of any description.

Wildlife Resources - Wild mammals, birds, and reptiles of every description, but not including other lower terrestrial forms of animal life.

Section III. That the President, under the authority granted him by Section 4, Chapter XVIII of the Acts passed by the Legislature of the Republic of Liberia during the session 1947-48 entitled "An act to Create a Department of Agriculture and Commerce" does hereby establish, and the action is hereby approved, a BUREAU OF FOREST CONSERVATION within the Department of Agriculture and Commerce.

Section IV. That the initial policies and primary objectives of the program to be carried out by the Bureau of Forest Conservation shall be to :

a. Establish a permanent forest estate, made up of reserved areas, upon which scientific forestry will be practiced;
b. Devote all publicly owned lands to their most productive use for the permanent good of the whole people considering both direct and indirect forest values;
c. Stop needless waste and destruction of forest and associated natural resources, and bring about the profitable harvesting of all forest products while assuring that supplies of these products are perpetuated;
d. Correlate forestry to all other land use and adjust the forests economy to the overall national economy;
e. Conduct essential research in conservation of forests and pattern action programs upon the results of such research;
f. Give training in the practices of forestry; offer technical assistance to all those engaged in forestry activities; and spread knowledge and acceptance of forestry and the conservation of natural resources throughout the country;
g. Conserve recreational, fish and wildlife resources of the country concurrently with the development of a forestry program.

Section V. That the functions of the Bureau of Forest Conservation shall be to :

a. Take all action necessary to permit the creation of Government Forest Reserves, Native Authority Forest Reserves, Communal Froests, and National Parks;
b. Administer all such reserved areas so as to best satisfy the policies and objectives set out in Section IV;
c. Enforce all laws and regulations for the conservation of our forests and the development of their resources with impartiality, industry, and dispatch;

d. Carry out a program for the wise use and perpetuation of the forest, recreational, fish and wildlife resources of the country.

Section VI. That the President is hereby empowered to create and establish GOVERNMENT FOREST RESERVES embracing any portion of the forests of the country, such reserves to be bounded and described at the time of their establishment and thereafter to be administered and protected as a permanent forest estate, in accordance with such rules and regulations as may be promulgated by the Secretary of Agriculture and Commerce for that purpose. All such Government Forest Reserves shall be created and established by presidential proclamation after all rights and claims of the original owners have been heard in a court settlement. Upon the adjudication of all such rights and claims and the proclamation of these reserves, all rights, title, and interest in them shall be vested in the Government.

Section VII. That the President is also hereby empowered to authorise the creation and establishment of NATIVE AUTHORITY FOREST RESERVES embracing forests lying in one or more tribal chiefdoms, such reserves to be bounded and described at the time of their establishment by presidential proclamation. Thereafter, such Native Authority Forest Reserves shall be protected as potential Government Forest Reserves in accordance with such rules and regulations as prescribed by the provision of this Act. The rules and regulations affecting reserves of this type shall be designed to minimize damage to the reserved forests and avoid unnecessary depletion of their resources pending the establishment of a Government Forest Reserve embracing the concerned area.

Section VIII. That the President is also hereby empowered to authorize the creation and establishment of Communal Forests to be administered by the concerned native authorities. Such forests shall be limited to small described forest areas immediately adjacent to one or more native villages, and use of these forests will be confined to the local population under such rules and regulations as prescribed by the provision of this Act. Said rules shall be designated to assure the perpetuation of such communal forests as a source of forest products for the private use of the local inhabitants and to prohibit any and all commercial use of forest products taken from these areas.

Section IX. That the President is also hereby empowered to appoint and commission, upon the recommendation of the Secretary of Agriculture and Comemrce, Forest Officers to perform all the duties and functions requisite to the accomplishment of the policies and objectives established for the Bureau of Forest Conservation. All such Forest Officers shall be authorized to exercise the powers necessary to fulfil their duties, and all shall be public servants in the employ of the Republic of Liberia. The categories of Forest Officers may include Chief, Bureau of Forest Conservation, Deputy Chief, Bureau of Froest Conservation, Forest Conservator, Assistant Forest Conservator, Forest Ranger, and Forest Guard. The salaries of all Forest Officers shall be fixed from time to time in the general budget in common with other public servants of the Republic.

Section X. That the President is also hereby empowered to create and establish NATIONAL PARKS embracing any areas of the country having such outstanding scenic, recreational, scientific or other pertinent values that it is deemed wise and expedient in the national interest to set aside as permanent parks to be retained insofar as is practicable in their existing condition. Such National Parks shall be created by presidential proclamation after all rights and claims of the original owners have been heard in a court of settlement. Upon the adjudication of all such rights and claims and the proclamation of these parks, all rights, title and interest in them shall be vested in the Government.

Section XI. That the Secretary of Agriculture and Commerce with the approval of the President, will describe and promulgate all rules and regulations as may be required to insure the accomplishment of all the purposes of the present Forest Conservation Act. The Secretary of Agriculture and Commerce shall also prescribe the form of all licenses, permits, agreements, and other instruments dealing with the use of forest, recreational, fish and wildlife resources; control the issuance of such instruments, and determine the conditions under which such instruments may be granted, exercised, produced, revoked, or returned. He shall also control the transportation or export of the products of forests by land, water or air, and he is responsible for the imposition and collection of all fees in connection with anything done under the rules and regulations cited above.

Section XII. That any person who contravenes any rule or regulation made under Section XI above, or fails to comply with the condition made a part of any license, permit, agreement, or other instrument issued or entered into under any such rule or regulation, for which no penalty is expressly provided in the rules and regulations, shall be deemed guilty of a misdemeanor and prosecuted according to law. Where any person is convicted of any offense against any of the rules and regulations promulgated under this Act, the Court may in addition to or in lieu of the imposition of any fine or term of imprisonment, order that the whole or any part of the forest products, in respect to which the offence was committed, be confiscated and forfeited to the Government, to be sold or otherwise disposed of in such manner as the Secretary of Agriculture and Commerce may prescribe.

Section XIII. That the law herein created and established shall become effective thirty days after it shall have been approved by the President and shall be published in handbills.

Any Law to the contrary Notwithstanding.

Approved April 17, 1953

SUPPLEMENTARY ACT

Approved February 28, 1957

STRUCTURE AND LIST OF PROVISIONS OF
THE SUPPLEMENTARY ACT

SUPPLEMENTARY ACT FOR THE CONSERVATION OF THE FORESTS
OF THE REPUBLIC OF LIBERIA

SUPPLEMENTARY ACT

Approved February 28, 1957

LIBERIA

Supplementary Act 1957

AN ACT SUPPLEMENTAL TO "AN ACT FOR THE CONSERVATION OF THE FOREST OF THE REPUBLIC OF LIBERIA", PASSED AND APPROVED APRIL 17, 1953

WHEREAS, it has become apparent that since the passage into Law of the Forest Conservation Act entitled "An Act for the Conservation of the Forests of the Republic of Liberia" approved April 17, 1953, for the conservation and utilization of our forests, animal wildlife and other natural resources, certain intricate problems and conditions, not previously envisaged nor contemplated, have arisen which demand the necessity for a supplementary legislation to implement the principles and objectives of the above cited Act; Therefore,

It is enacted by the Senate and House of Representatives of the Republic of Liberia in Legislature assembled:

PART I. TITLE

Section 1. That from and immediately after the passage of this Act entitled "An Act for the Conservation of the Forest of the Republic of Liberia" passed and approved April 17, 1953, be and the same is hereby amended and this Act shall be cited as the "SUPPLEMENTARY ACT FOR THE CONSERVATION OF THE FORESTS OF THE REPUBLIC OF LIBERIA".

PART II. DEFINITIONS

Section 1. Words used in the singular form in this Act shall be deemed to import the plural, and vice versa, as the case may demand. For the purpose of this Act the following words shall be construed, respectively, to mean:

Secretary - The Secretary of Agriculture and Commerce.

Bureau - The Bureau of Forest Conservation within the Department of Agriculture and Commerce.

Forest Officer - Any duly appointed officer of the Bureau of Forest Conservation.

Person - Any individual, firm, corporation, company, society, association, or other organized group of any of the foregoing.

Forest - All areas supporting woody vegetation other than planted or cultivated crops, regardless of the composition or age.

Reserve Forests - Forests within the boundaries of all publicly owned forests including those established by the National Government known as Government Forest Reserve, Native Authority Forest Reserves, National Parks and Communal Forests.

Granted Timber Areas - All forest areas covered by a timber concession agreement between the Government or other owners and a commercial timber operator.

Permittee - Person granted a permit to perform specified acts upon or in reserved forest.

Timber Sale Operator - Person or permittee authorized to harvest and utilize timber. Such authorization shall be covered by a timber sale agreement between the timber sale operator and the Government where the sale of National Forest timber is involved.

Commercial Use - Any use other than direct use for personal purposes, including uses involving barter sale, trade or any other disposition of forest products for which remuneration is received.

Take Timber - To cut down, cut the branches, girdle or otherwise injure any tree, or remove any timber from a tree.

Girth - Circumference of Tree outside the bark, at a point 4,5 feet above the average ground level or at a point 1 foot above the butt swell.

Diameter Breast High or DBH - Diameter or distance through a tree outside bark, at a point 4,5 foot above the average ground level or at a point 1 foot above the butt swell.

Protected Trees - A tree which may not be cut down, pruned, girdled, damaged in any way, or removed after felling without the permission of a responsible authority.

Seed Tree - A tree left uncut at the time of forest operations in order to provide natural regeneration.

Property Mark - A mark placed upon a tree or timber or log to denote ownership, usually made with a marking hammer or axe.

Soil - This term shall include all inorganic material such as sand, dirt, gravel, minerals and similar material.

Wildlife - Wild mammals, birds, fish and reptiles of every description, but not including other lower terrestial forms of life within the land area of the Republic of Liberia.

PART III. PERMITS

Section 1. A written permit must first be obtained before any act can be done that disturbs the vegetation, wildlife or soil on any Public Forest Reserve or National Park. This shall include a strip 200 feet wide outside the established boundary line.

Section 2. Authorized forest officers may issue permits for the following acts on Government Forest Reserves or National Parks. All permits shall be subject to the approval of the Chief, Bureau of Forest Conservation and the Secretary.

(a) Collect, convey, remove, or subject to any manufacturing process any forest products of plant origin.
(b) Cut down or damage any trees.
(c) Set fire to vegetation, logging slash, etc.
(d) Dig in the soil or prospect for minerals, coal or oil.
(e) Make a dam across a river, creek or other waterway.
(f) Reside, or build any structure necessary in connection with authorized forest harvesting operations.
(g) Build road, trails or railroads.

Section 3. Permission granted to perform any of the acts listed in Section 4 shall be subject to such written conditions as may be imposed by a Forest Officer, with the approval of the Chief, Bureau of Forest Conservation and the Secretary.

Section 4. Permits for the development of mineral resources on forest reserves will be issued by the Mining Board of the Republic of Liberia with the consent of the Secretary; and subject to such provisions as deemed necessary for the protection of surface uses as provided for in an Act to Revise the Mining Laws of the Republic of Liberia, Section V, Article 2 of March 26, 1952.

Section 5. Authorized Forest Officers may issue permits to perform any of the acts listed in Part III, Section 2 on or within Native Forest Reserves.

All permits shall be subject to such written conditions as may be imposed by the Forest Officer with the approval of the Chief, Bureau of Forest Conservation, and in consultation with the responsible Native Authority.

Section 6. Permits or authorizations issued to any person to cut or remove timber from Government Forests shall be in the form of a Timber Sale Agreement or Contract covering the following points :

(a) Name and address of permittee.
(b) Location and area covered.
(c) Cutting regulations including diameter or girth limits, etc.
(d) Amount and kind of timber to be removed.

(e) Method of designating trees to be cut.
(f) Method of measurement, whether by tree count, cubic foot or board foot.
(g) Amount of money to be paid per unit or measurement.
(h) Utilization requirements so as to cause the least waste.
(i) Waste disposal, sawdust, slabes, brush, etc.
(j) Responsibility of Permittee to prevent trespass, damage, fire, unauthorized use, etc. during the term of this contract.
(k) The Permittee shall be responsible to provide for the regeneration of tree species removed during cutting operations so that provision is made for the future maintenance of stand composition existing at the beginning of the operations covered by said timber sale agreement.
(l) Such additional clauses as may be required to cover special conditions pertaining to the particular area.

Section 7. Free use permits, issued without charge, shall be issued by the Chief, Bureau of Forest Conservation with approval of the Secretary of the Department of Agriculture and Commerce of the Republic of Liberia to take timber from reserved forests in public ownership for use in public works or the construction of public buildings. Such permits shall contain the same information required in Section 8.

Section 8. A Permittee or Grantee may not cut down any tree held to be a sacred tree or a medicine tree by the local population, nor may he cut down any tree growing within 100 yards of a village or market place, unless permission to do so is given by both the Forest Officer and the concerned tribal authorities.

Section 9. No person shall commit any of the following acts on Government Forest Reserves, National Park, Native Authority Forest, Communal Forests or Timber Concessions.

(a) Damage or destroy any Government property, including boundary markers, notices, corner posts, boundary trees, etc.
(b) Alter, deface, or obliterate any mark placed on a tree.
(c) Cut down tree or burn any area to make a farm or plantation.
(d) Fell and leave trees or other obstructions across any trail, road, railroad, or water.
(e) Enter a reserve or park with the intent to do any of the foregoing acts.

Section 10. No person shall make commercial use of forest products taken from a Communal Forest.

Section 11. No person shall allow sawdust, mill waste or other material that is harmful to fish, to enter any stream in Liberia.

Section 12. No person may possess, sell, buy, exchange, barter, or export any forest products which have been taken in violation of any of the forest Rules and Regulations.

PART IV. PROPERTY MARK

Section 1. No person may use a hammer or other tool to place property mark upon any tree unless the property mark has been properly registered with the Bureau, and the person using it authorized to do so.

Section 2. No person can register or use any of the following property marks:

RL (Republic of Liberia)
RLNP (Republic of Liberia National Park)
RLGR (Republic of Liberia Government Reserve)
RLNA (Republic of Liberia Native Authority)
RLCF (Republic of Liberia Communal Forest)

These property marks are reserved for the exclusive use of the Liberian Government and may not be used by anyone except an authorized Forest Officer.

Section 3. Each timber sale or concession agreement must specify the property mark or marks to be used on the area covered by the agreement and no property marks may be used outside the sale or granted area without specific authorization of a Forest Officer.

Section 4. When a property mark has been placed on a tree with a registered hammer, it cannot be altered in any way except with the consent of the owner of the mark and a Forest Officer.

Section 5. The stump and each log of every tree cut for commercial purposes shall be stamped with the registered property mark specified in the timber sale or concession agreement.

PART V. REVENUES AND RECEIPTS

Section 1. Any and all revenues accrued from Government Forest Reserves shall be deposited with the Bureau of Revenues, and a Flag Receipt obtained therefore, prior to the issuance of any permit.

Section 2. Any and all revenues obtained from Native Authority Forest Reserve will be divided equally between the Government of Liberia and the proper responsible Native Authority on whose land the operation takes place. The amount due to the Republic of Liberia will be deposited with the Collector of Revenues.

PART VI. GENERAL REGULATIONS

Section 1. All rights, title and interest, to any timber to which a permittee or grantee would otherwise be entitled, but not removed from the designated area, shall revert to the Government at the time of the expiration of the relative permit or agreement.

Section 2. No person shall, anywhere in Liberia unless it is otherwise stated in the timber sale or concession agreement or permit issued by the Chief, Bureau of Forest Conservation, and approved by the Secretary, cut or fell for the purpose of commercial conversion into sawn products any growing tree the girth of which is smaller than the following dimensions at a point one foot above the convergence of the buttress, if any, or at a point four and one half foot above average ground level where there are no such roots.

(a) In the case of the following species the girth of 8 feet or 3.5 inches in diameter.

Canarium schweinfurthii	Lovoa Klaineana
Entandrophragma sp.	Mimusops adjava
Khaya sp.	Mimusops heckelii

(b) In the case of the following species a girth of 7 feet or 27 inches in diameter

Anopyxis calaensis	Lophira procera
Chlorophora excelsa	Oldfieldia africana
Combretodendrum africana	Piptadenia africana
Erythrophloeum guineense	Sarcocephalus diderichii

(c) In the case of the following species a girth of 6 feet or 23 inches in diameter.

Berlinia sp.	Parinarium sp.
Mitragyna stipulosa	Saccoglottis gabonensis
Ochrocarpus africanus	Terminalia superba
Terminalia ivorensis	

(d) In the case of all other species a girth of 4,5 feet or 17 inches in diameter.

(e) If a grantee cuts for commercial purposes any tree below the above specified girth limit he will be assessed for each tree a penalty excise tax equal to three times the stumpage price set forth in the timber sale agreement. This penalty must be paid into the Bureau of Revenues by the grantee and upon his failure to do so within thirty (30) days he shall be prosecuted before any court of competent jurisdiction within the Republic for the recovery of the same.

PART VII. CONCESSION

Section 1. Any grantee, having received the rights from the Government to forest areas for the utilization of forest products, has the responsibility to protect these forest areas from destruction and encroachment by other persons. It shall also be the responsibility of the grantee to ensure that the forest areas under his agreement are maintained for forest production.

Section 2. The description of all boundaries located on the ground by the grantee shall be filed with the Bureau. Any disputes arising from such boundary location work will be handled according to existing laws of the Republic of Liberia.

Section 3. Grantees are wholly responsbile for any and all damages arising directly from their forest industry operations on land on which no valid timber rights have been obtained.

Section 4. A grantee operating a granted area is responsible for the determination and demarkation on the ground of the exterior boundaries of the gross area granted, as well as the coundaries of all interior trees not owned by the grantor, or otherwise excluded from the granted area. These boundary surveys shall be completed within 2½ years from the day the concession contract was approved by appropriate signatures.

Section 5. The grantee is responsible for the taking of adequate precautions to prevent the occurence and spread of fires in connection with any and all of his operations within or outside a granted timber area. Any damages resulting from fires caused by the grantee or anyone in his contract or employ, through carelessness or negligence shall be paid or satisfied by the grantee.

Section 6. It shall be the grantee's responsibility to assure that only trees, the taking of which is authorized by the concerned timber concession agreement, are cut and removed from any granted area.

Section 7. No unnecessary damage shall be done to young growth or to trees left standing on granted areas as a result of the grantee's operations. In all operations on a granted timber area, the grantee is responsible for the cutting of trees so as to cause the least practicable waste in stumps; to utilize all cut trees to as small a diameter in the tops as practicable; and to vary the length of logs so as to secure the greatest possible utilization of merchantile material.

Section 8. All roads and trails in or adjacent to a granted timber area shall be kept free of logs, brush, and other debris resulting from the grantee's operations. Any road, trail, or structure thereon not constructed by the grantee and damaged beyond ordinary wear and tear as a result of his operations shall within seven days be restored to its original condition by the grantee at his own proper cost and expense.

Section 9. All structure, improvements, or facilities constructed or operated by a grantee in connection with his forest industry operation within a granted timber area, shall be located and operated subject to conditions as may be imposed by an authorized Forest Officer to protect forest and water resources.

Section 10. In any area where the Government has granted rights for a specified time to an individual or a concern for the purpose of development, the following regulations will govern:

(a) After consultation and approval by a Forest Officer the grantee may cut any tree irrespective to the girth limits as long as it is used in the granted area for construction purposes. Cutting trees below the girth limit, however, shall be confined to a minimum and will be done only when the particular construction makes it necessary to use such smaller trees.

(b) Any timber product cut for commercial use and sold whether locally or for export will be assessed a stumpage price or excise tax in the sum of $3.00 per thousand board feet for lumber, planks timber or other parts manufactured products and $5.00 per thousand board feet for logs intended for export. Item 3 and under title III of the Export Schedule of the Republic of Liberia Customs Tariff Act of 1940 which assesses 2¢ per board foot of logs, timber and planks, be and the same is hereby repealed, and the above assessments substituted therefore.

(c) The grantee will be held responsible for furnishing the Chief, Bureau of Forest Conservation with monthly reports containing information on sale of products divided into the two classifications mentioned in paragraph (b).

(d) Authorized Forest Officer will make periodic inspections of grantee's operations for the purpose of assuring compliance with the Act.

Section 11. In any area where a permanent real estate deed has been granted the owner has the right to sell locally any timber product emanating from his property without paying a stumpage or excise tax. The Bureau of Forest Conservation will assist the owner in applying sustained yield forest management. However, if timber products are exported from Liberia from such property the owner must pay $3.00 per thousand board feet for lumber, planks timber or other partially manufactured products, or $5.00 per thousand board feet for logs.

PART VIII. WILDLIFE RESOURCES

Section 1. All regulations pertaining to wildlife resources shall originate from the office of the Secretary and after approval by the President shall be administered by the Bureau. Such regulations shall be binding on all lands of the Republic.

Section 2. Licenses - No person shall be permitted to sell, barter or trade or exchange Big Game, dead or alive, without having first obtained a commercial license issued by the Bureau:
(1) Big Game Hunting - $100.00 per annum.

Section 3. No permits will be issued to any person for the taking of live game animals out of the Republic except that the Secretary may issue special permits to take live game animals for medical or scientific purposes, and it shall be the responsibility of the applicant to give documentary and/or satisfactory evidence to that effect in the event where the said applicant is an agent of a Scientific or Medical Institution.

Section 4. Item 4 of Schedule A of the Act entitled: "An Act supplemental to an Act approved December 14, 1938" which Act was approved December 16, 1940, amending the Revenue Code of 1937 which imposes a license fee of $375.00 for Big Game Hunting, be it and the same is hereby repealed.

Section 5. Live Game Animals taken under Section 4 cannot be exported without first obtaining an export permit from the Secretary.

Section 6. Anyone having a live game animal in captivity at the time this Act becomes effective must within 90 days of this date register each animal with the Secretary.

Section 7. Any unregistered live game animal found in captivity after the specified time set out in Section 6 of this Act shall be confiscated and disposed of by the Bureau.

Section 8. No person shall hunt elephant without first having obtained a permit from the Secretary of Agriculture and Commerce. Foreigners shall be restricted to a limit of one elephant per year.

Section 9. Prohibited acts - All persons are prohibited from committing any of the following acts:

(a) Dynamiting or poisoning of water as a means of catching fish. Setting any animal traps, snares or similar devices in such a place as to endanger human beings. Adequate warning marks should be placed around such traps.
(b) Setting any gun, explosive bow and arrow traps.
(c) No chimpanzees shall be hunted, trapped or molested except as provided for in Section 3.
(d) No female monkey with sucklung young shall be shot or trapped within the forest reserve. Any person apprehended for violation of these Acts shall be taken before the District Commissioner or Justice of the Peace and upon conviction be fined in the sum of $5.00 for the first offence, and $25.00 for any repetition of the same offence.

Section 10. No hunting, trapping shooting or molesting of wild animals or commercial fishing shall be allowed in National Parks.

Section 11. No person without written permission of an authorized Forest Officer, may perform any of the following acts within a Government Forest Reserve:

(a) Hunt, shoot or set traps or snares for any wildlife.
(b) Fish, set traps or in any way catch fish.

Section 12. Any Big Game shot or killed must be reported to the Town chief of the nearest town within two days after the killing took place. The Town Chief will be required to report to the District Commissioner each month the total number of Big Game killed according to species. The Commissioner will submit monthly reports to the Bureau covering the number of Big Games species killed in his district.

Section 13. Portions, or all, of any National Forest within each of the provinces shall be proclaimed a Wildlife Refuge for the purpose of maintaining the wildlife found in Liberia. Absolutely no hunting shall be permitted on these refuges and it shall be the responsibility of the Bureau to maintain scientific control of the species population contained therein.

Section 14. An Advisory Conservation Committee shall be appointed within each county or province with the County Superintendent, District or County Commissioner as Chairman; The purpose of the Committee is to provide means of communicating to proper authorities the ideas, desires and opinions of the people on matters pertaining to forest and wildlife conservation and to exercise general supervision of the enforcement of Wildlife Regulations.

PART IX. FOREST OFFICERS' DUTIES AND RESPONSIBILITIES

Section 1. It is the duty and responsibility of the Chief, Bureau of Forest Conservation, to see that all the forest rules and regulations are carried out. It is also his duty and responsibility to see that every member of his organization is fully acquainted with these regulations and any changes that may be made.

Section 2. It is the duty and responsibility of every Forest Officer to impartially enforce these regulations and report every offence to his superior officer. Failure to do so will be considered just cause for disciplinary action.

Section 3. Every Forest Officer is hereby authorized and empowered to arrest any person whom he finds or reasonably suspects of violating any of the provisions of any regulation hereunder; and shall immediately take such person before the court of the County Superintendent, District Commissioner, Tribal Authority, Justice of the Peace or a court of competent judisdiction.

Section 4. It shall be discretionary with the Bureau of Forest Conservation to cancel or annul a certificate or permit issued in favour of any person or persons at any time when security or administrative reasons render it necessary.

Section 5. A Forest Officer may fell, cut, damage, tap or destroy trees within Government Forest Reserves, Native Authority Forest Reserves, Communal Forest, and National Parks, make clearings or remove timber therefrom for the purpose of planting trees, improving the growth of trees, or for the general better management of reserved forests.

Section 6. Any person convicted of any of the offences declared in this Act shall be fined a sum not less than five dollars ($5.00) nor more than three hundred dollars ($300.00) or imprisonment for a period not exceeding eighteen months nor less than fifteen days.

ANY LAW TO THE CONTRARY NOTWITHSTANDING

Approved February 28, 1957

TIMBER CONCESSION AGREEMENT STANDARD FORMATE

LIBERIA LIB/20

Timber Agreement

STRUCTURE AND LIST OF CLAUSES OF A TIMBER CONCESSION AGREEMENT

PREAMBLE

Article I. - THE TERMS OF THE CONCESSION
1. The Grant of Concession
2. Concession Area
3. Term of the Agreement; Periodic Review; Renewal of the Concession
4. Effective Date
5. Assignment of Concession
6. Surrender of All or Part of the Concession
7. Performance Bond and Required minimum Expenditure
8. Processing of Timber and other Forest Products
9. Government Inspection

Article II. - OPERATIONS OF THE CONCESSIONAIRE
1. Management Plan
2. Trees under Minimum Diameter not to be cut for Commercial Use
3. Selection and Exploitation of Timber Harvesting Tracts; Timing; Relinquishment
4. Measurement and Marking of Trees and Logs

Article III. - OTHER RIGHTS OF THE CONCESSIONAIRE
1. Occupation of Surface and Easements
2. Accessory Works and Installations
3. Right to Take and Use Water
4. Right to Take and Use Gravel, Sand, Clay and Stone
5. Agents or Independent Contractors

Article IV. - REPORTS, RECORDS, NOTICES AND COMMUNICATIONS
1. Reports concerning Surveys, Exploitation and Development
2. Other Reports to Government and Records to be Maintained
3. Local Resident Managers; Notices
4. Reports to be Confidential; Cost of Reports

Article V. - FISCAL OBLIGATIONS
1. Government Tax on Net Income; Accounting Principles
2. Surface Rent
3. Stumpage Fees
4. Duties and Exices

Article VI. - CAPITALIZATION; STOCK PURCHASES; DIRECTORS; FINANCE
1. Capitalization
2. Finance; Approved Indebtedness; Debt to Equity Ratio

STRUCTURE AND ARRANGEMENT OF SECTIONS OF A FOREST MANAGEMENT PLAN

1. INTRODUCTION

2. AREA UNDER MANAGEMENT
   - Forest Protection
   - Forest law
   - Violation

3. LOGGING
   - Method of Logging
   - Minimum diameter limits (Cut Limits)
   - Obligatory Species
   - Annual Coupe System : Determination of the Annual Coupe Allowance
     Permanent Forest
     Non-Permanent Forest
     Logging Plan
   - Entry & Re-Entry into the Felling Area :
     Entry
     Limitations of Felling Area
     Re-Entry into Logged-Over Areas
   - Penalties
   - Mensuration : Live Trees
     Felled Trees

4. SILVICULTURE
   - Reforestation : Reforestation by the Concessionaire
     Reforestation Fund
   - Experimental Plots

5. ROADS
   - Maximum Gradient
   - Other Obligations

6. MODIFICATIONS

APPENDIX I Minimum Diameter Limits
II Obligatory Species
III Future Obligatory Species
IV Procedure for Volume Assessment (Scaling)
V Definitions

ACT CREATING F D A
Approved November Ist, 1976

LIBERIA

FDA Act 1976

AN ACT CREATING THE FORESTRY DEVELOPMENT AUTHORITY

It is enacted by the Senate and House of Representatives of the Republic of Liberia, in Legislature Assembled :

Section 1. - Chapters 1 through 4 of the National Resources Law with respect to the Bureau of Forest Conservation, the Conservation of Forests, National Parks and the Conservation of Wild Life are hereby repealed,

Section 2. - Creation of Forestry Development Authority :
An authority to be known as the "Forestry Development Authority" is hereby created as a corporate body pursuant to the Public Authorities Law. As used in this Act, the term "Authority" means the authority hereby created.

Section 3. - Objects : The primary objectives of the Authority shall be to :

a) Establish a permanent forest estate made up of reserved areas upon which scientific forestry will be practised;
b) Devote all publicly owned forest lands to their most productive use for the permanent good of the whole people considering both direct and indirect values;
c) Stop needless waste and destruction of the forest and associated natural resources and bring about the profitable harvesting of all forest products while assuring that supplies of these products are perpetuated;
d) Correlate forestry to all other land use and adjust the forest economy to the overall national economy;
e) Conduct essential research in conservation of forest and pattern action programs upon the results of such research;
f) Give training in the practice of forestry; offer technical assistance to all those engaged in forestry activities; and spread knowledge of forestry and the acceptance of conservation of natural resources throughout the country;
g) Conserve recreational and wildlife resources of the country concurrently with the development of forestry program.

Section 4. - Powers : In addition to the powers conferred upon an authority by the Public Authorities Law, the Authority shall have the following powers :

a) To take all actions necessary to create and establish Government Forest Reserves, Native Authority Forest Reserves, Communal Forests, and national parks;

b) To administer all such reserved areas to fulfill the policies and objectives set out in Section 3 of this Act;

c) To enforce all laws and regulations for the conservation of forests and the development of their resources;

d) To assist the owners of timber land in applying sustained yield forest management;

e) To carry out a program for the wise use and perpetuation of the forest, recreational, and wildlife resources of the country except that regulations for the zoning of hunting grounds and the restriction of hunting to stated periods shall be promulgated by the President;

f) To prescribe the form of all licences, permits, agreements, and other instruments dealing with the use of forest resources;

g) To control the issuance of such instruments, and determine the conditions under which they may be granted, exercised, produced, revoked or returned;

h) To control the transportation or export of forest products by land, water, or air;

i) To be responsible for the collection of all fees payable under the rules and regulations promulgated under the authorization of this Section;

j) To promulgate rules and regulations required to insure the accomplishment of all the policies and objectives of the Authority;

k) To open and operate a main and subsidiary banking accounts, to receive and expend monies;

l) To continue existing services and to initiate new services, such as :
   (i) To establish a unit for market cost component analysis;
   (ii) To monitor real timber prices and production cost;
   (iii) To commission and carry out feasibility studies;
   (iv) To establish research, education and training facilities;
   (v) To levy cess on other components of the wood-using industries for purposes connected with the Authority's functions;

m) To negotiate, raise and make loans;

n) The power to issue, amend and rescind forestry regulations;

o) The power to engage in commercial undertakings as a principal or in conjunction with others, to enter into contracts, to sue and be sued;

p) As a principal or in conjunction with others to fell trees and prepare them for export or to have them processed locally, or both; to trade with such timber in the raw or processed state and to engage in all other operations directly or indirectly connected with the trade in forest products;

q) To make by-laws for its internal administration.

Section 5. - Penalty for violation of rules and regulations :
(omitted)

Section 6. - Board of Directors.

1. Composition. The policies of the Authority shall be formulated by the Board of Directors consisting of :
   a) The Minister of Agriculture as Chairman, the Minister of Finance, the Minister of Local Government, the Minister of Planning and Economic Affairs, the Minister of Commerce, Industry and Transportation and the President of the Liberian Bank of Development and Investment;
   b) The Managing Director of the Authority, to be appointed by the President.
   c) Two liberian nationals, one with experience in the field of law, and the other with experience in the field of business, to be appointed by the President.
2. Voting by Board Members without meeting. The Board may by regulation establish a procedure whereby the Managing Director, when he considers such action in the best interest of the Authority, may obtain a vote of the Board members on a special question without calling a meeting of the Board.
3. QUORUM. A quorum for any meeting of the Board shall be a majority of its members.
4. Financial interest. No Director or member of his immediate family shall hold any financial interest in a forestry concession or in any auxiliary undertaking in Liberia.
5. Reimbursement. The members of the Board, in their capacity as such, shall not receive salaries, but they may receive from the Authority a stipend for each meeting attended and reimbursement for all expenses they incur in discharging their duties to the Authority.

Section 7. - Officers. The Authority shall have a Managing Director and such other officers and staff as the Board may determine to be necessary or desirable for carrying out its lawful functions. The Managing Director shall be professionally qualified in forestry. He shall be responsible for the conduct of the general operation of the Authority and for that purpose shall exercise all powers delegated to him by the Board. Subject to the approval of the Board with regard to senior officers and staff, the Managing Director shall be responsible for the organization of the staff and the appointment and dismissal of the officers.

Section 8. - Appointment of Assistant Managing Director :
The President shall appoint an Assistant Managing Director for administration and finance who shall be a qualified financial controller and administrator rather than a professional forester.

Section 9. - Collection of revenues and expenditures :
The Authority shall be responsible for the collection of its revenues, settlement of its financial obligations and all other matters connected with the collection and disbursement of funds of the Authority.

Section 10. - Duration : The existence of the Authority shall continue until it shall be terminated by the legislature. Upon the termination of the existence of the Authority, all its rights and property shall rest in the Republic.

Section 11. - Audits : The Accounts of the Authority shall be subject to periodic audits by the Government. The accounts of the Authority shall also be audited annually by a firm of independent accountants appointed by the Board.

Section 12. - Reports : The Authority shall submit an annual report to the President and such other periodic reports as he may from time to time require. Such reports shall set out in detail facts describing the operation and fiscal transactions of the Authority during the preceding year; its financial condition and a statement to all receipts and disbursements during such year.

Section 13. - Power of Forest Officer to arrest Offenders : Every Forest Officer is hereby authorized and empowered to arrest any person whom he finds or reasonably suspects of violating any of the provisions of the statutes or regulations relating to conservation of forests. On arrest he shall immediately take such person before the Court of the County Superintendent, County Commissioner, Tribal Authority, or Justice of the Peace who shall immediately forward the matter to a court of competent jurisdiction in the County in which the reserve is located.

Section 14. - Power of Forest Officer with regard to trees : A Forest Officer may fell, cut, damage, tap, or destroy trees within Government Forest Reserves, Native Authority Forest Reserve,Communal Forests, and national parks, and make clearing or remove timber therefrom for the purpose of planting trees, improving the growth of trees, or for the general better management of reserved forests.

Section 15. - Advisory Committees: An Advisory Conservation Committee shall be appointed within each County with the County Superintendent or County Commissioner as Chairman. The purpose of the Committees shall be to provide means of communicating to the Forest Development Authority the ideas, desires and opinions of the people in matters pertaining to forest and wildlife conservation and to exercise general supervision of the enforcement of forest and wildlife regulations.

Section 16. - This Act shall take effect immediately upon publication in hand-bills.

Any law to the contrary notwithstanding.

APPROVED : NOVEMBER 1, 1976 MONROVIA

PUBLISHED BY AUTHORITY GOVERNMENT PRINTING OFFICE

MONROVIA, LIBERIA, DECEMBER 20, 1976

FORESTRY DEVELOPMENT AUTHORITY

REGULATION N° 5 OF 23 MARCH 1979

LIBERIA

FDA Regulation N°5

FORESTRY DEVELOPMENT AUTHORITY
REGULATION N° 5
ON ASSISTANCE TO OWNERS OF PRIVATE FOREST LANDS

WHEREAS, the Act creating the Forestry Development Authority (FDA ACT), approved November 1, 1976, published December 20, 1976, obligates FDA to offer technical assistance to all those engaged in forestry activities and assist the owners of timber land in applying sustained yield forest management (sections 3 (f) and 4 (c); and

WHEREAS, the Act Adopting a New Revenue and Finance Law of the Republic of Liberia, approved May 24, 1977, published June 20, 1977, in its Chapter 20 requires any person who on privately owned land cuts timber for commercial use to pay certain forest fees and taxes; and

WHEREAS, said FDA Act charges the FDA with the responsibility for the collection of the forest revenues (Sections 4 (i) and 9); and

WHEREAS, said FDA Act has conferred upon the FDA the power to promulgate, issue, amend and rescind forestry Rules and Regulations to assure and accomplishment of all the policies and objectives of the FDA (Section 4 (j) and (n); and

NOW, THEREFORE, the Forestry Development Authority does hereby rule and regulate :

PART I : DEFINITIONS

Section 1. - Definitions

In this Regulation the following words have the meaning indicated unless the context otherwise requires :

a) FDA - Forestry Development Authority

b) Owner - Anyone, individual, group or Corporation, who owns real property in the Republic of Liberia under a valid Deed according to the laws of General Application;

c) Commercial Use - Any use other than direct use for personnel purpose including uses involving barter, trade or any other disposition of forest products for which remuneration is received;

d) Operations'Plan - An outline of the sequence of activities intended;

e) Concessionaire - Any Company that has concluded a valid Forest Concession Agreement with the Government of Liberia;

f) Operator - Any individual or Company without a valid Forest Concession Agreement with the Government of Liberia intending to engage in or already engaged in tree felling and harvesting for remuneration;

g) FDA Regulation N° 4 - Control of Non Concession Forest Operators - Approved March 21, 1979, published in "FDA News Letter", Vol. II, N° 2 of March 23, 1979.

PART II : GENERAL RULES

Section 2. - Notification of FDA

Anyone who intends to make Commercial Use of his forest land shall first notify FDA in writing explaining his plans and intentions so that FDA can determine the appropriate assistance to the owner.

Section 3. - Assistance by FDA

1. Upon receipt of a notification according to Section 2 above, FDA shall in writing and/or meetings clarify and determine with the Owner the scope, requirements, finances, advantages and disadvantages of the Owner's plans and intentions with a view to jointly decide whether :

a) Management of the Owner's forest land in order to achieve, sustained or continous yield and income from it (see Part III of this Regulation); or
b) Clear-cutting of all merchantable trees (salvage operation) (see Part IV of this Regulation).

is in the Owner's best interest.

2. Subject to the laws of General Application, the Owner shall be free to implement the FDA's considered expertise and recommendations.

Section 4. - Owner's Title

The FDA shall especially make the Owner understand that he can plan forest operations only if his title and deed to his forest land are in order under the laws of General Application and that no act or assistance rendered by FDA to him under this Regulation shall be understood or construed to be a legal confirmation by or opinion of FDA regarding the Owner's property rights.

Section 5. Owner's Boundaries

Subject to Section 4 above, the FDA shall have the right to request the Owner to clearcut and demarcate the boundaries of the forest land prior to the commencement of any operations by the Owner in order to permit reasonable planning.

PART III : FOREST MANAGEMENT

Section 6. - Management Plan

In the event the Owner decides according to Section 3 (a) above to manage his forest land in order to achieve sustained or continous yield and income from it, FDA shall assist him to jointly draw up a Forest Management Plan suitable for the Owner's forest land, if possible.

Section 7. - Forestry Consultancy

1. Upon the request of the Owner, FDA shall, if possible, also render consultancy services to the Owner in all other aspects of the management and utilization of the Owner's forest land and forest resources, including but not limited to, the hiring of qualified staff, the Organization of Operations, the acquiring, use and maintenance of machinery, the processing and marketing of forest products.

2. Consultancy Services to be rendered according to Section 7.1 above shall be clearly identified and defined in a Consultancy Agreement which shall be concluded between FDA and the Owner.

3. FDA shall have the right to charge the owner for FDA's Consultancy Services.

PART IV : FOREST CLEARCUTTING (SALVAGE)

Section 8. - Operations

1. In the event the owner decides according to Section 3 (b) above to clearcut (salvage) all merchantable trees on his forest land, he shall submit to FDA an Operations'Plan for FDA's prior approval.

2. FDA shall assist the owner in the drawing up of the Operations' Plan or in its improvement, as the case may be.

Section 9. - Owner's Operations

1. In the event the Owner intends to implement the Operations' Plan himself, he shall, prior to any operations apply for and acquire the Permit required according to FDA Regulation Nº 4 "On Registration of Non Concession Forest Operators".

2. Upon receipt of the Owner's application for said Permit, FDA shall together with the Owner and the Ministry of Finance determine an arrangement, which shall take into account the Owner's individual circumstances, for the Owner's payment of the forest fees and taxes according to the Revenue and Finance law and said arrangement shall be condition and part of the said Permit.

Section 10. - Contract with a Concessionaire

1. In the event the Owner intends not to implement the Operations's Plan by himself, he shall first offer the Contract to a Concessionaire in the environs of the Owner's forest land and both parties shall negotiate in good faith but shall be free to accept or refuse the offer of the other.

2. The Owner shall, prior to any operations by a Concessionaire, notify FDA of any contract offer made, accepted or rejected and shall file a copy of any Contract, for FDA's prior approval.

3. FDA shall have the right to refuse approval of the Owner's Contract with a Concessionaire in, but not limited to, the event that the Concessionaire has defaulted in his Contractual obligations towards Government, especially is in arrears with any payments of forest fees, taxes or land rentals.

Section 11. - Contract with Operator

1. In the event the Owner cannot conclude a Contract with any Concessionaire to implement his Operations' Plan, the Owner shall have the right to offer the Contract to an Operator, provided, however, that such Operator shall have a valid Permit according to FDA Regulation Nº 4 "On Registration of Non Concession Forest Operators".

2. The Owner shall, prior to any operations by an Operator, notify FDA of any Contract offer made, accepted or rejected and shall file a copy of any Contract for FDA's prior approval.

3. Upon receipt of a Contract for approval according to Section 11.2 above, FDA shall together with the Operator, the Owner and the Ministry of Finance determine an arrangement, which shall take into account the individual circumstances, for the payment of the forest fees and taxes according to the Revenue and Finance Law, and said arrangement shall form an integral part of the Contract between Owner and Operator.

Section 12. - Duties of Owner, Concessionaire and Operator

1. The owner, Concessionaire or Operator, as the case may be, shall implement the Operations'Plan approved by FDA as laid down in Section 8 above according to best forestry and logging practices and standards.

2. Whoever implements the approved Operations'Plan shall immediately notify FDA of the actual commencement and the site of operations as well as of any temporary suspension or cessation thereof.

Section 13. - Duties of FDA

Upon notification of commencement of operations according to Section 12.2 above, FDA shall nominate and send an FDA Scaler to the site of operations to undertake the assessment of the production as required by the Revenue and Finance Law and the FDA Act.

PART V : REPEALERS AND EFFECTIVE DATE

Section 14. - Repealers

All prior Rules and Regulations concerning Commercial Use of privately owned forest lands are hereby repealed.

Section 15. - Effective Date

This Regulation shall become effective on April 15, 1979 and shall be announced in the public media and be published in the FDA News Letter.

Monrovia, LIBERIA March 23, 1979

ACTING MANAGING DIRECTOR

FORESTRY DEVELOPMENT AUTHORITY

LIBERIA

REGULATION N° 6 OF SEPTEMBER 1st 1979

FDA Regulation N° 6

FORESTRY DEVELOPMENT AUTHORITY

REGULATION N° 6

" ON EXPLOITATION PERMITS FOR NON CONCESSION PUBLIC FOREST LAND"

WHEREAS, the Liberian publicly owned forest land since 1959 only small areas, usually less than 100 000 acres remain free for new granting and new investment; and

WHEREAS, the present model Liberian Concession Agreement, applied since 1973, its Management Rules, Investment and Processing requirements as well as its Investment Incentive having been designed for Forest Concession of 100 000 or more acres is evidently not being applicable for the special circumstances of the remaining small-sized parcels of publicly owned forest land; and

WHEREAS, the Forestry Development Authority according to the Act creating the Forestry Development Authority, approved November 1, 1976, published December 20, 1976, is charged, according to Section 3 (b), to devote ALL publicly owned forest lands to their most productive use for the permanent good of the whole people considering both direct and indirect values; and

WHEREAS, said FDA Act has conferred upon the Forestry Development Authority the power to prescribe the form of all Licences, Permits, Agreements and other instruments dealing with the use of forest resources, as well as the power to control the issuance of such instruments, and determine the conditions under which they may be granted, exercised, produced, revoked, or returned through Rules and Regulations (Sections 4 (f), (j) and (n),

NOW, THEREFORE, the Forestry Development Authority does hereby rule and regulate :

PART I : DEFINITIONS

Section 1. - Definitions

a) FDA - Forestry Development Authority;

b) Forest Exploitation Area - publicly owned Forest Land of 100 000 or less acres.

PART II : APPLICATION

Section 2. - Application for FDA Forest Exploitation Permit

Anyone desiring and able to invest in Forest Exploitation Area may apply in writing to the FDA, Sinkor Old Road, P.O.Box 3010, Monrovia, Liberia, for the granting of an FDA Forest Exploitation Permit.

Section 3. - Attachments

The application according to Section 2 above shall be accompanied by :

a) a rough sketch/map of the envisaged area;
b) a rough feasibility study based on sustained yield management of the area;
c) a description of the technical know-how, the equipment and financial resources of the Applicant including Bank references.

Section 4. - Evaluation

FDA shall have the right and duty, if necessary to :

a) request clarification as well as further information regarding Section 3 above from the Applicant; and
b) make a field investigation of the requested area.

PART III : FDA'S RESERVATIONS

Section 5. - Reservations

Upon receipt of an application for FDA Forest Exploitation Permit, FDA reserves the right to refuse the application if one of the following forms of management of the requested area promise a more productive use of the parcel of publicly owned forest land in question.

a) offer the area publicly for competitive bidding;
b) grant it to an existing neighbouring Concession with proven technical know-how, superior equipment, management and financial resources;
c) manage the ara itself, as a principal or in conjunction with others, as provided for in said FDA Act, Section (4) (o) and (p).

PART IV : FDA FOREST SURVEY PERMIT

Section 6. - FDA Survey permit

In the event that FDA is satisfied that application and Applicant meet the requirements and standards according to Sections 3 and 4, FDA shall first issue an FDA Forest Survey Permit.

Section 7. - Contents of FDA Forest Survey permit

The FDA Forest Survey permit shall require, including but not limited to, the following :

a) Surveying and Drawing of precise Maps;
b) Demarcate the boundaries on the ground;
c) a 5% sample Enumeration of the Survey Area;
d) The term of the FDA Survey permit which shall be six (6) months.

PART V : FDA FOREST EXPLOITATION PERMIT

Section 8. - Grant of FDA Forest Exploitation Permit

In the event that the conditions of the FDA Forest Survey Permit have been met on time, FDA shall issue an FDA Forest Exploitation Permit to the Applicant.

Section 9. - Contents of FDA Forest Exploitation Permit

The FDA Forest Exploitation Permit shall require including but not limited to :

a) The deposit of a sum of money, the amount to be determined by the FDA, as an advance payment of the anticipated forest fees and taxes;
b) The payment of land rental at the regular rate in force;
c) The elaboration and determination with FDA of the number of years of the felling cycle and onsequently of the annual coupe considering the special size of the specific forest area;
d) The submission of an operational plan, based on the result of the enumeration (Section 7 (c) and the felling cycle plus annual coupe (c) above;
e) The determination of the percentages/quotas of logs that shall be sold on the local market for local processing and the percentages/quotas of logs that may exported;
f) The term of the Exploitation Permit which shall not be less than 5 years and not be more than 10 years depending on the size of the Forest Exploitation Area.

Section 10. - Incentives

The FDA Exploitation Permit does not grant any exemptions from the Liberian laws of General Application (Investment Incentives) but the FDA may or may not recommend the granting of Investment Incentives by the proper Authority under the Law depending on the circumstances of the individual FDA Forest Exploitation Permit.

PART V : EFFECTIVE DATE

Section 11. - Effective Date

This Regulation shall become effective on September 4, 1979, and shall be announced in the public media and be published in the FDA News Letter.

Monrovia, LIBERIA, September 1, 1979

MANAGING DIRECTOR

An ACT ADOPTING A
REVENUE AND FINANCIAL LAW
APPROVED MAY 24, 1977

LIBERIA

Revenue Law 1977

CHAPTER 20 : STUMPAGE AND FOREST PRODUCTS FEE

Section 20.1 - Definitions

In this chapter the following words and symbols have the meaning indicated unless the context otherwise requires :

"Commercial use" : Any use other than direct use for personal purpose, including uses involving barter, trade or any other disposition of forest products for which remuneration is received;

"Forests" : All areas supporting woody vegetation other than planted or cultivated crops regardless of the composition or age;

"Reserved forests" : Forests within the boundaries of all publicly owned areas including those established by the national government known as Government Forest Reserves, Native Authority Forest Reserves, Communal Forests and national parks;

"Unprocessed forest products" include (but not limited to) the following :

a) Trees, which include seedlings, saplings, brushwood, palms and canes;
b) Felled trees, stumps and sections of felled trees commonly referred to as logs;
c) Wild rubber, wood oil, resin and gums;
d) Leaves, flowers, fruits, seeds and all other parts of trees not hereinbefore mentioned;
e) Plants other than trees (including grass, vines, reeds, and moss) growing in the forest which are not cultivated for agricultural purposes, and all parts and products of nut plants.

"Processed forest products" : Materials yielded by forests and parts thereof which have been worked upon by machinery or by hand including sawing and dividing, and which are considered by the market to be different and new products, but with the exclusion of cleaning, debranching and debarking.

Section 20.2 - Local Stumpage Fee

A fee shall be paid on the severance of timber cut for commercial use in reserved forests and on privately owned forest land by any person who cuts or causes trees to be cut for commercial use.

The rate and basis of assessment of the local stumpage fee shall be determined by the Forestry Development Authority together with the Ministry of Finance. However, the amount of the local stumpage fee shall in no case be lower than $ 1.50 (one dollar and fifty cents) per cubic meter.

Section 20.3 - Industrialization Incentive Fee
A fee shall be paid by any person who ships or causes to be shipped unprocessed round logs for processing abroad, in order to encourage the establishment of a timber industry in the Republic of Liberia.

The Forestry Development Authority together with the ministry of Finance shall determine and regulate from time to time the rates and basis of assessment of the Industrialization Incentive Fee for unprocessed round logs.

Section 20.4 - Forest Products Fee
A fee shall be paid by any person who ships or causes to be shipped abroad lumber, planks, sawdust, bark, slabs, edgings, trimmings, and other similar products.

The rate of the Forest products' fee shall be not lower than 5% and not higher than 15% of the F.O.B. market value as determined by the Forestry Development Authority.

Section 20.5 - Reforestation Fee
A fee of not less than $ 1.50 and not higher than $ 4.50 per cubic meter shall be paid by any person who harvests a tree from a forest for commercial use. However, any person may be exempt from paying said fee if he undertakes to carry out his own reforestation in accordance with recognized methods of reforestation as promulgated and amended from time to time by the Forestry Development Authority upon the written consent of the Government.

Section 2.6 - Assessment
The fees according to Sections 20.2, 20.3, 20.4 and 20.5 shall be assessed by the competent forest authority.

APPROVED MAY 24, 1977

PUBLISHED JUNE 20, 1977

EFFECTIVE JULY 1, 1977

FDA Reg. N°7 OF DEC. 7, 1979

LIBERIA

FDA Reg. N°7

FORESTRY DEVELOPMENT AUTHORITY REGULATION N°7

Regulation on revised Forest Fees and Taxes

WHEREAS, the Act Adopting a Revenue and Finance Law, approved May 24, 1977, published June 20, 1977, in Chapter 20 on Stumpage and Forest Products Fee, charges the Forestry Development Authority and the Ministry of Finance to determine and regulate from time to time the rates and basis of the Liberian forest fees and taxes (Sections 20.2, 20.3, 20.4 and 20.5 respectively); and

WHEREAS, since August 29, 1977, when the last Regulation on Revised Forest Fees and Taxes (Revised Stumpage Fees - Bulletin N°5A) was promulgated, the tropical red woods are being depleted in the whole of West Africa and round logs prices are continuously rising on the international markets; and

WHEREAS, the changed circumstances including the statistics and the priority policy for accelerated industrialization, the generation of employment and value-added in the Liberian forest sector have been in detail discussed and accepted by the meeting of the FDA Board of Directors on September 1979, and in an information meeting with the forest concessionaires on November 21, 1979;

NOW, THEREFORE, the Forestry Development Authority together with the Ministry of Finance do hereby rule and regulate :

Section 1. - Severance Fee

1) All trees cut for commercial purposes on all lands within the Republic of Liberia shall be assessed at $ 1.50 per cubic meter of Full Tree Length irrespective of grade and intended use.

2) Full Tree Length is the log after Felling, Topping, removal of Defects, if any, but before any crosscutting, with Felling, Topping and Defects as prescribed by FDA's forest management rules and regulations from time to time.

Section 2. - Industrialization Incentive Fee :

All round logs intended for export shall be assessed according to species at the following rates per cubic meter (m^3) :

			$/$m^3$
1.	Entandrophragma utile	Sipo/Utile	75
2.	Tieghemella heckelli	Makore	40
3.	Entandrophragma cylindricum	Sapele	35
4.	" candollei	Kosipo	35
5.	" angolense	Tiama/Edinam	35
6.	Khaya ivorensis/anthotheca	Acajou/Khaya	35
7.	Lovoa trichiliides	Dibetou/Lovoa	35
8.	Heritiera utilis	Niangon	35
9.	Guarea cedrata	Bosse/Guarea	17
10.	Chlorophora spp.	Iroko	17
11.	Mansonia altissima	Bete/Mansonia	17
12.	Guibourtia ehie	Amazakoue	15
13.	Triplochiton scleroxylon	Wawa/Obeche	15
14.	Terminalia ivorensis	Framire	15
15.	Aningeria robusta	Aningre	15
16.	Terminalia superba	Frake	5
17.	Erythrophleum spp.	Tali	5
18.	Nesogordonia papaverifera	Danta/Kotibe	5
19.	Brachstegia leonensis	Naga	3
20.	Pycnanthus angolensis	Ilomba	3
21.	Afzelia spp.	Doussie	3
22.	Tetraberlinia tubmaniana	Sikon	3
23.	Distemonanthus benthamianus	Movingui	3
24.	Pterygota macrocarpa	Koto	3
25.	Nauclea spp.	Kusia/Bilinga	3
26.	Canarium schweinfurthii	Aiele	3
27.	Lophira alata	Azobe/Ekki	2
28.	All other species	N.A.	2

Section 3. - Forest Products Fee :

1) The wood products defined herebelow intended for export shall be assessed according to species at the following rates per m^3.

2) Sawn T&T (Sawn Through and Through) includes :
   a) a log which is sawn through and through without the individual lumber having been further processed but which is rebundled to its original round log form (Boule), and
   b) the individual lumber of a log sawn through and through without the lumber having been further processed.

3) Square-edged on four sides (undressed in the rough) is lumber with four sides square edged but not planed.

4) Square-edged dressed four sides is lumber with four sides square edged and planed including moldings but excluding veneer and plywwod.

		Sawn T&T	Square edged four sides (undressed in the rough)	Square edged dressed four sides
Entandrophragma utile	Sipo/Utile	60/m³	15/m³	2/m³
Tieghemella heckelli	Makore	34	10	2
Entandrophragma cylind.	Sapele	30	10	2
" cand.	Kosipo	30	10	2
" ang.	Tiama/Edinam	30	10	2
Khaya spp.	Acajou/Khaya	30	10	2
Lovoa trichiliodes	Dibetou/Lovoa	30	10	2
Heritiera utilis	Niangon	30	10	2
Guarea cedrata	Bosse/Gharea	15	5	1
Chlorophora spp.	Iroko	15	5	1
Mansonia altissima	Bete/Mansonia	15	5	1
Guibourtia ehie	Amazakoue	13	5	1
Triplochiton scleroxylon	Wawa/Obeche	13	5	1
Terminalia ivor.	Framire	13	5	1
Aningeria robusta	Aningre	4	2.50	1
Terminalia superba	Frake	4	2.50	1
Erythrophleum spp.	Tali	4	2.50	1
Nesogordonia papaverifera	Danta/Kotibe	4	2.50	1
Brachystegia leon.	Naga	2	1.50	1
Pycnanthus ang.	Ilomba	2	1.50	1
Afzelia spp.	Doussie	2	1.50	1
Tetraberlinia tubm.	Sikon	2	1.50	1
Distemonanthus benth.	Movingui	2	1.50	1
Pterygota mac.	Koto	2	1.50	1
Nauclea spp.	Kusia/Bilinga	2	1.50	1
Canarium sch.	Aiele	2	1.50	1
Lophira alata	Azobe/Ekki	2	1.50	1
All other species	N.A	1.50	1.50	1

Section 4. - Reforestation Fee

1) All trees cut for commercial purposes on all lands within the Republic of Liberia shall be assessed at $3.00 per cubic meter (m³) of Full Tree Length.

2) Full Tree Length is the log after Felling, Topping and removal of Defects, if any, but before any crosscutting, with Felling Topping and Defects as prescribed by FDA's forest management rules and regulations from time to time.

Section 5. - Repeal :
The Revised Stumpage Fees - Bulletin N°5A of August 29, 1977, effective November 1, 1977, is hereby repealed.

Section 6. - Effective Date :
This regulation shall become effective on January 1, 1980, and shall be announced in the public media and be published in the FDA-Newsletter.

Monrovia, Liberia December 7, 1979

Minister of Finance

Managing Director
Forestry Development Authority

FORESTRY DEVELOPMENT AUTHORITY
REGULATION N° 10 OF NOVEMBER 8, 1982

LIBERIA

FDA Regulation N°10

REGULATION ENABLING A FURTHER REDUCTION OF CERTAIN FOREST FEES

WHEREAS, the Act Adopting a Revenue and Finance Law Approved May 24, 1977, published June 20, 1977, in Chapter 20 on Stumpage and Forest Products Fee, charges the Forestry Development Authority and the Ministry of Finance to determine and regulate from time to time the rates and basis of the Liberian Forest Fees and Taxes (Section 20.2, 20.3, 20.4 and 20.5 respectively); and

WHEREAS, in 1981 and 1982, because of a decline in international demand and trade in logs with a consequent reduction of prices, the Government of Liberia in order to assist the timber trade during a period of depression, did lower the Industrialization Incentive Fee payable on the export of logs and lumber of certain species; and

WHEREAS, the depressed condition of the timber trade still continues, causing a further slackening of demand for certain species; and

WHEREAS, the Government of Liberia still continues to allow the unhindered and free export of logs of all species; and

WHEREAS, some further sharing of the burden of the recession is required in the form of a reduction of certain forest fees without destroying the structure of the scale of fees and the reasonable production of the proceeds of trade due to the Government of Liberia.

NOW, THEREFORE, the Forestry Development Authority together with the Ministry of Finance do hereby rule and regulate that a further reduction of 20% be made from the present rate of Industrialization Incentive Fees for certain species as shown in Section 1.

Section 1. - Industrialization Incentive Fee

All logs intended for export shall be assessed according to species at the following rates per cubic meter (M^3)

	BOTANICAL NAMES	TRADE NAMES	RATES $
1.	Entandrophragma utile	Sipo	75.00
2.	Tieghemelia heckellii	Makore	32.00
3.	Entandrophragma cylindricum	Sapele	28.00
4.	" candollei	Kosipo	20.16
5.	" angolense	Tiama	20.16
6.	Khaya ivorensis/anthotheca	Acajou/Khaya	20.16
7.	Lovoa trichilioides	Dibetou/Lovoa	20.16
8.	Heritiera utilis	Niangon	20.16
9.	Guarea cedrata	Bosse/Guarea	9.79
10.	Chlorophora spp.	Iroko	9.79
11.	Mansonia altissima	Bete/Mansonia	9.79
12.	Guibouritia ehie	Amazakoue	8.64

13.	Triplochiton scleroxylon	Wawa/Obeche	8.64
14.	Terminalia ivorensis	Framire	8.64
15.	Aningeria robusta	Aningeria	8.64
16.	Terminalia superba	Frake	3.60
17.	Erythrophleum spp.	Tali	3.60
18.	Nesogordonia papaverifera	Danta/Kotibe	3.60
19.	Brachystegia leonensis	Naga	2.70
20.	Pycnanthus angolensis	Ilomba	2.70
21.	Afzelia spp.	Doussie	2.70
22.	Tetraberlinia tubmaniana	Sikon	2.70
23.	Distemonanthus benthamianus	Movingui	2.70
24.	Pterygota macrocarpa	Koto	2.70
25.	Nauclea spp.	Kusia/Bilinga	2.70
26.	Canarium schweinfurthii	Aiele	1.80
27.	Lophira alata	Azobe/Ekki	1.80
28.	All other species		1.80

Section 2. Severance Fee

All trees cut for commercial purposes on all lands within the Republic of Liberia shall be assessed at $1.50 per cubic meter of Full Tree Length irrespective of grade and intended use.

Section 3. - Reforestation Fee

1. All trees cut for commercial purposes on all lands within the Republic of Liberia shall be assessed at $3.00 per cubic meter ($M^3$) of Full Tree Length.

2. Full Tree Length is the log after Felling, Topping, and Removal of Defects, if any, but before any cross-cutting, with Felling, Topping and Defects as prescribed by FDA's Forest Management Rules and Regulations from time to time.

Section 4. - Forest Products Fee

1. The wood products defined herebelow intended for export shall be assessed according to species at the following rates per M^3.

	Botanical Names	Trade Names	Sawn T&T $/M³	Square edged four sides (undressed) in the rough $/M³	Square edged dressed four sides $/M³
1.	Entandrophragma utile	Sipo/Utile	51.00	15.00	2
2.	Tieghemella heckelli	Makore	22.95	10.00	2
3.	Entandrophragma cylindricum	Sapele	20.40	10.00	2
4.	" candollei	Kosipo	18.70	10.00	2
5.	" angolense	Tiama/Edinam	18.70	10.00	2
6.	Khaya spp.	Acajou/Khaya	18.70	10.00	2
7.	Lovoa trichilioides	Dibetou/Lovoa	18.70	10.00	2
8.	Heritiera utilis	Niangon	18.70	10.00	2
9.	Guarea cedrata	Bosse/Guarea	8.50	5.00	1
10.	Chlorophora spp.	Iroko	8.50	5.00	1
11.	Mansonia altissima	Bete/Mansonia	8.50	5.00	1
12.	Guibourtia ehie	Amazakoue	7.75	5.00	1

13. Triplochiton scleroxylon	Wawa/Obeche	7.65	5.00	1
14. Terminalia ivorensis	Framire	7.65	5.00	1
15. Aningeria robusta	Aningeria	3.00	2.50	1
16. Terminalia superba	Frake	3.00	2.50	1
17. Erythrophleum spp.	Tali	3.00	2.50	1
18. Nesogordonia papaverifera	Danta/Kotibe	3.00	2.50	1
19. Brachystegia leonensis	Naga	2.00	1.50	1
20. Pycnanthus angolensis	Ilomba	2.00	1.50	1
21. Afzelia spp.	Doussie	2.00	1.50	1
22. Tetraberlinia tubmaniana	Sikon	2.00	1.50	1
23. Distemonanthus benthamianus	Movingui	2.00	1.50	1
24. Pterygota macrocarpa	Koto	2.00	1.50	1
25. Nauclea spp.	Kusia/Bilinga	2.00	1.50	1
26. Canarium schweinfurthii	Aiele	1.50	1.25	1
27. Lophira alata	Azobe/Ekki	-	-	-
28. Other species	N . A	1.50	1.25	1

2. Sawn T & T - (Sawn Through and Through) including :

   a) A log which is sawn through and through without the individual lumber having been further processed but which is rebundled to its original round log form (Boule); and

   b) The individual lumber of a log sawn througn and through without the lumber having been further processed.

   For wood products sawn through and through, the rates have been adjusted to encourage further proceesing in respect of the present reduction in the rates for round logs intended for export.

3. Square Edged on Four Sides (Undressed in the Rough) - is the lumber with four sides square edged but not planed.

4. Square Edged Dressed Four Sides - is the lumber with four sides square edged and planed including mouldings but excluding veneer and plywood.

Section 5. - Repeal

Regulation N°9 on Special Trade Depression Allowance on Certain Forest Fees effective April 1, 1982 is hereby repealed.

Section 6. - Effective Date

This Regulation shall become effective on December 1, 1982 and be announced in the public media and published in the FDA-NEWSLETTER.

MANAGING DIRECTOR, FDA

Monrovia, LIBERIA, November 8, 1982

MINISTER OF FINANCE

FORESTRY ORDINANCE
N.R. 1960
Northern Region

NIGERIA

Forestry Ordinance

1. - This Ordinance may be cited as the Forestry Ordinance

PART I : PRELIMINARY

Interpretation

2. - Cattel
Communal forestry area
Communal lands
Enclave
Export
Forests
Forest growth
Forestry officer
Forest produce
Forestry property
Forest reserve
Girth
Inquiry
Lands at the disposal of the Government
Minor forest produce
Native community
Native lands
Protected forests
Protected tree
Resident
Timber
To take minor forest produce
To take timber
Tree
Working plan

Appointment of forestry officers

3. - The Minister, or a native authority or a local government council with the approval of the Minister, may appoint such officers as may be considered necessary for the purpose of giving effect to the provisions of this Ordinance.

PART II : SPECIAL PROVISIONS RELATING TO GOVERNMENT FOREST RESERVES AND GOVERNMENT PROTECTED FORESTS

Governor may constitute reserves and protected forests

4. - 1) It shall be lawful for a Governor to constitute as Government forest reserves any of the following lands within his Region :

a) lands at the disposal of Government or native lands;
b) any lands in respect of which it appears to a Governor in Council on the advice of the Chief Conservator of Forests of the Region that the forest growth on such lands should be protected or reserved or forest growth be established.

2) A governor may by notice in the Regional Gazette declare any such lands to be a Government protected forest.

Preliminary Procedure for Constitution Government Forest Reserves

Notification of intention to create a reserve and appointment of reserve settlement officer

5. - 1) Before constituting any lands a Government forest reserve a notice shall be published by the Governor in the Regional Gazette :

a) specifying as nearly as may be the situation and limits of the lands;
b) declaring that the lands now form a protected forest;
c) declaring whether the lands are at the disposal of the Government or are native lands or are lands coming within section 4 (1) (b);
d) declaring that it is intended to constitute such lands a Government forest reserve, either for the general use and benefit, wholly or in part, of any class of persons or for the benefit of any native community or authority or a local government council;
e) appointing an officer, hereinafter referred to as the reserve settlement officer, to inquire into and determine the existence, nature and extent of any rights claimed by or alleged to exist in favour of any persons or communities or brought to the knowledge of the said officer affecting the lands or any other rights in or over the lands which it is proposed to constitute a Government forest reserve.

2) If for any reason the reserve settlement officer appointed under this section is unable to perform his duties, the Governor may, by notice in the Regional Gazette, appoint any person to act on his behalf or as his successor; such notice may have retrospective effect for a period not exceeding one month.

Notice of inquiry

6. - Upon publication of the notice aforesaid the reserve settlement officer shall :

a) immediately cause the particulars contained therein to be made known in the district or districts in which the lands are situated by causing the same to be read and interpreted in the local native language in every native court in the said district or districts and also by, as far as he considers essential, informing the chiefs of the communities dwelling on, and the native authorities or local government councils having jurisdiction over, the lands aforesaid in writing; and

b) fix and, in the manner aforesaid, make known a period within which and a place to or at which any person or community claiming any right or rights in or over or affecting the lands which it is proposed to constitute a forest reserve shall either send in a written statement of claims to him or appear before him and state orally the nature and extent of their alleged rights.

The Inquiry

Duty of reserve settlement officer at inquiry

7. - 1) As soon as possible after the expiration of the period fixed by the reserve settlement officer he shall :
a) inquire into and determine the limits of the lands specified in the notice aforesaid; and
b) determine the nature and extent of any claims or alleged rights affecting the land which have been preferred or brought to his notice.

2) The reserve settlement officer shall keep a record in writing of :
a) all such claims and alleged rights;
b) all objections which may be made to such claims or alleged rights; and
c) any evidence in support of or in opposition to any claim or alleged right.

Reserve settlement officer to have juridical powers

8. - For the purposes of the inquiry the reserve settlement officer shall have all the powers conferred upon a magistrate.

Reserve settlement officer may sever or join claims

9. - The reserve settlement officer may at any time during the inquiry join any number of claims or sever any claims joined and in his judgment may give a decision which may join any number of claims or sever any claims which were formerly joined.

Judgment

Publication of judgment on completion of the inquiry

10. - 1) Upon the completion of the inquiry, the reserve settlement officer shall deliver his judgment, describing the limits of the land specified in the notice aforesaid and setting forth, with all such particulars as may be necessary to define their nature, duration, incidence and extend, all claims and alleged rights preferred or brought to his knowledge in respect of the lands and admitting or rejecting the same wholly or in part, and shall file it at the Land Registry.

2) A notice shall be published in the Regional Gazette specifying the land which it is intended to reserve, the privileges conceded in respect of such land and stating the special conditions intended to govern the reservation thereof. In addition such notice shall be made known so far as may be practicable to every person who, and the head of any community which, preferred any claim, or in respect of which any claim was brought to the knowledge of the reserve settlement officer.

Appeal

Appeals from judgment of reserve settlement officer

11. - 1) Any person who has made a claim on his own behalf, or where a claim has been made on behalf of a community that person or the representative of that community may, within three months of the date of delivery of the judgment, appeal to a District Judge against that portion of the reserve settlement officer's judgment which affects his claim or the claim made on behalf of the community which he represents.

2) An appeal shall lie to the High Court within whose area of jurisdiction the forest reserve is situated from a decision of a District Judge on appeal against the judgment of the reserve settlement officer within thirty days of the date of the decision appealed against. The decision of the High Court shall be final.

Constitution of the Forest Reserve

Order of the Governor constituting the forest reserve

12. - 1) The Governor may make an order constituting the lands in respect of which an inquiry has been held a Government forest reserve at the expiration of three months from the date of publication of the notice in the Regional Gazette under section 10.

Provided that if an appeal has been made against the judgment of the reserve settlement officer no such order shall be made until such appeal has been determined or until the expiration of the time within which an appeal may be laid.

2) Such order shall, subject to the provisions of subsection (3), set forth :

a) the limits of the lands which constitute the reserve; and
b) all rights affecting the same as set forth in the judgment of the reserve settlement officer or established by the court upon appeal from such judgment; and
c) such additional rights as the Governor shall consider it just and equitable to allow notwithstanding that such rights have not been allowed in the judgment of the reserve settlement officer.

3) Such order shall not include therein such rights as may have been allowed by the reserve settlement officer but which have been subsequently modified or extinguished as hereinafter provided and where the boundaries of the reserve have been modified as hereinafter provided the order shall set forth the boundaries as subsequently determined.

4) Such order shall be published in the Regional Gazette and made known in the same manner as was the judgment of the reserve settlement officer.

5) From the date of the publication of such order in the Regional Gazette such lands shall be a Government forest reserve.

6) Any order made under this section may be revised or modified by the Governor and such revision or modification may be given retrospective effect.

7) In any revision or modification made by the Governor under subsection (6) the Governor may, after such inquiry, if any, as he shall in his discretion think fit :

a) exercise the powers conferred upon him by subparagraph (ii) of section 13; and
b) add such additional rights as he shall consider it just and equitable to allow notwithstanding that such rights had not been allowed in the judgment of the reserve settlement officer.

Modification of the Reserve Settlement Officer's Judgment

Rights may be extinguished or modified by Minister

13. - Where the reserve settlement officer had admitted wholly or in part any right or claim and in the opinion of the Minister the exercise of such right or claim or any part thereof :

a) would stultify the objects of any forest reserve;
b) would seriously hinder the efficient working of any forest reserve; or
c) would do serious damage to any forest reserve ;

the Minister :

(i) may extinguish any such rights or claims and shall either give monetary compensation, or grant in exchange similar rights on other similarly situated land either within or without the final boundaries of the forest reserve; or
(ii) may confine or restrict any rights or claims to certain areas either within or without the forest reserve or the exercise of such rights to certain times of the year; or
(iii) may adopt wholly or in part any combination of the above methods of dealing with the matter.

Reserve may be made to exclude areas over which claims are admitted

14. - If the reserve settlement officer has admitted wholly or in part any rights on any area which in the opinion of the Chief Conservator of Forests of the Region could be exercised from the reserve without materially altering or stultifying the objects of the reserve the Governor may so amend the boundaries of the reserve that such areas are excluded from the reserve or he may create such excluded areas or enclaves within the boundaries of the reserve:

Provided that in altering the external boundaries of the reserve the officer shall not include any area which lies outside the original boundaries set out in the notice of the proposed reserve published in accordance with the provisions of section 5.

Further Provisions affecting Rights

Minister may close existing rights of way and watercourses if alternatives exist

15. - In any Government forest reserve the Minister of the Region may close any right of way or watercourse :

Provided that in his opinion another right of way or watercourse equally convenient already exists or is provided.

Extinguishment and revival of rights

16. - Every right in or over land in respect of which no claim shall have been made to the reserve settlement officer, or of which no knowledge shall have been acquired by that officer before delivery of his judgment shall be extinguished :

Provided that if any person shall, within one year of the date of delivery of the final judgment, claim and satisfy the Minister that he was possessed of a right in respect of which he might have made a claim, and that through ignorance of the fact that an inquiry was being held or for other sufficient reason he failed to make such claim, the Minister may direct :

a) that such right shall be revived; or
b) that such claim shall be modified or extinguished in accordance with the provisions of section 13.

Non-exercise of rights

17. - If any right within a forest reserve shall not have been exercised for a period of ten years it shall be deemed to be extinguished.

Rights in reserves may not be alienated without consent of the Minister

18. - It shall not be lawful for any person to alienate any right, affecting land included in a forest reserve, which has been established before the reserve settlement officer or before a court under section 11 or revived under section 16, by sale, mortgage or transfer without the consent of the Minister of the Region first had and obtained. Any such sale, mortgage or transfer effected without such consent shall be null and void.

No new rights to be acquired in land to be constituted a reserve except with approval

19. - During the period between the dates of the publication by a Governor of the notice of his intention to create a Government forest reserve and of the order of that Governor constituting the reserve :

a) no right shall be acquired in or over the land comprised within such notice otherwise than by succession or under a grant or contract in writing entered into with the approval of the Governor; and
b) save as hereinafter provided :
   (i) no new house shall be built;
   (ii) no tree shall be cut; and
   (iii) no forest produce shall be removed :

Provided that nothing in this section shall be deemed to prohibit any act done with the permission in writing of the district officer or a forestry officer of a rank not below that of assistant conservator of forests.

Miscellaneous

Powers of native to grant land absolutely to Government and to grant licences and permits

20. - Any native, and the chief or head of any native community on behalf of such community, notwithstanding any native law or custom to the contrary, shall be entitled to enter into any agreement :

a) to grant and convey absolutely to the Government any lands and any rights in and over any lands, owned by him or them which it is proposed to constitute a forest reserve under the provisions of this Ordinance;

b) to grant, subject to the provisions of this Ordinance, any licence or permit for the taking of forest produce owned by him or them.

Power to de-reserve

21. - A Governor may by order direct that from a date named therein any lands within his Region or any part thereof constituted a Government forest reserve under section 12 shall cease to be a Government forest reserve or a part of such reserve and thereupon from such a date such lands shall cease to be a Government forest reserve or a part of such reserve :

Provided that the rights, if any, which may have been extinguished therein shall not revive in consequence of such cessation.

PART III : SPECIAL PROVISIONS RELATING TO NATIVE AUTHORITY OR LOCAL GOVERNMENT COUNCIL FOREST RESERVES

Constitution of Reserves

Constitution of native authority or local government council forest reserve

22. - It shall be lawful for a native authority or local government council, by order made with the approval of the Minister of the Region, to constitute as a native authority or local government council forest reserve any land lying within the area of its jurisdiction.

Procedure

Action precedent to the constitution of a native authority or local government council forest reserve

23. - 1) No lands shall be constituted a native authority or local government council forest reserve under section 22 unless and until :

a) the intention to constitute such lands a native authority or local government council forest reserve has been announced by the native authority or local government council in a manner approved by the Resident; and

b) the existence, nature and extent of any rights claimed by or alleged to exist in favour of any persons or communities affecting the lands which it is proposed to constitute a native authority or local government council forest reserve has been inquired into and determined by or under the direction of the native authority or local government council.

2) The announcement in this section referred to shall require all claimants to put forward their claims within three months.

3) All claims not put forward within three months of the announcement shall be invalid :

Provided that the provisions of the proviso to section 16 shall apply to native authority or local government council forest reserves, mutatis mutandis, and that the powers vested thereunder in the Minister shall be vested in the Resident in respect thereto.

4) If the inquiry made in accordance with paragraph (b) of subsection (1) discloses the existence of any rights in any area which in the opinion of the Chief Conservator of Forests of the Region could be excluded in whole or in part from the reserve without materially altering or stultifying the objects of the reserve, the Resident may so amend the boundaries of the reserve that such areas are excluded from the reserve or he may create such excluded areas as enclaves within the boundaries of the reserve :

Provided that in altering the external boundaries of the reserve the Resident shall not include any area which lies outside the boundaries of the proposed reserve as announced in accordance with paragraph (a) of subsection (1).

5) At any time either before or after an order has been made under section 22, the Resident, after reference to the Chief Conservator of Forests of the Region, may cause a further inquiry to be held by or under the direction of the native authority or local government council or by any other person for the better determination of the rights affecting the land which it is proposed to constitute or which has been constituted a native authority or local government council forest reserve.

Order constituting native authority or local government council forest reserves

24. - Every order under section 22 constituting a native authority or local government council forest reserve shall after approval by the Minister, be published in the Regional Gazette and shall set forth the limits, situation and approximate area of the lands which constitute the native authority or local government council forest reserve and all rights affecting the same as determined under the provisions of the last preceding section.

Approval of Minister

25. - No order under section 22 constituting a native authority or local government council forest reserve shall be published in the Regional Gazette or be of any effect unless and until it has received the approval in writing of the Minister, which approval may be granted or withheld or granted on such conditions as the Minister may in his absolute discretion consider necessary.

Power to Native Authority or local government council to revise or modify order

26. - 1) A native authority or local government council by order published in the Regional Gazette may revise or modify any order made under section 22 so that the rights affecting the land set forth in the order constituting the reserve shall accord with the determinations of any further inquiry held under the provisions of subsection (5) of section 23 and any rights declared invalid by subsection (3) of section 23 may be revived by an order under this section.

2) The provisions of section 25 shall apply to an order made under this section in like manner as if to an order under section 22.

Provisions Affecting Rights in Native Authority or Local Government Council Reserves

Extinguishment of rights in native authority or local government council forest reserve

27. - Every right in or over land within an area constituted a native authority or local government council forest reserve under section 22, other than the rights set forth in the order constituting such reserve, shall be extinguished upon the coming into operation of the order, save as provided in section 23.

Control of alienation of rights in native authority or local government council forest reserve

28. - It shall not be lawful for any person to alienate any right in or over land within an area constituted a native authority or local government council forest reserve under section 22 by sale, mortgage of transfer without the consent of the native authority or local government council which constituted such native authority or local government council forest reserve or within whose jurisdiction it is situated. Any such sale, mortgage or transfer effected without such consent shall be null and void.

Power to de-reserve

29. - A native authority or local government council may by order made with the approval of the Minister of the Region direct that from a date named therein any lands or any part thereof constituted a native authority or local government council forest reserve under section 22 shall cease to be a native authority or local government council forest reserve or a part of such reserve and thereupon from such date such lands shall cease to be a native authority or local government council forest reserve or a part of such reserve :

Provided that the rights, if any, which may have been extinguished therein shall not revive in consequence of such cessation.

PART IV : GOVERNMENT FOREST RESERVES CONVERTED TO NATIVE AUTHORITY OR LOCAL GOVERNMENT COUNCIL FOREST RESERVE

Power to convert existing forest reserves into native authority or local government council forest reserves

30. - It shall be lawful for a Governor by order published in the Regional Gazette to declare that any Government forest reserve within his Region constituted under section 4 shall from a date to be specified in the order become a native authority or local government council forest reserve or be divided up into two or more forest reserves, all or any of which shall become native authority or local government council forest reserves. Thereupon all the preceding provisions of this Ordinance relating to native authority or local government council forest reserves constituted under section 22 shall so far as possible apply to any so declared native authority or local government council forest reserve as if it had been constituted a native authority or local government forest reserve by the native authority or local government council within whose jurisdiction it is situated :

Provided that the rights recognized in the order under section 12 by which such forest reserve was constituted a Government forest reserve shall continue to be recognized together with all other lawful existing rights:

Provided further that no rights acquired by the Government under any of the provisions of this Ordinance or of any Ordinance repealed by this Ordinance shall be thereby extinguished.

PART V : NATIVE AUTHORITY OR LOCAL GOVERNMENT COUNCIL PROTECTED FORESTS

Protected forests

31. - 1) Any native authority or local government council with the approval of the Minister of the Region may, from time to time by notice, declare any land within its jurisdiction a native authority or local government council protected forest.

2) Every such notice shall be made public in the area to which it relates in the same manner as are orders of the native authority or local government council for that area.

Power to withdraw protection

32. - A native authority or local government council with the approval of the Minister of the Region may declare by notice in the Regional Gazette that from the date named therein any land which has been declared a protected forest under section 31 shall cease to be a protected forest and thereupon such land shall cease to be a protected forest :

Provided that the rights, if any, which may have been extinguished therein shall not revive in consequence of such cessation.

PART VI : ADMINISTRATION OF NATIVE AUTHORITY OR LOCAL GOVERNMENT COUNCIL FOREST RESERVES

Management of native authority or local government council forest reserves

33. - 1) The protection, control and management of a native authority or local government council forest reserve shall be undertaken by the native authority or local government council constituting it, or within whose jurisdiction it is situated, subject to the supervision and control of the Resident, exercised with the advice of the Chief Conservator of Forests of the Region.

2) Such protection, control and management may, upon a notification to that effect being published in the Regional Gazette, be placed temporarily under the guidance and direction of the Chief Conservator of Forests of the Region either at the request of the native authority or local government council concerned, or upon the instructions of the Minister if he is of the opinion that such a step is necessary or expedient for ensuring the proper and sufficient protection, control and management of such native authority or local government council forest reserve.

3) Any native authority or local government council forest reserve placed temporarily under the guidance and direction of the Chief Conservator of Forests of the Region in pursuance of the provisions of subsection (2) shall be protected, controlled and managed on behalf and for the benefit of the native authority or local government council concerned.

PART VII : COMMUNAL FORESTRY AREAS

Power of native authority or local government council to declare lands communal forestry areas

34. - Any native authority or local government council at the request of any native community within the area of its jurisdiction may, with the approval of the Resident, declare any lands within the area occupied by such native community a communal forestry area.

Notification of declaration

35. - Such declaration shall be made known in the same manner as native authority or local government council orders are made known to natives ordinarily subject to the jurisdiction of a native court and also by posting a notice setting forth the situation, extent and limits of the communal forestry area outside the office or other meeting place of the native authority or local government council.

Management and control

36. - A communal forestry area shall be managed and controlled by the native community acting on the advice of the native authority or local government council and the forestry officer.

Power of native authority or local government council to make rules

37. - 1) A native authority or local government council with the approval of the Resident may make rules for the protection and management of communal forestry areas within the area of its jurisdiction for all or any of the following purposes :

a) prescribing the duties of native communities;
b) prohibiting or regulating the taking, free or on payment of forest produce or of any specified kind of forest produce;
c) prohibiting the sale or export of forest produce or of any specified kind of forest produce by any person other than specified persons or classes of persons;
d) prohibiting the destruction of, or any act which may tend to the destruction of or cause injury to, any forest produce or forest growth or forest property;
e) providing for the seizure, detention, and disposal of timber or forest produce taken, collected, prepared, sold, purchased, possessed or not marked, in contravention of the rules or in respect of which any offence against the rules has been committed;

f) providing for the extablishment of nurseries and for the afforestation of lands, the preservation and production of forest produce and the introduction of forest produce and the introduction of new species of trees or other forest produce;

g) providing for the management, utilization and protection of the areas;

h) protection the forest produce in the areas by :

(i) prescribing the time at which and the manner in which the rights recognized by native law and custom may be exercised ;

(ii) prohibiting the taking or destruction of any specified kind of timber or minor forest produce in the exercise of such rights as aforesaid; and

(iii) prohibiting the exercise of all or any of such rights as aforesaid in any specified part of the area;

i) generally for giving effect to the objects and purposes for which the areas are established.

2) Rules made under this section shall be made known in the same manner as declarations made under section 34 and subject to any exception specified in such rules, all rules made under this section shall apply to all persons who are subject to the jurisdiction of the native authority or local government council, and may be made to apply in whole or in part to such other persons as may be specified in the said rules.

Power to vary or cancel declaration

38. - A native authority or local government council may, with the approval of the Resident, declare that from a specified date lands or any part thereof declared to be a communal forestry area shall cease to be a communal forestry area or a part of such area and thereafter from such date such lands shall cease to be a communal forestry area or a part of such area. Such declaration shall be made known in the same manner as a declaration declaring an area a communal forestry area.

PART VIII : GENERAL PROVISIONS

Entry upon lands

Marking of boundaries

39. - Any person required so to do by the Resident or the Conservator of Forests with necessary workmen may enter upon any land for the purpose of erecting any beacons or demarcating or cutting any boundary lines within and around any land which it is proposed to constitute a forest reserve or a protected forest and around any portions of land included as enclaves.

Improvement of forest generally

40. - Any forestry officer not below the rank of assistant conservator of forests may enter upon any land and may cut out and destroy any diseased, dead or dying tree likely to cause damage to any forestry property or to life or property.

Prevention of offence

41. - It shall be lawful for any forestry officer, administrative officer or police officer to prevent the commission of any forestry offence.

Power to exempt certain classes and districts

42. - In any Region the Governor may by notice in the Regional Gazette withdraw from the operation of all or any of the provisions of this Ordinance any class of persons or any tribe or part of a tribe, or any area specified therein, either for the period mentioned in the notice or without period assigned.

Miscellaneous

Forest produce required for public purposes from native authority or local government council forest reserves

43. - It shall be lawful for a Minister to take from any native authority forest reserve or native authority or local government council protected forest within his Region any forest produce which may be required for public purposes upon payment of a fair and reasonable price therefor which price shall not exceed such fees and royalties as may be specified in regulation and are generally applicable in respect of the reserve or protected forest whence the forest produce was taken.

Disposal of fees and royalties

44. - 1) All fees received under this Ordinance shall be paid by the officer receiving them into the Treasury of the Region save that fees payable in respect of any native authority forest reserve or any other area in a Region which the Governor of the Region may by notice in the Regional Gazette prescribe as an area in respect of which fees are payable to the native authority, shall be paid by the officer receiving them direct into the native treasury of such native authority administering the same.

2) Royalties shall be paid to the person or persons entitled to receive them

3) All fees payable in respect of any local government council forest reserve shall be paid by the officer receiving them direct to the local government council concerned.

Saving of power to acquire land in native authority or local government council forest reserves for public purposes

45. - Nothing in this Ordinance shall be construed so as to prevent the acquisition under any Ordinance relating to the acquisition of lands for public purposes of land included in the area of a native authority or local government council forest reserve or protected forests.

Regulations

Power of Governor in Council to make regulations

46. - A Governor in council may make regulations for his Region for all or any of the purposes following and may specify the area or area to which all or any regulations shall apply :

1) prohibiting or regulating the taking of forest produce or of any specified kind of forest produce on lands at the disposal of Government or on native lands or communal lands;

2) prohibiting the sale and purchase of forest produce or of any specified kind of forest produce by any person other than the holders of licences and permits granted under this Ordinance, or by any other persons or by any classes of persons specified in the regulations;

3)prohibiting the sale, purchase and possession of forest produce taken, collected or prepared in contravention of this Ordinance;

4) prohibiting the destruction of, or any act which may tend to the destruction of or cause injury to, any forest produce or forest growth or forest property in any forest reserve or on lands at the disposal of the Government or on native lands or communal lands;

5) (deleted by Order 47 of 1951)

6) regulating the grant and precribing the form that any licences or permits may take in any particular case :
a) to take forest produce in forest reserves or on lands at the disposal of Government, or on native lands or communal lands, and
b) to sell and purchase forest produce;

7) precribing the procedure for fixing and making known to the public the fees to be paid on the application for and the grant of any licence or permit and the royalties and fees to be paid by the holders thereof;

8) prescribing the persons who may declare any specified kind of tree to be a protected tree and any specified kind of minor forest produce to be protected minor forest produce under this Ordinance, and the procedure therefor;

9) (deleted by Order 47 of 1951)

10) providing for the collection, payment and disposal of fees, royalties, tolls and costs of survey and demarcation;

11) providing for the survey and demarcation of forest reserves and forests;

12) regulating the marking of timber and the manufacture, use and possession of marking instruments;

13) regulating the taking, collection and preparation of forest produce;

14) providing for the seizure, detention and disposal of timber or forest produce taken, collected, prepared, sold, purchased, possessed or not marked, in contravention of this Ordinance or in respect of which any of offence against this Ordinance has been committed;

15) requiring the holders of licences and permits to render returns and accounts and to submit their books for inspection;

16) providing for the termination, revocation and forfeiture of licences and permits;

17) regulating the transit of forest produce by land and water and by different means and classes of transport;

18) regulating the salving and disposal of drift timber;

19) prohibiting any act which might cause the obstruction of any waterway or cause danger to navigation;

20) providing for the establishment and maintenance of nurseries and for the afforestation of lands, the preservation and production of forest produce and the introduction of new species of trees or other forest produce;

21) providing for the management, utilization and protection of forest reserves;

21A) providing for the appointment and prescribing the powers and duties of forestry officers and providing for the maintenance of discipline

22) (deleted by Order 47 of 1951)

23) authorising the payment of grants and bonuses out of the public revenue for the encouragement of forestry;

24) the protection of forest produce in forest reserves by :

a) prescribing the time at which and the manner in which the rights reserved or recognized by the reserve settlement officer may be exercised;
b) prohibiting the taking or destruction of any specified kind of timber or minor forest produce in the exercise of such rights as aforesaid; and
c) prohibiting the exercise of all or any of such rights as aforesaid in any specified part of a forest reserve;

25) the control of protected forests and the protection of forest produce in a protected forest mutatis mutandis as if such protected forest were a forest reserve;

26) regulating the kindling of fires for any purpose within a protected forest or a forest reserve and prescribing the persons who may allocate the period during which fire may or may not be allowed for any purpose;

27) providing for the remission or reduction of any royalty, fee or toll charged or payable under the provisions of this Ordinance;

28) generally for giving effect to the objects and purposes of this Ordinance.

Power of Governor-General in Council to make regualtions

46 A.- The Governor-General in Council may make regulations for all or any of the purposes following :

1) prohibiting the export of forest produce or of any specified kind of forest produce by any person other than the holders of licences and permits granted under this Ordinance, or by any other persons or by any classes of persons specified in the regulations;

2) prohibiting the export of forest produce taken, collected or prepared in contravention of this Ordinance;

3) prohibiting or regulating the export of any specified kind of forest produce;

4) regulating the grant and prescribing the form that any licences or permits may take in any particular case to export forest produce;

5)imposing tolls on forest produce or any kind of forest produce conveyed on any inland waterways opened or improved by the Government;

6) (deleted by L.N. 131 of 1951)

47. - The Governor-General may by notice in the Gazette exclude from the operation of any regulations made under section 46A of this Ordinance any area prescribed in such notice and a Governor may by notice in the Regional Gazette exclude from the operation of any regulation made under section 46 of this Ordinance any area within his Region prescribed in such notice.

Power of native authority or local government council to make rules

48. - 1) A native authority or local government council, with the approval of the Governor of the Region, may make rules for any of the purposes prescribed in section 46 for the general protection and management of forests and forest produce, exclusive of Government forest reserves, in the area within its jurisdiction.

2) Subject to any exception specified in such rules, all rules made under this section shall apply to all persons who are subject to the jurisdiction of the native authority or local government council and may be made to apply in whole or in part to such other persons as may be specified in the said rules.

Offences and Legal Procedure

49. - Acts prohibited in a forest reserve (omitted)
50. - Saving in respect of section 49 (omitted)
51. - Offences in protected forests (omitted
52. - Seizure of certain forest produce (omitted)
53. - Offences (omitted)
54. - Arrest of suspected persons (omitted)
55. - Innocent possession (omitted)
56. - Onus of proof (omitted)
57. - Authority to compound offences (omitted)
58. - General penalty (omitted)
59. - Additional penalty (omitted)
60. - Institution and conduct of legal proceedings (omitted)

PART IX : REPEAL AND SAVING

Saving as to existing forest reserves

61. - All forest reserves existing at the time of the coming into force of this Ordinance shall be deemed to have been constituted under and in accordance with the provisions of this Ordinance.

Effect of certain things done under Chapter 95 in the 1923 edition

62. - a) Any notification, appointment or notice made or published; or

b) any inquiry or the decision or judgment or any person holding such inquiry, held or given whether at the inquiry or on appeal,

under the provisions of the Forestry Ordinance specified in the first column in the Schedule shall be deemed to have been made or published, given or held under the provisions of this Ordinance set out in the corresponding line of the second column in the Schedule.

63. - (omitted)

FORESTRY (AMENDMENT) EDICT 1969
OF DECEMBER 12, 1969
(Date of Commencement August 1st 1970)
WESTERN STATE

NIGERIA

Forestry Amendment
Edict 1969

SECTION 4 AND SECTION 18

Section 1 to 3 OMITTED

Section 4. - Insertion of new Part 1A in Cap. 38

The principal Law is hereby amended by inserting immediately after section 3 thereof a new Part as follows :

"PART 1A

ESTABLISHMENT, CONSTITUTION AND FUNCTIONS OF FORESTRY ADVISORY COMMISSION

Establishment and constitution of Forestry Advisory Commission

3A.- 1) There shall be establish a body to be known as the Forestry Advisory Commission, the functions of which shall be as hereinafter prescribed.

2) The Commission shall consist of the following ten members who shall be appointed by the Military Governor :

a) a Chairman, being one appearing to the Military Governor to be an outstanding person in respect of experience or knowledge in forestry matters;
b) two Members, being persons appearing to the Military Governor to represent the interests of the timber trade and industry;
c) two Members (whether or not Obas or Chiefs) being persons appearing to the Military Governor to represent the interests of communal owners of forests in the State;
d) four Members two of whom at least shall be persons appearing to the military Governor to possess special knowledge or experience in forestry matters;
e) the Chief Conservator of Forests as an ex officio Member.

3) The allowances to be paid to the Chairman and the other Members of the Commission, except the ex officio Member, in relation to their functions, shall be at such rate as the Military Governor may prescribe, and different rates of allocances may be prescribed according as the circumstances of the Chairman or other Members differ one from the other.

First Schedule

4) The provisions contained in the First Schedule hereto shall have effect with respect to the constitution and proceedings of the Commission.

5) There shall be assigned by the authority having power so to do, an officer in the public service of the State as Secretary to the Commission who shall discharge secretarial functions for the Commission and also such other functions of an administrative nature as the Chairman of the Commission or the Commission may from time to time direct; and in the discharge

of its functions the Commission shall be assisted from time to time for that purpose by the authority having power so to do.

Functions of the Commission

3B.- The Commission shall have the following functions :

a) advising the Commissioner, the Executive Council or the Military Governor, as the case may be, with respect to -

(i) the formulation of both short-term and long-term policies on planned forestry protection, control and management including the whole range of forestry activities such as the areas to be exploited or to be regenerated from time to time; the manner of such exploitation or regeneration; the establishment of forestry industries and the furtherance of forestry trade;

(ii) the ways and means whereby finances can be raised for the effective implementation of the forestry policies of the Government from time to time;

b) exercising such other powers and discharging such other duties relating to forestry matters as may be conferred or imposed upon the Commisison by the Forestry Law or any other law, or as it may be required to exercise or discharge from time to time by the Commissioner, the Executive Council or the Military Governor. "

Section 5 to 17 OMITTED

Section 18. - Repeal and substitution of Parts V and VI comprising ss.31 to 33 of Cap. 38.

Part V of the principal Law comprising sections 31 and 32 and Part VI comprising section 33 thereof are hereby repealed and the following new Parts substituted therefor respectively :

"PART V
FORESTRY TRUST FUND

Establishment and administration of Forestry Trust Fund

31.- 1) There shall be established a fund to be known as the "Forestry Trust Fund" into which shall be paid such proportion of any such fees received under this Law and in such manner as the Executive Council may from time to time direct.

2) The said fund shall be applied for the purposes of regeneration or afforestation of particular forest reserves or generally for forest regeneration and afforestation in the State, according as the Executive Council may from time to time deem necessary or expedient

3) Subject to the provisions of this section, disbursements from the fund shall be made in accordance with such special or general directions as the Commissioner for Finance, acting after consultation with the Commissioner, may give from time to time.

PART VI

TRANSFER OF LOCAL GOVERNMENT FORESTRY STAFF AND PROPERTY

Provisions with respect to transfer of former forestry staff of councils, and joint boards to the public service of the State

32.- 1) There may be appointed on transfer to the public service of the State, and subject to the agreement of each in his own behalf, all such employees in the forestry service of all local government councils and joint boards immediately before the coming into operation of this section, as the authority having power to make such appointments may consider suitable for the same.

2) Where any employee in the forestry service of any local government council or joint board is not considered suitable for appointment on transfer to the public service of the State, his case shall be dealt with by such council (including a contributing council to the joint board) with a view either to assigning him to an alternative post or bringing his services with the council to an end whether by way of retirement or of termination appointment or otherwise, as the council may consider appropriate in his particular circumstances, due regard being had to the provisions of any existing law and practice relating to such matters.

Transfer of forestry property, etc., of councils and joint board to the Government

33.- 1) Subject to the provisions of this law, the Commissioner, after consultation with the Commissioner for Local Government and Chieftaincy Affairs, may give directions for the transfer to the Government of the forestry property of any local government council or joint board subject to such terms and conditions as to compensation, if any, or otherwise, as the Commissioner may decide.

2) The Commissioner may, after such consultation as aforesaid by the same or subsequent directions make supplemental provisions with respect to administrative and other arrangements as may appear to him necessary for the purpose of giving better effect to the provisions of this Law, and without prejudice to the generality of the foregoing may, by such directions -

a) provide for the transfer to, or for any adjustments with respect to, the assets and liabilities of any local government council;
b) provide for the transfer of any records, documents, or other things relating to functions hitherto discharged by any local government council or joint board, with respect of forestry matters. "

Section 19 to 37 OMITTED

"FIRST SCHEDULE (Section 3A (4))
PROVISIONS RELATING TO THE CONSTITUTION AND PROCEEDINGS OF THE FORESTRY ADVISORY COMMISSION

Tenure of office of members

1. Every Member of the Commission (other than an ex officio member) shall, subject to the provisions of this Schedule, hold office for a period of three years from the date of his appointment.

Resignation

2. Any member of the Commission may, at any time by writing under his hand addressed to the Military Governor, resign his office.

Eligibility for re-appointment

3. A Member of the Commisison who has ceased to be such Member shall be eligible for re-appointment.

Vacation of office

4. If the Military Governor is satisfied that a Member of the Commission -

a) has been absent from two consecutive meetings of the Commission without the permission of the Military Governor in the case of the Chairman, or of the Chairman in the case of any other Member; or
b) has become bankrupt or made an arrangement with his creditors; or
c) is incapacitated by physical or mental illness; or
d) is otherwise unable or unfit to discharge the functions of a Member,

the Military Governor may declare his office as a Member of the Commission to be vacant and shall notify the fact in such manner as he thinks fit, and upon such declaration such Member shall vacate his office.

Vacancies

5. No act or proceedings of the Commission shall be questioned on account of any vacancy among its Members or on account of the appointment of any member having been defective.

Temporary membership

6. Where any member of the Commisison is temporarily incapacitated by illness or is temporarily absent from Nigeria, the Military Governor may appoint any person to hold temporarily the office held by such incapacitated or absent Member during the period of such incapacity or absence and all the powers and duties of such Member under this Law shall devolve upon the person so temporarily appointed.

Co-option of persons

7. Where upon any special occasion the Commission desire to obtain the advice of any person on any particular matter, the Commission may co-opt such a person to be a Member for such meeting or meetings as may be required, and such person whilst so co-opted shall have all rights and privileges of a Member save that he shall not be entitled to vote on any question.

Ordinary and special meetings

8. The Commission shall ordinarily meet at least four times in every financial year at such times and places as its Chairman may direct :

Provided that no less than half of the Members of the Commission, by notice in writing signed by them, may request the Chairman to summon a special meeting of the Commission for such purposes as shall be specified in the notice, and upon receipt of such notice, the Chairman shall, at the earliest convenient date, summon a special meeting for the purposes set out therein.

Quorum

9. One-half of the Members (including the Chairman or other member presiding) shall form a quorum at any meeting of the Commission.

Presiding and voting

10. 1) At every meeting of the Commisison, the Chairman, if present, shall preside, but in his absence the Commission, subject to the provisions of paragraph 6 of this Schedule, may appoint one of its Members present to preside.

2) Subject to the provisions of this law, any question which falls to be determined by the Commission at any of its meetings shall be decided by a majority of the Members present and voting on the same.

3) The Chairman of the commission or any other Member presiding in his absence shall have an original vote, and in the event of equality of votes, a casting vote.

Disqualification of Members from voting etc., on account of interest and penalty

11. 1) If any member or co-opted member of the Commission has any pecuniary interest, direct or indirect, in any contract or proposed contract or other matter and is present at a meeting of the Commission at which the contract or other matter is the subject of consideration, he shall at the meeting disclose the fact and shall not take part in the consideration or discussion of, or vote on any question with respect to, the contract or other matter, and if the person presiding so directs, he shall withdraw from the meeting during such consideration or discussion :

Provided that this section shall not apply to an interest in a contract or other matter which a Member of the Commission may have as an inhabitant of any area.

2) For the purposes of this section and without prejudice to the generality of the provisions of subparagraph (1) hereof, a person shall be regarded as having indirectly a pecuniary interest in a contract or other matter if -

a) He or any nominee of his is a member of a company or other body with which the contract is made or is proposed to be made, or which has a direct pecuniary interest in the contract or other matter under consideration; or
b) he is a partner, or is in the employment, of a person with whom the contract is made or is proposed to be made, or who has a direct pecuniary interest in the contract or other matter under consideration :

Provided that :

(i) this sub-paragraph shall not apply to membership of, or employment under, any public body;

(ii) a member of a company or other body shall not, by reason only of his membership, be treated as being so interested if he has no beneficial interest in any shares or stock of that company or other body.

3)In the case of married persons living together, the interest of one spouse shall, if known to the other, be deemed for the purposes of this paragraph to be also an interest of the other spouse.

4) Any person who contravenes any of the provisions of this paragraph shall be guilty of an offence and shall be liable on conviction to imprisonment for three years.

Standing Orders

12. Subject to the provisions of this Law, the Commission may make Standing orders for the purpose of regulating its own proceedings, and without prejudice to the generality of the foregoing, for the purpose of regulating procedure with regard to the holding of meetings, the proceedings thereat and the keeping of minutes. "

FORESTRY REGULATIONS OF 1943
with consequential amendments
WESTERN REGION

NIGERIA

Forestry
Regulations

STRUCTURE AND LIST OF PROVISIONS OF
THE FOREST REGULATIONS
(Section 48 of the Forestry Law)

FORESTRY REGULATIONS
OD.S.L.N. 5 OF 1981

Forestry Tariffs Review Notice 1980

NIGERIA

Forestry
Regulations
(Fees and Royalties)

FORESTRY TARIFFS REVIEW NOTICE 1980

Date of Commencement : 28th October 1980

In exercise of the powers conferred by regulations 33(1) and 41 of the Forestry Regulations and by virtue of all other powers enabling in that behalf, the following notice is hereby given by the Chief Conservator of Forests with the approval of the Commissioner for Agriculture and Rural Development, Ondo State :

(i) The fees and royalties set out in Schedules "A", "B" and "C" hereto are now applicable to the issue of permits in respect of Forestry Reserves in Ondo State of Nigeria to which the Forestry Regulations apply.

(ii) The Tariff set out in Schedule "A" hereto is now applicable to the issue of permits in respect of protected trees in areas outside forest reserves to which the Forestry Regulations apply.

(iii) The installation and renewal fees set out in Schedule "D" hereto are now applicable to the installations of sawmills and renewal of Sawmill Licenses throughout the State.

2. - This Notice may be cited as Forestry Tariffs Review Notice 1980 and shall be deemed to have come into force on 1st November, 1980.

Given at Akure this 24th of October 1980

CHIEF CONSERVATOR OF FORESTS

Approved this 28th of October 1980

COMMISSIONER FOR AGRICULTURE AND RURAL DEVELOPMENT

SCHEDULE "A"

TARIFF OF STUMPAGE RATES

Serial Nº	Species	Local Name	Minimum girth at breast height in metres	Approved rates per Tree ₦
1.	Chlorophora excelsa	Iroko	2.13	30.00
2.	Periscopis elata	-	1.82	40.00
3.	Entandrophragma species	Ijebo	2.13	30.00
4.	Khaya, all species	Oganwo	2.13	30.00
5.	Tectona grandis	Teak	1.52	30.00
6.	Lovoa trichiliodes	Apopo	2.13	30.00
7.	Guibourtia, all species	-	1.82	30.00
8.	Afzelia, all species	Apa	2.13	25.00
9.	Triplochiton scleroxylon	Arere	1.82	20.00
10.	Nauclea diderrichii	Opepe	2.13	25.00
11.	Guarea, all species	Olofun	2.13	20.00
12.	Gossweilerodendron balsamiferum	Agba	2.13	20.00
13.	Terminalia ivorensis	Idigbo	1.82	25.00
14.	Cordia, all species	Omo	1.82	20.00
15.	Piptadeniastrum africanum	Agboin	2.13	20.00
16.	Nesogordonia papaverifera	Ole, Danta	1.82	20.00
17.	Mimusops, all species	-	2.13	15.00
18.	Terminalia superba	Afara	1.82	25.00
19.	Brachystegia spp.	Eku	2.13	20.00
20	Distemonanthus benthamianus	Ayan	2.13	15.00
21.	Daniellia ogea	Ogea	1.82	10.00
22.	Lophira alata	Ekki	2.13	10.00
23.	Pterygota, all species	Oporoporo	1.82	20.00
24.	Mansonia altissima	Ofun	1.82	10.00
25.	Antiaris africana	Oriro	2.13	20.00
26.	Antrocaryon poluneurum	-	1.82	10.00
27.	Canarium schweinfurthii	-	2.13	10.00
28.	Cylicodiscus gabunensis	-	2.13	5.00
29.	Mitragyna stipulosa	Abura	1.52	10.00
30.	Alstonia, all species	Ahun	1.82	6.00
31.	Berlinia, all species	-	1.82	6.00
32.	Bombax, all species	-	2.13	8.00
33.	Ceiba pentandra	Araba	2.13	8.00
34.	Casearia, all species	-	1.82	5.00
35.	Erythrophyleum spp.	Eru	1.82	5.00
36.	Poga oleosa	-	2.13	5.00
37.	Ricinodendron heudelotii	Erinmade	1.82	5.00
38.	Sterculia oblonga	Aye, Koko Igbo	1.52	10.00
39.	Albizia, all spp.	Ayunre	1.82	10.00
40.	Others	-	-	5.00

SCHEDULE "B"

TIMBER FEES BASED ON YIELD

S/N°	Name of Forest Reserve 1/	Approved rate per hectare ₦ 2/
1.	Akure	250
2.	Akure Ofosu	250
3.	Ikere	200
4.	Ogbese	200
5.	Oluwa OA2	250
6.	Idanre	250
7.	Owo	250
8.	Oluwa OA3	250
9.	Ifon/Okeluse	200
10.	Ise	250
11.	Ala North and South	200
12.	Oluwa OA1	250
13.	Little Ose	200
14.	Aramoko	200
15.	Otu	200
16.	Onishere	200
17.	Irele	200
18.	Ipele/Idoani	200
19	Ogotun Series	200

Note :

1/ The rates are in addition to the Afforestation levy (i.e. funds contributed towards replacement of Forests now being cut over) which is currently fixed at ₦ 2,000.00 per square kilometre or ₦ 20 per hectare.

2/ Salvaging of area felled previously may be charged at 25% rebate of the rate shown for particular reserve.

SCHEDULE "C"

TARIFF OF OUT-TURN VOLUME RATES

Species	Local	Current rate per cubic metre (M^3)	Rate per cubic foot conversion factor (0.028)
		₦	k
Tectona grandis	Teak	32.00	90
Entrandrophragma cylindricum	Ijebo	15.00	42
" utile	Ijebo	15.00	42
Khaya, all species	Oganwo	15.00	42
Periscopsis elata	-	15.00	42
Lovoa trichilioides	Apopo	15.00	42
Guibourtia species	-	15.00	42
Triplochiton scleroxylon	Arere	15.00	42
Entandrophragma all species except utile and cylindricum	Ijebo	25.00	42
Mimusops, all species	-	15.00	42
Chlorophora excelsa	Iroko	15.00	42
Mansonia altissima	Ofun	15.00	42
Afzelia all species	Apa	15.00	42
Nauclea diderrichii	Opepe	15.00	42
Guarea all species	Olofun	15.00	42
Nesogordonia papaverifera	Ole, Danta	15.00	42
Cordia all species	Omo	15.00	42
Terminalia ivorensis	Idigbo	15.00	42
Diospyros mesiformis	-	15.00	42
Piptadeniastrum africanum	Agboin	15.00	42
Gossweilerodendron balsamiferum	Agba	15.00	42
Terminalia superba	Afara	15.00	42
Antiaris africana	Oriro	12.00	34
Brachystegia all species	Eku	12.00	34
Lophira alata	Ekki	15.00	42
Mitragyna all species	Abura	15.00	42
Sterculia oblonga	Ayo, Koko Igbo	15.00	42
Erythrophyleum all species	Eru	15.00	42
Sterculia all species	Aye	15.00	42
Pterygota	Oporopo	10.00	28
Daniellia ogea	Ogea	15.00	42
Berlinia all species	-	14.00	39
Canarium schweinfurthii	-	10.00	28
Cylicodiscus gabunensis	-	10.00	28
Other	-	10.00	28

SCHEDULE "D"

SAWMILLS LICENCES FEE (O m i t t e d)

STRUCTURE AND LIST OF PROVISIONS OF THE PROPOSED FOREST ORDINANCE 1982

PART V : PROTECTION OF FORESTS AND FOREST PRODUCE

Sec. 15. - Restrictions and Prohibitions within State Forests
Sec. 15A.- Incorporation of Director of Beekeeping
Sec. 16. - Public to assist in protection state forests from fire
Sec. 17. - National Trees
Sec. 18. - Restrictions over the use of trees
Sec. 18A.- Restrictions over the movement of Forest Produce
Sec. 18B.- Assessment of Natural Forest Resources

PART VI : LICENCES

Sec. 19. - Grant of licences
Sec. 19A.- Prohibition of Forest Officers to have timber dealings
Sec. 19B.- Restrictions over the felling of young trees (undersize)

PART VII : POWERS OF OFFICERS

Sec. 20. - Power to demand licence, seizure and arrest
Sec. 20A.- Power to enter private land
Sec. 20B.- Power to inspect forest based industries
Sec. 21. - Power to compound offences

PART VIII : MISCELLANEOUS OFFENCES

Sec. 22. - Unlawful possession of forest produce
Sec. 22A.- Contravention of conditions given in a licence
Sec. 23. - Counterfeiting and similar offences
Sec. 24. - Interference with or obstruction of officers of the Forest Division

Part IX : GENERAL PROVISIONS

Sec. 25. - Trees or forest produce presumed to be property of the Government
Sec. 26. - Penalties
Sec. 26A.- Court Order to forfeit property
Sec. 27. - Operation of other laws not barred
Sec. 28. - Saving of common law
Sec. 29. - Power to reward informers
Sec. 29A.- Power of Forest Officers to prosecute
Sec. 29B.- Protection of Forest Officers

PART X : RULES, EXEMPTIONS AND SAVINGS

Sec. 30. - Power to make rules
Sec. 31. - Power to grant exemptions
Sec. 32. - Saving of mining rights and amendment of the Mining Ordinance
Sec. 33. - Repeal of Chapter 132

TANZANIA

Forest Ordinance 1982
- Text -

PART I : PRELIMINARY

Short title and commencement

1. - This Ordinance may be cited as the Forest Ordinance Amendment which came into operation on the ....

Inter-pretation

2. - In this Ordinance unless the context otherwise requires :

"Administrative Officer" includes Ward and Divisional Secretaries.

"Court" means a court having jurisdiction to try a charge of any offence against this Ordinance.

"Director of Forestry" means the person for the time being performing the duties of the Director of Forestry as defined in the executive of the Government.

"District Forest Reserves" - A forest Reserve under district council.

"Disease" includes fungus and other Pathogens.

"Firewood" includes parts of trees made up into bundles or loads, or cut wood for burning, and refuse wood generally, but does not include logs or poles.

"Forest Officer" means any Officer of the Forest Division of or above the ranks of forest attendants and includes an honorary forest officer appointed under section 4.

" Forest produce" includes :

(i) trees, timber, firewood, charcoal, sawdust, withies, bark, bast, roots, fibres, resins, gums, latex, sap, galls, leaves, fruits, and seeds; and

(ii) within State Forest and Local Authority Forest only, vegetation of any kind, litter, soil, sand, rock, peat, honey, wax and wild silk; and

(iii) such other things as the Minister may from time to time by noticepublished in the Gazette declare forest produce, either generally or within any forest reserve.

"Leasehold land" does not include land held under a mining claim granted under the Mining Ordinance.

"Licence" means a valid licence granted by the Director of Forestry or any person duly authorized by him in that behalf or by a local authority, under section 19.

"Livestock" includes cattle, sheep, goats, pigs, horses, donkeys, mules and all other domesticated animals and their young.

"Local Authority" includes a District Council, City Council, Municipal Council, Town Council, Local Council and Village Council.

"Local Authority Forest Reserve" means an area declared to be a local authority forest under section 5.

"Log" means the stem of a tree or a length of stem or branch after felling, cross-cuttings and trimming, but does not include a pole.

"Owner" in relation to any land means any person having a freehold, leasehold estate (or any interest deemed under any law for the time being in force to be a freehold or leasehold estate) in, or a right of occupancy in respect of such land.

"Permit" means a permit in writing issued by the Director of Forestry under section 8.

"Provisional Forest Reserve" is an intermediary forest reserve or Reclaimation areas.

"Pole" means a tree or part of a tree of suitable size for use in the round as a telegraph, telephone, power transmission or building pole or for similar purposes.

"Police Officer" means a police officer as defined in section 2 of the Police Force Ordinance.

"Public highway" means a public highway as defined in the Public Highway Ordinance.

"National tree" means any tree declared by order made under section 17 to be a National tree.

"Senior Forest Officer" means any officer of the Forest Division of or above the rank of Assistant Forest Officer III.

"State Forest Reserve" means an area of land declared to be a state forest reserve under section 5.

"Timber" means any tree or part of a tree which has fallen or been felled or cut off, and all wood, whether or not sawn, split, hewn or otherwise cut up or fashioned, but does not include firewood or poles.

"Tree" includes palms, bamboos, canes, shrubs, bushes, plants, poles, climbers, seedlings, saplings, and the regrowth thereof of all ages and all kinds and any part thereof.

"Unreserved Forest Land" means forest land not situated within a state forest reserve or Local Authority Forest Reserve which is not freehold or leasehold land (or not deemed to be freehold or leasehold land under any law for the time being in force) or land occupied under a right of occupancy granted under the provisions of section 6 of the Land Ordinance.

"Village Forest Reserve" means forest owned by a village council.

"Forest Land" means area of land covered with trees, grass, and other vegetation including bare land.

"Provisional State Forest Reserve" means an area of land declared to be a Provisional State Forest reserve under section 5 (1A)

"Forest" means all land bearing a vegetative association dominated by trees of any size, exploitable or not, capable of producing wood or other products, of exercing an influence on the climate or on the water regime or providing shelter for life-stock and wildlife.

Administration of Ordinance

3. - The Director of Forestry and the officers and staff of the Forest Division shall be responsible for the administration of this Ordinance.

Honorary forest officers

4. - The Minister may by notice published in the Gazette appoint any person he deems fit to be an honorary forest officer for the purposes of this Ordinance. Any such appointment shall be for such period as may be specified in the said notice. The Minister may in his discretion at any time revoke such appointment.

PART II : CREATION OF STATE FOREST RESERVES AND LOCAL AUTHORITY FOREST RESERVES

Declaration of State Forest Reserve

5. - 1) Subject to the provisions of section 6, the Minister may by order published in the Gazette declare any area of state controlled forest reserve or provisional state forest reserve to be a State Forest Reserve as from the date and may at any time vary or revoke such order.

Declaration of provisional State Forest Reserve

1A) The Director of Forestry may by order published in the Gazette declare any maltreated land or land liable to deterioration by maltreatment to be a Provisional State Forest Reserve for the purpose of land improvement or land reclaimation or forest development.

Authority for exchange of State Forest Reserve

1B) The Minister may authorize the exchange of any part of a State Forest Reserve with any other land and may receive or pay money for assets on the land to equalize the exchange, and the land so exchanged shall be deemed to have been declared a State Forest Reserve while the exchanged part of the State Forest Reserve shall thereafter cease to be a part of the State Forest Reserve.

Declaration of L.A. Forest Reserve to become a State Forest Reserve

1C) The Minister may by order published in the Gazette declare any Local Authority Forest Reserve to be a State Forest when in his opinion such Local Authority Forest has been abandoned by a dissolution of the Authority and no other Authority or corporate body formed to take over the forest as one of its assets and such forest shall upon such declaration cease to be a Local Authority Forest.

Amalgamation and fragmentation of State Forest Reserve

1D) The Minister may by order published in the Gazette amalgamate State Forest Reserves into one or divide a State Forest Reserve into two or more State Forest Reserves or assign or change the Area of State Forest Reserve.

Demarcation of State Forest Reserve boundaries

2) As soon as practicable after the publication of the order made under the provisions of subsections (1), (1A), (1B), (1C) and (1D), the Director of Forestry shall cause the boundaries of such forest to be visibly demarcated on the ground.

Creation of Local Authority Forest Reserve

Declaration Local Authority Forest Reserve

5.A- 1) Local Authorities may in consultation with the Director of Forestry declare by order published in the Gazette any area in state controlled Forest Reserves or any area within their jurisdiction to be a Local Authority Forest Reserve as from the date specified in the order and may at any time vary or revoke such order provided that :

a) A map of the area has been certified by the District Land Officer;

b) The District Land Officer in the area has ascertained that no other rights are claimed over the area under consideration; or

c) Where there are such rights being claimed, these rights are dealt with in accordance with the law of the Country.

2) Except as herein before provided nothing in subsection (1) of this section shall be deemed to transfer to or vest in a Local Authority any right, title or interest what so ever in or over any land declared to be a Local Authority Forest.

3) As soon as practicable after the publication of the order made under the provisions of subsection (1) of this section, the Local Authority concerned shall cause the boundaries of such forest to be visibly demarcated on the ground.

Mapping of State & L. A. Forest Reserves

5.B- Maps of all state Forests, provisional Forests and Local Authority Forests must be certified and registered by the Director of Surveys and shall be deposited in the Head Office of the Forest Division before the declaration of such forests.

Restrictions on creation of new rights in the State & L.A. Forest Reserves

5.C- Any State Forest Reserve declared under this Ordinance shall be deemed to be permanend and save, and except as shall be detailed elsewhere in this Ordinance, such State Forest Reserve and forest produce contained therein shall not be dealt with except in conformity with good forestry.

Requirement for declaration of area as Forest Reserve

6. - 1) Before the Minister makes any order under section 5. declaring any area of land to be or form part of a forest reserve he shall :

a) Ascertain that the Director of Forestry has given not less than ninety days notice in writing of the proposed declaration of the said area as a forest reserve, which notice shall describe the proposed boundaries of the forest reserve and that such notice has been published in the Gazette and exhibited at the office of every Area Commissioner within whose district any part of the said area is situated, and in such other manner as may be customary in the area concerned ;

b) Take into consideration any grounds of objection that may be notified in accordance with subsection (2);

c) Satisfy himself :

(i) That all claims to rights in relation to land or forest produce notified in accordance with subsection (2) have been investigated and determined in accordance with the provision of subsection (4)

(ii) that all rights so claimed which have been determined to be lawfully exercisable by any person or group of persons within the said area have been recorded in accordance with the provisions of section 7 or have been voluntarily surrendered,

(iii) that in the case of voluntary surrender such compensation (if any) as may be attributable to the loss of the said rights has been assessed in accordance with the provisions of subsection (9) and has been or will be duly paid.

SUB-SECTIONS 2 to 11 : OMITTED

" This part of Section 6. deals in considerable detail with the legal and administrative procedures for constituting forest reserves. It covers in particular the following points :

2) Notification of objections and claimed rights

3) Report to the Minister

4) nvestigation on stated claims to rights

5) Rights arising subsequent to public announcement of proposed declaration

6) Appeal to a Court of a Resident Magistrate

7) Abolition of unclaimed rights

8) urrender of rights

9) Payment of compensation for surrendered rights

10) Payments from governmental funds and/or funds of Local Authorities

11) Refund of payments made by Local Authorities "

Recording of rights in relation to land or forest produce

7. - 1) Any rights in relation to land or forest produce, which have been determined under section 6 to be lawfully exercisable within any area declared to be a forest reserve, shall if they are not voluntarily surrendered be recorded within such time and in such manner as may be prescribed and the Director of Forestry shall subject to the provisions of section 8. , permit the exercise of such rights.

2) A copy of any record made as aforesaid concerning rights in relation to land or forest produce which is certified by or on behalf of the Director of Forestry, or such other person or officer as the Minister may appoint for the purpose, as the case may be, shall be prima facie evidence for all purposes of the possession of such rights as may be therein set forth by such person or group of person respectively shown therein as possessing such rights.

Restriction on creation of new rights in area of intended Reserve

9. - OMITTED

"This section determines in detail the rights prior to the public announcement of the proposed declaration as well as the restrictions on the creation of new rights after the public announcement. "

PART III : MANAGEMENT OF LOCAL AUTHORITY FORESTS

Management and Control of Local Authority Forests Reserve

10. - Each Local Authority shall be responsible for the maintenance, control and management of all the Local Authority Forests under its jurisdiction, meeting the costs of management, and any revenue derived from fees for forest produce charged or issued in respect of such Local Authority Forest Reserve shall form part of the revenue of the Local Authority.

Manager of Local Authority Forest

11. - Each Local Authority shall appoint a forest officer to manage the Local Authority Forest, declared by the Local Authority, under the direction of the Local Authority.

Control of Director of Forestry over Management of L.A. Forest

12. - 1) A Local Authority Forest shall be managed by the Local Authority owning it in accordance with the advice of Director of Forestry.

2) The Director of Forestry shall be entitled to make such written representations as he thinks fit to the Local Authority concerned regarding the management of a Local Authority Forest and shall be entitled upon making a written request to such effect to appear before the Local Authority personally or by his representative for the purpose of making such representations orally.

3) If the Minister after considering a report from the Director of Forestry is satisfied that owing to miss-management of any Local Authority Forest by the Local Authority concerned, it is in the public interest so that such Local Authority Forest shall cease to be managed by such Local Authority; he may by order published in the Gazette direct that such Local Authority Forest be managed by the Director of Forestry and thereupon the Director of Forestry shall exercise all and any powers conferred on the Local Authority under this Ordinance and such Local Authority shall cease to exercise such powers.

4) The Director of Forestry shall manage any Local Authority Forest which he is directed to manage according to provisions of subsection (3) on behalf of and for the benefit of the Local Authority concerned and the net profits of management (if any) shall after deduction of the costs of management and development, be deemed to be part of the revenue of the said Local Authority, which shall likewise bear any loss incurred.

5) For the purpose of the preceeding subsection the net profits of management and development shall be such sums as the Director of Forestry, with the prior approval of the Minister, shall notify in writing to the Local Authority.

6) The Director of Forestry shall have access to plans and reports drawn for the management of the Local Authority Forests as he may from time to time require.

Cancellation of declaration in respect of L.A. Forest Reserve

13. - No Local Authority Forest shall be revoked or varied after declaration save and except where a written approval of the Director of Forestry has been given.

PART III A : MANAGEMENT OF STATE FOREST RESERVES

Preparation of Management and Working Plans

10.A - 1) The Director of Forestry shall cause the preparation and implementation of working or management plans to manage, develop, establish or otherwise use any State Forest Reserve or Provisional Forest Reserve to achieve the objective or objectives detailed in the declaration or other-otherwise stated by the Director of Forestry in furtherance of sound forestry.

2) The Director of Forestry may cause improvement of any State Forest by constructing or improving houses or buildings or roads bridges or any other permanent or temporary structure for use consistant with sound management of the State Forest.

Regulations of leasing of part of a State Forest for construction work

11.A - 1) The Director of Forestry may in the cause of managing any State Forest, lease a part or parts of the State Forest to a person or a corporate body a site or sites for a factory or factories and buildings and houses related to the factory or factories or other buildings or houses if in his opinion such lease will be to the interests and furtherance of Forestry provided that :

(i) The lease of the site shall be consistent with the requirements of the Land Ordinance;

(ii) The lease holder of the site or sites in a State Forest shall abide to such terms and such conditions as the Director of Forestry may detail and from time to time amend, provided that such terms and conditions are not in consistent with the interests of the factory or factories or the purposes of erecting such buildings or houses or structures

2) The Director of Forestry may lease, part or parts of a State Forest Reserve for recreational or other purposes which in his opinion are of public interest, provided that such lease shall be in accordance with the Land Ordinance, and shall not be deemed to have excised the part or parts of the State Forest Reserve in question.

PART IV : FORESTRY DEDICATION CONVENANTS

Forestry dedication convenants

14. - 1) The Director of Forestry may enter into a convenant with any person or institutions owning land to the effect that such land or any part thereof shall not be used otherwise than for the growing, in accordance with rules of practice of good Forestry, of trees for the commercial production of forest produce or water conservation.

2) Where the owner of any land enters into a convenant with Director of Forestry in Accordance with subsection (1) of this section, such convenant shall, subject to the provisions of subsection (1), be enforceable against the convenator and, subject to the intention expressed in such covenants, against his successors in title and all persons deriving title under him or them.

3) As respects the enforcement of any such convenant the Director of Forestry shall have the like rights, as if he has at all material time been the absolute owner in possession of ascertained land adjacent to the land, in respect of which the convenant is sought to be enforced and capable of being benefited by the convenant and the convenant had been expressed to be for the benefit of that adjacent land.

4) Nothing in this section shall render enforceable any convenant entered into under subsection (2) where the use of such land in accordance with such convenant contravenes the provisions of any laws for the time being inforce or inconsistent with any other prior convenant relating the use of such land and binding on such owner and his successors in title and persons deriving title under him or them.

5) The Minister may subject to the approval of Treasury give by way of grants under terms and such conditions as will be agreed upon, funds to colleges and other public institutions to establish and manage the land if he is satisfied that such grant is of public interest.

PART V : PROTECTION OF FORESTS AND FOREST PRODUCE

Restrictions

15. - 1) <u>OMITTED</u>
"This subsection determines the activities in State Forest Reserves, which constitute an offence if undertaken without licence or other lawful authority."

2) If any person, without lawful excuse, the burden of proof of which shall be on him, within, or in the vicinity of, any State Forest has in his possession any implement for cutting, taking, working or rendering any forest produce, he shall be guilty of an offence against this Ordinance.

3) The Director of Forestry may by notice published in the Gazette exempt the whole or any part of any State Forest Reserve from the application of the provisions of any of the paragraphs of subsection (1), or of subsection (2), subject to such conditions and limitations as he may think fit.

4) If any livestock are found grazing, or depastured in, or entering any State Forest such livestock shall be presumed, unless the contrary is shown, to have been grazed, depastured or allowed to enter by the authority of the owner and of the person, if any, actually in charge of such livestock.

Incorporation of Director of Beekeeping

15.A - Notwithstanding the provisions of this section, the Director of Beekeeping may in consultation with the Director of Forestry make Beekeeping Development Laws to be used within State Forest provided that nothing in the said law shall be prejudicial to good forestry.

Public to assist in protecting State Forests from fire

16. - 1) It shall be lawfull for any Forest Officer, Beekeeping Officer, Game Officer, Police Officer, Administrative Officer, or any Local Authority, Fire Brigade Officer to require any person who is within a reasonable distance of any State Forest to assist in averting or extinguishing any fire in or threatening to enter or effect such State Forest, or in securing any property within the State Forest from loss or damage arising from fire:

Provided that no such person shall be required to do anything which may reasonably be expected to expose him to the risk of death or serious injury.

2) Any person other than a Govt. Officer or an employee of Institution or an employee of Local Authority who has been required by one of persons mentioned in subsection (1) to assist in averting or extinguishing any fire or in securing any property from loss or damage by fire, may receive payment for such rates as a Senior Forest Officer may determine regard being had in the amount of work done by the person in averting or extinguishing the fire or securing the property from loss or damage by the fire.

National Trees

17. - The Minister may by Order published in the Gazette declare any tree or class of trees to be National Trees.

Restrictions over the use of trees

18. - 1) Any person who without a licence or other lawfull authority fells, cuts, damages, or removes any national Tree in any state controlled forest, lease hold or free hold land, shall be guilty of an offence against this Ordinance.

2) Any person who without a licence or other lawful authority fells, cuts damages or removes any tree on any state controlled forest for the commercial undertaking shall be guilty of an offence against this Ordinance.

3) Any person who without a licence or other lawful authority sets fire in state controlled forests shall be guilty of an offence against this Ordinance.

Restrictions over the movement of forest produce

18.A - 1) The Director of Forestry may by order published in the Gazette restrict the movement of any trees or forest produce or forest insect from one part of country to another or the importation of any trees or forest produce or insects if in his opinion such movement or importation will spread or introduce a disease or diseases disastrous to forests.

2) Notwithstanding the provisions of subsection (1) of this section the Director of Forestry may require by order published in the Gazette all imported trees or imported forest produce to be subject to state quarantine regulations.

3) Any person who fails to observe the provision of subsection (1) and (2) of this section shall be guilty of an offence against this Ordinance.

Assessment of Natural Forest Resources

18.B - The Director of Forestry shall assess Natural Forest Resources, especially in water catchment areas on private owned land to determine the extent and contents and shall thereafter prepare a plan to be implemented by the owner to use the forest for public interest.

PART VI : LICENCES

Grant of licences

19. - 1) The Director of Forestry or any person authorised by him in that behalf may grant licences for all or any of the purposes of this Ordinance. Every such licence shall be subject to such conditions as may be specified therein, and there shall be payable in respect thereof such fee as may be prescribed.

2) A Local Authority may grant licences for all or any of the purposes of this Ordinance in respect of any Local Authority Forest which it maintains or controls. Every such licence shall be subject to such conditions as may be specified therein, and there shall be payable in respect thereof such fee as may be prescribed.

3) The Director of Forestry, or any person authorised by him in that behalf, or a Local Authority as the case may be, may at any time cancel or suspend any licence granted by or on behalf of the Director of Forestry or such Local Authority.

3 A) The Director of Forestry or any person authorised by him in that behalf or a Local Authority as the case may be, at any time refuse to grant a licence to a person or persons whom he has evidence of that he has repeatedly committed offences against this Ordinance or has repeatedly infringed conditions of licences granted to him.

Prohibitions of Forest Officers to have timber dealings

19.A - Forest Officers are prohibited by this section to have licences issued under this Ordinance, to have shares or partnership with a person or private corporations operating business or industry on the bases of licence or licences issued under this Ordinance except where such a licence is granted for forest produce for domestic use.

19.B - Any tree felling licence issued under this Ordinance shall not authorise the holder to remove trees which have not reached the minimum breast height diameter stated for that tree or group of trees except where it is stated so in the licence.

PART VII : POWERS OF OFFICERS

Power to demand licence or authority for acts done; seizure and arrest

20. - OMITTED

"This section determines the powers of administrative officers, forest officers, game officers, fisheries and police officers."

Power to enter private land

20.A - The director of Forestry or any person duly authorised by him may enter private lands to assess the forest thereon or to investigate or check any forest disease or any forest phenomenon which he considers to be curious provided that the owner shall be notified in writing of the intentions and date to enter the land and provided further that due respect shall be paid to the property of the owner of the land.

Power to inspect forest based industries

20.B - A Senior Forest Officer may enter the premises of any forest based industry or forest produce dealers to inspect any forest produce placed or found within the premises to satisfy himself that the dealers abide to the provisions of this Ordinance.

Power to compound offences

21. - OMITTED

"This section determines the power of Senior Forest Officers to compound forest offences under certain specific conditions."

PART VIII : MISCELLANEOUS OFFENCES

Unlawful possession of forest produce

22. - OMITTED

Contravention of conditions given in a licence

22.A - OMITTED

Counter feiting and similar offences

23. - OMITTED

Interference or obstruction of officers, etc. of the Forest Division

24. - OMITTED

PART IX : GENERAL PROVISIONS - STATE CONTROLLED FOREST

Trees or forest produce presumed to be property of the Government

25. - When in any proceedings under this Ordinance a question arises as to whether any tree or forest produce is the property of the Government or of a Local Authority, or whether any land State Controlled Forest, such tree or forest produce shall be presumed to be the property of the Government or of the Local Authority, as the case may be, and such forest shall be deemed to be State Controlled Forest until the contrary is proved.

Penalties

26. - OMITTED

"This section determines a general penalty; forfeiture of a licence; payment of compensation for damages; removal of unauthorised installations; and disposal of forfeited forest produce."

Court order to forfeit property

26.A - OMITTED

Operation of other laws not barred

27. - OMITTED

Saving of common law rights

28. - OMITTED

Power to reward informers — 29. - OMITTED

Powers of Forest Officers to prosecute — 29.A - OMITTED

Protection of Forest Officers — 29.B - OMITTED

PART X : RULES, EXEMPTIONS AND SAVING

Power to make rules — 30. - 1) The Minister may from time to time make rules either of general application or in respect of any particular State Forest or in respect of any forest produce for any or all of the following purposes :

a) Regulating the felling, working and removal of forest produce;
b) Prescribing any areas of State Controlled Forest in which all or any forest produce may or may not be cut or removed;
c) Prohibiting or regulating the use and occupation of land in State Forests for residential, cultivation, commercial or industrial purpose or grazing;
d) Prohibiting or regulating the use of land in State Forest for camping or any other purpose of such nature;
e) Prescribing the time and manner of recording and publicising rights in relation to land or forest produce within State Forest;
f) Regulating the manner and circumstances in which licences or permits may be applied for, granted, varied, refused or c ncelled, providing for the conditions and terms subject to which they may be granted, prescribing the fees payable for any licence or authority and providing for the exemption of any person from payment of such fees and any conditions and limitations relating to exemption;
g) Regulating the sale and disposal forest produce by tender public auction, private treaty or otherwise, and matters incidental thereto, fixing the price of forest produce by assessment within prescribed limits by specified persons or otherwise, prescribing the fees to be paid for the cutting or removal of forest produce and providing for the remission of all or any part of any such fee, either generally or in individual cases;
h) Prohibiting or regulating any act liable to cause damage to forest or forest produce;

i) Prohibiting or controlling the entry of persons, animals or vehicles into any State Forest or part thereof, and regulating the period during which such persons, animals or vehicles may remain therein and providing for the conditions subject to which they may do so.
j) Providing for the declaration of insect and fungal pests dangerous to forest produce and prescribing measures to be taken to control or eradicate such notified pests;
k) Prescribing the names to be applied to forest produce in order to promote its better utilization and marketing and providing for the manner in which any list of names made hereunder may be from time to time amended or varied;
l) Providing for the compulsory use of property marks by timber dealers licenced to take timber under this Ordinance or any rules made thereunder, and the registration of such marks;
m) Providing for the compulsory use of property marks by Local Authorities and owners of private woodland for the purpose of dentifying timber sold from Local Authority forests and private woodland;
n) Providing for the prohibition of the use of marks not registered under the provisions of rules made under this section;
o) Prohibiting or regulating the use of roads other than public highways within State Forests and providing for the repair of roads, tracks or bridges in a forest reserve by any person damaging the same;
p) Prohibiting or regulating within State Forests the lighting of fires, smoking, or the carrying, kindling or throwing of any fire or light or inflammable material
q) Providing for the registration of stamps and marks for use by the Forest Division for marking forest produce or indicating State Forests;
r) Providing for the registration and use of such Forest Division brands, tags, or other devices for marking livestock as may be necessary to identify livestock licenced to graze in State Forests;
s) Prohibiting or regulating the export from the Republic or from any area of the Republic of forest produce;
t) Prescribing the form of forestry dedication convenants;
t1) Prohibiting or regulating the establishment of forest based industries;
t2) Prohibiting or regulating the use of maltreated land or land liable to deterioration;
u) Providing generally for the carrying out of the purposes and provisions of this Ordinance prescribing anything which may be provided for this Ordinance.

2) Any rule made under the provisions of this section may require acts or things to be performed or done to the satisfaction of a prescribed person, and may empower a prescribed person to issue orders to be performed or done, imposing conditions and prescribing periods and dates upon,within or before which, such acts or things shall be performed or done or such conditions shall be fulfilled.

3) A Local Authority may, with the approval of the Minister for Local Government and Administration, make rules applicable to any Local Authority Forest which it maintains and controls prescribing for such forest any or all of the matters which the Minister may prescribe or regulate under the provisions of paragraph (a), (c), (d), (f), (g), (h), (i), (o), and (p) or subsection (t) and specifying the officers who may act on its behalf in administering the provisions of this Ordinance.

4) The Minister or a Local Authority may in making a rule under this section prescribe for a breach thereof a fine not exceeding three thousands shillings or imprisonment for a term not exceeding six months, or both such fine and such imprisonment.

Power to grant exemptions

31. - The Minister may, by notice published in the Gazette exempt any person or class of persons or any land or class of lands from any or all of the provisions of this Ordinance or any rules orders or notices made thereunder, subject to such conditions and limitations as may be specified in such notice.

Saving of mining rights and amendment of the Mining Ordinance

32. - 1) OMITTED

Right of leasee and claim holder to take timber

2)

(i) A leasee may, on the lands included within the area of his lease, cut, take and use any tree when necessary in the course of mining operations or when required for mining or domestic purposes :

Provided that he shall be liable for any fees and royalties which may be payable under the law relating to forestry;

(ii) A claim holder may on the lands included within the area of this claim, cut, take and use any tree when necessary in the course of mining operations or when required for mining or domestic purposes : Provided that he shall not cut, take or use :

a) Any tree within a State Forest; or
b) Any tree situated outside a State Forest which has been declared under section 17 of the Forest Ordinance, to be a National Tree,

Unless licensed in that behalf under the provisions of that Ordinance; and provided further that he shall be liable for any fees and royalties which may be payable under that Ordinance.

Repeal of CAP. 132

33. - The Forest Ordinance is hereby repealed : Provided that, notwithstanding such repeal :

a) (i) All forest reserves other than Local Authority Forest Reserves existing at the time of commencement of this Ordinance shall be deemed to have been declared as National Forest Reserves under the provisions of this Ordinance;

(ii) All Local Authority Forest Reserves existing at the time of commencement of this Ordinance shall be deemed to have been declared as Local Authority Forest Reserves under the provisions of this Ordinance;

(iii) All Local Authorities who have been nominated under the provisions of the Ordinance so repealed to exercise control over such Local Authority Forest, shall be deemed to be responsible, respectively to have been declared under the provisions of this Ordinance to be responsible for the maintenance and control of the same as Local Authority Forests Reserve;

(iv) All licences granted under the provisions of the Ordinance so repealed shall be deemed to have been made or granted under this Ordinance;

(v) All the rules made under the provisions of the Ordinance so repealed which are specified in the Schedule to this Ordinance shall, to the extent that the same remain in force at the commencement of this Ordinance, be deemed to have been made under the provisions of this Ordinance.

b) All declarations, licences and rules as are referred to in paragraph (a) of this provision may be varied or amended in accordance with the provisions of this Ordinance and, if not previously expired, shall remain in force until cancelled or revoked hereunder.

STRUCTURE ET LISTE DES DISPOSITIONS DU DECRET DU 11 AVRIL 1949 SUR LE REGIME FORESTIER

Section 2 : Modes d'Exploitation

Section 3 : De la Licence d'Achat

Section 4 : Reboisement et Régénération des Forêts

Section 5 : Réglementation - Restriction

TITRE C : FORETS PRIVEES

TITRE D : PENALITES

DECRET DU 11 AVRIL 1949 SUR LE REGIME FORESTIER

TITRE A : GENERALITES

1. - Le régime forestier est l'ensemble des règles spéciales régissant l'administration, l'aménagement, l'exploitation, la surveillance et la police des forêts.

Au terme du présent décret, on entend par forêts :

a) Les terrains recouverts d'une formation végétale à base d'arbres et d'arbustes :
 1° capables de produire du bois ou des produits forestiers;
 2° ou exerçant un effet indirect sur le climat, le régime des eaux ou le sol;

b) Les terrains qui étaient recouverts de forêts récemment coupées à blanc ou incendiées, mais qui seront soumis à la régénération naturelle ou reboisées artificiellement.

Par produit forestier, il faut entendre notamment : les écorces, les fruits, le latex, les résines, les gommes et tous autres végétaux ne constituant pas un produit agricole.

Par extension, sont compris dans l'acception du terme forêts, les terrains réservés pour être couverts d'essences ligneuses soit pour la production de bois, soit pour la régénération ou la protection du sol.

TITRE B : DU REGIME FORESTIER

Chapitre I : Des forêts soumises au régime forestier

2. - Sont soumises au régime forestier et administrées conformément aux dispositions du présent décret :

1° les forêts qui font partie du domaine de la Colonie;

2° les forêts indigènes, c'est à dire les terrains recouverts d'une formation végétale spontanée d'arbres ou d'arbustes, sur lesquels un droit d'occupation au profit des indigènes est établi à la suite d'une constatation de l'Administrateur territorial. Des instructions administratives préciseront la procédure de cette constatation;

3° (Omis)

4° Les forêts concédées, sans préjudice des conditions spéciales qui peuvent être stipulées au contrat de concession.

3. - Les forêts faisant l'objet d'un titre de propriété aux termes du décret du 30 juin 1913 ainsi que les boisements effectués par les indigènes et par les non-indigènes ne sont pas soumises au régime forestier. Il appartiendra toutefois aux propriétaires de se conformer à ce qui sera spécifié à leur égard à l'Art.37 du présent décret.

Section 1 : Classification des forêts

4. - Les forêts domaniales et indigènes sont soumises à la gestion directe du Service des Eaux et Forêts; elles sont réparties en deux catégories :
1° les forêts classées;
2° les forêts protégées.

5. - Sont classées :

1° Les réserves forestières de l'Etat, existant à la date de promulgation du présent décret ou qui seront constituées dans l'avenir par ordonnance du Gouverneur Général.

2° Les parties de terrain nu ou insuffisamment boisé dont la protection aura été déclarée nécessaire par ordonnance du Gouverneur Général, notamment :
a) pour leur afforestation ou leur restauration;
b) pour la protection des pentes contre l'érosion;
c) pour la protection des sources et des cours d'eau;
d) pour l'exécution de travaux présentant un caractère d'utilité ou de salubrité publique.

3° Les blocs forestiers dans lesquels les indigènes ont entrepris des travaux d'aménagement.

Les forêts citées aux 1° et 2° constituent le domaine classé de la Colonie; celles citées au 3°, le domaine classé des indigènes.

6. - Sont forêts protégées, toutes les autres forêts, non cédées, n'ayant pas fait l'objet d'une ordonnance de classement.

Section 2 : Aliénation

7. - Les forêts ne pourront être aliénées en totalité ou en partie qu'après déclassement par ordonnance du Gouverneur Général, le Service des Eaux et Forêts entendu.

Chapitre II : Des usages indigènes

Section 1 : Usages coutumiers et exploitation à caractère commercial

8. - Les droits coutumiers, sauf ceux reconnus par les ordonnances de classement, ne peuvent être exercés dans les forêts classées.

9. - Sous réserve des règlements ou mesures concernant l'exercice de ces droits, les indigènes exercent leurs droits coutumiers dans les forêts protégées indigènes ou domaniales.

10. - L'exploitation commerciale par les indigènes des produits forestiers qu'il récoltent selon leurs usages coutumiers est libre dans les forêts protégées domaniales.

Le Gouverneur Général pourra réglementer ou interdire la récolte de tel produit forestier dont il jugera utile de contrôler l'exploitation.

Section 2 : Culture en terrain forestier

11. - A l'exclusion des travaux de sylviculture, les travaux agricoles sont interdits dans les forêts classées. Toutefois, le Gouverneur de Province pourra autoriser des cultures temporaires placées sous le contrôle du Service des Eaux et Forêts qui proposera l'emplacement, la durée et les modalités d'exécution.

Dans les forêts protégées, les cultures sont autorisées; elles pourront être défendues par le Gouverneur de Province là où la rareté, l'état de dégradation ou l'intérêt futur du massif forestier nécessiteront cette mesure.

Chapitre III : Exploitation des bois

Section 1 : Généralités

12. - Dans les forêts protégées domaniales, toute personne peut librement couper ou faire couper le bois de chauffage pour son usage domestique et y ramasser ou y faire ramasser le bois mort destiné au même usage.

Tout indigène de la Colonie, non soumis à l'impôt personnel peut, en outre, couper librement dans les forêts protégées domaniales, et sous réserve des droits coutumiers, dans les forêts protégées indigènes, le bois d'oeuvre nécessaire à ses besoins, à l'exercice de son métier ou de son industrie coutumière; il peut également y ramasser le bois mort ou y couper librement le bois de chauffage pour son usage ou pour la vente.

13. - Les indigènes soumis à l'impôt personnel et les non-indigènes peuvent, dans les forêts protégées domaniales, moyennant un permis qui leur sera délivré gratuitement par l'Administrateur du Territoire ou son délégué, couper, faire couper, ramasser ou faire ramasser le bois nécessaire, soit à la construction de leurs établissements, soit au chauffage des fours à brique destinés à cette construction, soit à la fabrication de leur mobilier.

Le permis indique l'endroit, la durée et les conditions spéciales de l'exploitation, ainsi que la nature et la quantité de bois que le titulaire est autorisé à couper. Cette quantité pourra atteindre trente mètres cubes de bois d'oeuvre et mille stères de bois de chauffage.

Une même personne ne peut obtenir simultanément ou successivement, plusieurs permis valables pour le même territoire, que par décision du Gouverneur de Province.

Le titulaire sera tenu au paiement des redevances proportionnelles frappant l'exploitation, sous le couvert du permis de coupe de bois prévu à l'Art.18 du présent décret, suivant les modalités arrêtées pour le paiement de ces redevances.

14. - L'application des dispositions des Art.12 et 13 peut être limitée ou suspendue par le Gouverneur de Province dans les régions qu'il détermine, si cette application est de nature à y compromettre l'existence des forêts ou de certaines essences.

Toutefois, le Gouverneur de Province pourra déterminer dans ces régions des parties de forêts dans lesquelles les indigènes non soumis à l'impôt personnel pourront couper ou faire couper du bois dans les conditions prévues à l'Art.12.

15. - Dès que l'inventaire des forêts soumises au régime forestier permettra d'en déterminer la POSSIBILITE, le Gouverneur Général fixera par voie d'ordonnance, l'aménagement qui y sera mis en vigueur.

A défaut d'un inventaire complet des peuplements et lorsque l'épuisement prématuré des forêts est à craindre, une POSSIBILITE provisoire pourra être fixée.

Toute dérogation ou modification à un aménagement ne peut être autorisée que par ordonnance du Gouverneur Général sous réserve des droits acquis.

Section 2 : Modes d'exploitation

16. - L'exploitation des forêts soumises au régime forestier ne peut se faire que sous le couvert d'un permis, en régie ou à la suite de vente de coupes par adjudication publique.

17. - Dans les forêts classées et dans les forêts indigènes, l'exploitation ne peut avoir lieu qu'en régie ou à la suite d'achat de coupes en adjudication publique. Toutefois, des dérogations pourront être admises par le Gouverneur Général sous réserve des droits acquis.

Dans les forêts protégées domaniales, l'exploitation par l'indigène soumis à l'impôt personnel et par le non-indigène, pour d'autres usages que ceux prévus à l'alinéa ler de L'Art.12 et de l'Art.13, se fera, en vertu d'un permis de coupe de bois délivré par le Gouverneur de Province.

La Colonie se réserve le droit de recourir à la vente par adjudication ou à l'exploitation en régie, pour tel bloc de forêt protégée spécialement désigné, sous réserve des droits acquis.

1. - Du permis de coupe de bois

18. - Le permis de coupe de bois est délivré pour une durée minimum d'un an. Il peut être renouvelé et le renouvellement ne portera effet qu'à partir du moment où le permis en cours sera périmé.

Le permis de coupe porte sur une superficie maximum de mille hectares. L'étendue accordée sera fonction de la POSSIBILITE de la forêt et des moyens de production dont dispose l'exploitant au moment de la demande.

Sauf les coupes de bois de chauffage, le Gouverneur de Province peut, après avis du Service des Eaux et Forêts, subordonner la délivrance d'un permis sur une superficie égale ou supérieure à 100 hectares, à l'obligation pour l'exploitant de posséder une scierie de capacité définie dans chaque cas particulier ou d'user de tout autre procédé mécanique d'exploitation.

Dans ce cas seulement, il pourra, sur demande du requérant et dans la mesure du possible, lui garantir, pour un maximum de 4 années consécutives, le renouvellement de son permis pour une coupe de superficie égale à la coupe initiale. La reconduction de cette garantie pourra être sollicitée annuellement par l'exploitant.

Pour les coupes de superficie inférieure à 100 hectares, le permis ne donnera lieu qu'à délivrance d'une seule coupe sans garantie de renouvelelemnt.

19. - Le permis de coupe est strictement personnel. Il ne peut être cédé ni transféré sans autorisation préalable du Gouverneur de Province.

20. - Le permis de coupe de bois n'est délivré qu'après examen d'une demande introduite par le requérant auprès du Gouverneur de Province. Cette demande doit contenir un exposé complet du projet d'exploitation et des moyens matériels mis en oeuvre. Elle sera accompagnée d'un plan du terrain sur lequel le permis est sollicité.

Le Gouverneur de Province déterminera, pour chaque permis, l'ordre des coupes auquel sera soumise l'exploitation du bloc délivré, ainsi que les conditions spéciales d'exploitation.

Si l'exploitation cesse avant la date d'expiration du permis, le titulaire de celui-ci est tenu de notifier cette cessation, endéans le mois, au pouvoir concédant. A défaut de ce faire, il sera tenu pour responsable des dégâts occasionnés, même par des tiers, jusqu'à la date d'expiration de son permis.

21. - La délivrance du permis de coupe ou son renouvellement sont subordonnés au paiement d'une taxe qui ne sera pas inférieure à 1 500 francs par 100 hectares et fractions de 100 hectares.

Le détenteur d'un permis de coupe sera tenu, en outre, de payer les redevances proportionnelles qui seront fixées par ordonnance du Gouverneur Général.

22. - Pour le service de tous bateaux à vapeur, le Gouverneur de Province délivre un permis annuel, valable pour toutes les forêts domaniales protégées de la Colonie, donnant le droit de couper, faire couper, ramasser, faire ramasser, et acquérir de quelque personne que ce soit, le bois de chauffage.

Le Gouverneur de Province pourra désigner les forêts ou partie de forêts dans lesquelles la coupe pour le service des bateaux est interdite.

La délivrance du permis est subordonnée au paiement d'une taxe forfaitaire dont le montant sera fixé par ordonnance du Gouverneur Général et qui ne pourra être inférieure aux taux ci-après :
- Bateaux dont la jauge totale est de 50 mètres cubes au moins d'après le certificat de jaugeage : 1 000 francs;
- Bateaux d'une jauge totale de plus de 50 mètres cubes jusqu'à 250 mètres cubes : 3 000 francs;
- Bateaux d'une jauge totale de plus de 250 mètres cubes jusqu'à 500 mètres cubes : 6 000 francs;
- Bateaux d'une jauge totale de plus de 500 mètres cubes jusqu'à 1 000 mètres cubes : 10 000 francs;
- Bateaux d'une jauge totale de plus de 1 000 mètres cubes : 15 000 francs.

Par dérogation aux dispositions qui précèdent, le prix des permis délivrés pour les bateaux servant exclusivement au remorquage ou au touage est établi d'après le nombre de mètres cubes de jauge vide, indiqué au certificat de jaugeage.

Les permis destinés aux bateaux des associations scientifiques, religieuses ou philanthropiques sont délivrés gratuitement, à la condition que celles-ci n'effectuent pas de transport dans un but lucratif.

La durée de validité du permis pour bateaux expire le 31 décembre de l'année de sa délivrance, quelle qu'ait été la date de celle-ci.

23. - Par dérogation aux dispositions des Art.18, 19, 20 et 21, les concessionnaires de mines et les titulaires de permis de traitement, peuvent couper ou faire couper du bois dans les forêts protégées domaniales, pour leurs besoins, suivant des conditions spéciales déterminées par ordonnance du Gouverneur Général.

Ils auront à payer les redevances :
1° sur les bois destinés à l'exploitation des mines, au traitement des minerais et aux installations industrielles;
2° sur le bois destiné au chauffage, à la construction des habitations de leurs employés non-indigènes et travailleurs indigènes, ainsi qu'aux autres usages prévus aux Art.12 et 13 du présent décret.

Les redevances prévues par les alinéas précédents seront déterminés par le Gouverneur Général, sans qu'elle puissent être inférieures aux taux suivants :
1° 20 francs par mètre cube de bois destiné à l'exploitation, au traitement et aux installations industrielles;

2° annuellement 100 francs par employé non-indigène et 30 francs par travailleur indigène au service des employeurs redevables pour le bois destiné au chauffage, à la construction des habitations de ces employés et travailleurs, ainsi qu'aux usages prévus aux Art.12 et 13.

Le montant annuel des redevances prévues aux 1° et 2° ci-dessus pourra être déterminé forfaitairement, d'après la consommation normale des redevables.

24. - (Omis)

25. - Les concessionaires de mines et les titulaires de permis de traitement ou leurs préposés doivent faire, aux époques déterminées par le Gouverneur Général, une déclaration quant à l'importance des bois coupés pour leurs besoins industriels et quant au nombre des employés ou travailleurs à leur service.

2. - De l'adjudication

26. - La mise en adjudication publique d'une coupe de bois est soumise à décision du Gouverneur de Province.

Les coupes à mettre en adjudication publique sont proposées par le Service des Eaux et Forêts qui en effectue l'estimation et fixe la mise à prix.

Les cahiers des charges spéciaux à chaque adjudication sont établis par le Service des Eaux et Forêts et soumis à l'approbation du Gouverneur de Province. Ils spécifient les conditions de l'adjudication ainsi que les règles auxquelles est soumise l'exploitation.

3. - De l'exploitation en régie

27. - L'exploitation en régie par la Colonie, les circonscriptions indigènes ainsi que par les organismes qui se constitueraient conformément à des règles édictées par arrêté, pourra être faite :
1° dans le but de satisfaire les besoins en bois des services de la Colonie;
2° en vue de l'aménagement d'une forêt déterminée.

Dans le premier cas, l'exploitation en régie sera soumise à la décision du Gouverneur de Province. Dans le second cas, celle-ci sera décidée par ordonnance du Gouverneur Général qui en fixera les modalités.

Les redevances proportionnelles à verser par ces régies seront celles auxquelles sont soumis les titulaires de permis de coupe, sauf exceptions admises par le Gouverneur Général.

Section 3 : De la licence d'achat

28. - }
29. - } Abrogés par Ordonnance N° 52/273 du 10.9.1956

Section 4 : Reboisements et régénération des forêts

30. - Quel que soit le titre couvrant l'exploitation forestière, le titulaire de celui-ci sera tenu de payer, indépendamment des redevances proportionnelles, une taxe de reboisement. Cette taxe, dont le montant sera fixé par ordonnance du Gouverneur Général, ne sera pas inférieure à 20% des redevances forestières.

Cette taxe de reboisement sera appliquée également aux bois achetés en vertu de la licence d'achat de bois prévue à l'Art.28. Pour les bois coupés, ramassés ou achetés par les titulaires d'un permis pour bateau, ou coupés par les concessionnaires de mines et les titulaires de permis de traitement, cette taxe pourra être déterminée en fonction de la taxe forfaitaire payée par les redevables.

Les sommes perçues par les Comités par application de la taxe de reboisement seront acquises aux Comités, à charge de les affecter intégralement à un programme de reboisement arrêté par la Colonie et sous le contrôle du Service des Eaux et Forêts.

Section 5 : Réglementation - Restriction

31. - Les taxes et redevances prévues par le présent décret sont payables dans les délais et conditions déterminées par le Gouverneur Général...

Les taxes auxquelle est subordonnée la délivrance du permis pour bateau et de la licence d'achat de bois sont réduites de moitié lorsque ces permis et licences seront délivrés après le 30 juin.

32. - (O.L. du 22 février 1951)
Les revenus de l'exploitation des forêts indigènes seront versés aux propriétaires de la forêt ou à défaut de pouvoir les déterminer, à la caisse administrative de la circonscription indigène dans le ressort de laquelle se trouve la forêt, sous déduction d'une quote-part fixée par ordonnance du Gouverneur, laquelle sera attribuée au Trésor,...
en contre-partie des dépenses occasionnées par la gérance des dites forêts.

Cette disposition sort ses effets le 15 mai 1950.

33. - Le droit de couper ou de ramasser le bois n'est consenti que sous réserve de droits de tiers, indigènes et non-indigènes.

34. - Le Gouverneur Général détermine les règles à suivre dans les coupes de bois

35. - Il peut interdire ou réglementer les coupes des forêts ou des essences forestières qu'il y a lieu de protéger, ordonner les opérations culturales jugées nécessaires, prescrire les mesures pour empêcher la disparition d'arbres servant à la délimitation ou au mesurage des terres et prendre toutes mesures utiles à la préservation ou à la conservation des forêts.

Ces interdictions, restrictions et réglementations, peuvent s'appliquer à tous bois et forêts, quelle que soit la nature du droit qui grève le fonds, à l'exception des boisements et reboisements faits spontanément par le propriétaire ou le concessionnaire.

36. - (Omis)

TITRE C : FORETS PRIVEES

37. - Les propriétaires de forêts, spécifiés à l'Art.3 du présent décret ne peuvent pratiquer le déboisement sur des pentes dont l'inclinaison dépasse 20 degrés, ainsi que dans un rayon de 75 mètres autour des sources sauf autorisation du Gouverneur de Province qui déterminera, éventuellement, sous quelles conditions.

Les exploitations abusives ou toutes autres pratiques qui auraient pour conséquence d'entraîner la destruction de la forêt sont assimilées à des déboisements.

TITRE D : PENALITES

(Art. 38 à 44 omis)

45. - A la date de mise en vigueur du présent décret, le décret du 4 avril 1934 modifié par le décret du 13 juin 1936, sur les coupes de bois dans les forêts domaniales est abrogé.

Toutes dispositions contraires au présent décret sont abrogées.

Les ordonnances et réglements d'exécution des décrets précités restent en vigueur tant qu'ils n'auront pas été remplacés par des dispositions nouvelles rendues en exécution du présent décret, et restent applicables à toutes les forêts soumises au régime forestier établi par le présent décret.

46. - Le présent décret entrera en vigueur à la date qui sera fixée par ordonnance du Gouverneur Général. Par mesure transitoire, les permis et licences en cours à la date de la mise en vigueur du présent décret, continueront à sortir leurs effets jusqu'à la date d'expiration normale de leur validité.

ARRETE INTERDEPARTEMENTAL N° 01059 DU 22 OCTOBRE 1975
Portant réglementation sur l'exportation des grumes

VU la constitution

VU le Décret du 11 Avril 1949, sur le régime forestier

VU l'Ordonnance N° 41/131 du 14 avril 1948, modifiée par les Ordonnances N° 41/130 du 25 mai 1957 et N° 41/2 du 3 janvier 1958;

VU l'Ordonnance N° 52/371 du 28 octobre 1950 modifiée par l'Ordonnance N° 52/207 du 19 juin 1952;

VU l'Ordonnance N° 52/348 du 13 novembre 1949 remplacée par l'Ordonnance N° 52/189 du 29 août 1955;

VU l'Arrêté ministériel N° 08/CAB/MA/68 du 15 janvier 1968 portant nouvelles dispositions en matière d'octroi de permis de coupe de bois;

ATTENDU qu'il y a lieu de favoriser l'équipement pour le développement de l'industrie nationale du bois;

ATTENDU qu'il y a nécessité de protéger notre marché international en grumes exportables;

ARRETENT

Article 1. - La mesure interdisant l'exportation des bois sous forme de grumes est levée.

Article 2. - Le Département de l'Agriculture détermine annuellement pour chaque entreprise forestière, un quota de grumes exportables variant entre 15 et 45% calculé sur base des critères repris sur le tableau en annexe.

Article 3. - Le volume des bois Limba (Terminalia superba), le Wenge (Milletia laurentii) et l'Afrormosia (Afrormosia elata) exportables en grumes ne peut dépasser pour chaque essence 10% de l'ensemble du volume de grumes exportables autorisées.

Article 4. - Le pourcentage total des grumes exportables est calculé par région et par entreprise ou succursales d'entreprise.

Article 5. - Les exploitants forestiers ne disposant pas de matériel de transformation sont autorisés à exporter 600 m^3 de bois grumes par an renouvelables sur base d'un rapport établi par l'office National du Bois.

Toutefois ces exploitants doivent répondre aux conditions reprises à l'Art. 18 du Décret du 11 avril 1949. Les dispositions de l'Art.3 ci-dessus sont aussi applicables au présent article.

Article 6. - L'exportation des essences à promouvoir ci-après ne peut dépasser 5 000 m^3 par an et par entreprise. Il s'agit de Limbali (Gilbertiodendron dewevrei), Tchitola (Oxystigma oxyphyllum), Dibetou (Lovoa trichilioides), Mukulungu (Autranella congolensis).

Article 7. - L'exportation des bois en grumes (avec ou sans aubier) et équarris est soumise aux conditions suivantes : les bois en grumes et les bois équarris doivent être droits, exempts de piqûres, pourriture, échauffement, coeur mou, ainsi que de fentes, de bosses et de noeuds vicieux.

La tolérance de fentes, en additionnant celles des deux bouts, ne pourra dépasser 15% de la longueur de la grume. Par bois équarri, il faut entendre : bois rond transformé en une pièce de section carrée ou rectangulaire et renfermant au centre le coeur de l'arbre.

Article 8. - Toute entreprise forestière qui sera reconnue coupable d'abus aux dispositions du présent Arrêté sera passible selon la gravité du cas des peines suivantes :
- Paiement d'une amende dont le montant sera déterminé par le Département de l'Agriculture;
- Retrait d'autorisation d'exportation du bois;
- Suspension de l'entreprise.

Article 9. - Les autres dispositions prévues au Titre D Article 38 du Régime Forestier de 1955 restent d'application.

Article 10. - La note N° 02623/CAB/AGRI/74 du 23 septembre 1975 du Commissaire d'Etat à l'Agriculture est abrogée.

Article 11. - Le présent Arrêté entre en vigueur à la date de sa signature.

Fait à Kinshasa, le 22 octobre 1975

LE COMMISSAIRE D'ETAT A L'AGRICULTURE

LE COMMISSAIRE D'ETAT AU COMMERCE

ANNEXE

TABLEAU DE CALCUL DU QUOTA DE GRUMES POUR LES EXPORTATIONS

Distance port d'Exportation	% Volume Bois Distance au Port d'Exportation	Implantation Coupe et Usine dans la même Région	Volume du Bois traité l'année précédente	TOTAL % Grumes Exportables
0 à 500 Km	4	5	6	15
501 à 1 000 Km	8	5	12	25
1 001 à 1 500 Km	15	5	15	35
1 501 Km et plus	20	5	20	45

ORDONNANCE N° 244/79 DU 16 OCTOBRE 1979 ZAIRE ZAI/14

fixant les Taux et règles "d'Assiette" et de Recouvrement des Taxes et Redevances en matière Administrative Judiciaire et Domaniale perçues à l'initiative du DEPARTEMENT DE L'ENVIRONNEMENT, CONSERVATION DE LA NATURE ET TOURISME

Ordonnance N°244/79

LE PRESIDENT-FONDATEUR DU MOUVEMENT POPULAIRE DE LA REVOLUTION, PRESIDENT DE LA REPUBLIQUE

VU la Constitution;

VU la Loi N° 78-014 du 11 juillet 1978 portant statut des Agences de Voyage en République du Zaire;

VU la Loi N° 78-015 du 11 juillet 1978 portant statut des établissements hôteliers au Zaire;

VU la Loi N° 79-004 du 11 juillet 1979 portant réglementation de l'assiette, du taux et des modalités de recouvrement des taxes et redevances ou titres de recettes administratives, judiciaires et domaniales;

VU le Décret du 24 avril 1937 sur la Chasse et la Pêche tel que modifié à ce jour;

REVU les Ordonnances N° 76-003 et 76-005 du 8 janvier 1976 fixant les taux et règles d'assiette et de recouvrement des taxes et redevances en matière administrative, judiciaire et domaniale perçues à l'initiative des Départements de l'Agriculture et de l'Economie Nationale;

VU l'Ordonnance N° 76-126 du 16 juillet 1976 modifiant et complétant l'Ordonnance N° 41-48 du 12 février 1953 relative aux Etablissements dangereux, insalubres et incommodes;

VU l'Ordonnance N° 77-023 du 22 février 1977 portant actualisation des taxes et redevances sur l'exploitation forestière au Zaire;

SUR proposition du Commissaire d'Etat à l'Environnement, Conservation de la Nature et Tourisme;
Le Conseil Exécutif entendu,
ORDONNE :

TITRE I : DES GENERALITES

Article 1. - Les recettes perçues à l'initiative du Département de l'Environnement, Conservation de la Nature et Tourisme proviennent:
- du produit de l'exploitation des forêts;
- des redevances perçues lors de la délivrance des permis de chasse et de pêche;
- de la taxe sur l'implantation et sur les permis d'exploitation des établisseemnts dangereux, insalubres ou incommodes;
- de la taxe sur l'implantation et l'exploitation des hôtels et restaurants;
- de la taxe sur l'exploitation des agences de voyage.

Article 2. - Ces recettes sont déterminées comme suit :

A. Exploitation des forêts

- Permis de coupe de bois;
- Taxe de superficie;
- Licence d'achat et de vente de bois de feu et de charbon de bois;
- Permis de récolte des menus produits forestiers;
- Redevance proportionnelles sur la récolte de bois;
- Taxe de validation de contrat de vente;
- Produits des transactions résultant des infractions à la législation forestière.

B. Chasse et capture des animaux

- Permis de chasse, licence de guides de chasse et licence spéciale de séjour dans les domaines ou réserves de chasse;
- Taxes d'abattage et de capture;
- Produits des transactions résultant des infractions à la législation sur la chasse.

C. Pêche

- Permis de pêche;
- Taxes sur les embarcations et filets utilisés pour la pêche;
- Produits des transactions résultant des infractions à la législation sur la pêche.

D. Etablissements dangereux, insalubres ou incommodes

OMIS

E. Installation et exploitation des établissements hôteliers

OMIS

F. Agences de Voyage

OMIS

TITRE II : DE L'EXPLOITATION FORESTIERE

Article 3. - Permis de coupe de bois

Le permis de coupe de bois est délivré par le Commissaire d'Etat ayant la Conservation de la Nature dans ses attributions ou son délégué.

La délivrance est soumise au paiement d'une taxe de 90,00 Zaïres par cent hectares ou fraction de cent hectares.

Article 4. - Taxe de superficie

Une taxe de superficie de 10 Zaïres pour cent hectares est acquitée annuellement par le bénéficiaire d'une garantie d'approvisionnement.

Article 5. - Licence d'achat

L'achat de bois de feu ou de charbon de bois est soumis à l'obtention préalable d'une licence d'achat délivrée par le Département ayant l'Environnement, Conservation de la Nature et Tourisme dans ses attributions.

La licence détermine le nombre de stères de bois ou de tonnes de charbon de bois que le titulaire de la licence peut acheter ainsi que la localisation du lieu d'achat.

La licence est délivrée après paiement d'une taxe fixée à 1 Zaïre le stère de bois et 6 Zaïres la tonne de charbon de bois.

Article 6. - Récolte et exportation des menus produits forestiers

La délivrance d'un permis de récolte des menus produits forestiers, en ce compris notamment les plantes médicinales, gommes, résines, copal, laques, est soumise au paiement préalable d'une taxe déterminée de la manière suivante :
- Rauwolfia : 20 Zaïres / Tonne
- Voacanga, digitale : 70 Zaïres / tonne
- Gommes, laque, résine, copal et autres menus produits forestiers : : 50 Zaïres / Tonne

L'exportation de tout menu produit forestier doit être constatée par un permis d'exportation délivré par le Département ayant l'Environnement, Conservation de la Nature et Tourisme dans ses attributions.

La délivrance du permis d'exportation visé par le paragraphe précédent est soumise au paiement d'une taxe fixée comme suit :
- Rauwolfia : 50 Zaïres / Tonne
- Voacanga, digitale : 50 Zaïres / Tonne
- Gommes, laque, résine, copal et autres menus produits forestiers : : 20 Zaïres / Tonne

Article 7. - Redevances proportionnelles

Le titulaire d'une licence paie une redevance trimestrielle calculée d'après les quantités de bois ou de charbon de bois acquises suivant le barème ci-après :

a) Bois d'oeuvre et de construction : 4,50 Zaïres le m³

b) Bois de mines, rondins et perches :
- catégorie 1 : de 0 à 0,10 m de diamètre : 3 K le m. courant
- catégorie 2 : de 0,10 à 0,30 m " " : 15 K "
- catégorie 3 : de 0,30 à 0,50 m " " : 45 K "

c) Bois de feu : 0,40 Z. le stère
d) Charbon de bois : 6,00 Z. la tonne
e) Bois de briqueterie : 0,90 Z. pour mille briques

Toutefois pour le bois d'oeuvre et de construction, les sommes dues subissent les correctifs suivants selon la classe des essences :
- Classe 1 : aucune réduction
- Classe 2 : réduction de 5%
- Classe 3 : réduction de 10%

Les forêts sont réparties en deux zones suivant les difficultés d'exploitation et de transport. Ceci, en vue de l'application des réductions ci-après sur les tarifs appliqués en vertu de l'alinéa précédent.

Zone I :	Région du Bas-Zaïre	: aucune réduction
	Région de Bandundu	: "
	Région de l'Equateur	: "
	Région du Haut-Zaïre	: "
Zone II :	Région du Kasaï Oriental	: 10% de réduction
	Région du Kasaï Occidental	: "
	Région du Shaba	: "
	Région du Kivu	: "

Article 8. - Le Commissaire d'Etat ayant la Conservation de la Nature dans ses attributions peut majorer les montants des taxes prévus à l'art. 7 lorsque la récolte de tout produit forestier proviennent des forêts aménagées par l'Etat.

Article 9. - Les contrats de vente de tout produit forestier à l'extérieur du territoire national doivent être validés par le Commissaire d'Etat ayant l'Environnement, la Conservation de la Nature et le Tourisme dans ses attributions ou son délégué.

La validation est soumise au paiement d'une taxe fixe comme suit :
- Grume : 30 Z / m³
- Placage : 15 Z / m³
- Sciage : 5 Z / m³
- Contreplaqué et autres produits finis : 5 Z / m³

Article 10. - Transactions

Toute récolte de produits et menus produits forestiers effectuée dans des conditions illégales donné lieu à mesurage.

Le contrevenant est tenu d'acquitter sans délai le montant quintuplé des permis de coupe, licences, taxes de superficie, permis de récolte, redevances forestières, sans préjudice des sanctions pénales.

Article 11. - Les montants des taxes et redevances prévues par les dispositions de l'art. 9 peuvent être modifiés au besoin par arrêté du Commissaire d'Etat ayant l'Environnement, Conservation de la Nature et Tourisme dans ses attributions.

TITRE III : DE LA CHASSE ET DE LA PECHE
(Art. 12 à 25 Omis)

TITRE IV : LES ETABLISSEMENTS DANGEREUX, INSALUBRES OU INCOMMODES
(Art. 26 à 29 Omis)

TITRE V : HOTELS ET RESTAURANTS
(Art. 30 à 37 Omis)

TITRE VI : EXPLOITATION DES AGENCES DE VOYAGE
(Art. 38 à 41 Omis)

TITRE VII : DISPOSITIONS FINALES

Article 42. - Les Commissaires d'Etat à l'Environnement, Conservation de la Nature et Tourisme et aux Finances sont chargés, chacun en ce qui le concerne, de l'exécution dela présente Ordonnance, qui entre en vigueur à la date de sa signature.

Fait à Lubumbashi, le 16 octobre 1979

ARRETE N°OOOI/CCE/ADRE/83 DU 26 JANVIER 1983

ZAIRE

Portant Modification de certains Taux des Taxes et Redevances prévues par l'Ordonnance N°79-244 du 16 octobre 1979 et perçues à l'initiative du DEPARTEMENT DE L'ENVIRONNEMENT, CONSERVATION DE LA NATURE ET TOURISME

Arrêté N°OOOI/83

VU la Constitution;

VU la Loi N° 82-002 du 28 mai 1982 portant réglementation de la Chasse;

VU l'Ordonnance N° 79-244 du 16 octobre 1979 fixant le taux et règles d'assiette et de recouvrement des taxes et redevances en matière administrative, judiciaire et domaniale perçues à l'initiative du Département de l'Environnement, Conservation de la Nature et Tourisme;

Considérant la nécessité de maximiser les recettes de l'Etat notamment par la perception des taxes et redevances;

ARRETE :

TITRE I : DE L'EXPLOITATION FORESTIERE

Article 1. - La taxe sur le permis de coupe de bois prévue par l'art. 3 de l'Ordonnance N° 79-244 du 16 octobre 1979 est de 150,00 Z. par cent hectares ou fraction de cent hectares.

Article 2. - La taxe de superficie prévue à l'art. 4 de l'ordonnance précitée est de 20,00 Z. pour cent hectares.

Article 3. - La taxe sur la licence d'achat de bois de feu ou de charbon, telle que prévue par l'art. 5 de l'ordonnance précitée est de 3,00 Z. le stère de bois et 30,00 Z. la tonne de charbon de bois.

Article 4. - La taxe sur le permis de récolte des menus produits forestiers, en ce comprises notamment les plantes médicinales, prévue par l'art. 6 de l'Ordonnance précitée est de :
- 40 Z./ Tonne pour la rauwolfia;
- 140 Z./ Tonne pour le voacanga, digitale;
- 100 Z./ Tonne pour les gommes, laque, résine, copal et autres menus produits forestiers.

- La taxe sur le permis d'exportation est de :
- 150 Z./ Tonne pour la rauwolfia;
- 150 Z./ Tonne pour le voacanga, digitale;
- 60 Z./ Tonne pour les gommes, laque, résine, copal et autres menus produits forestiers.

Article 5. - Les redevances proportionnelles prévues par l'art. 7 de l'Ordonnance sont dues comme suit :

a) Bois d'oeuvre et de construction : 13,50 Z. le m^3

b) Bois de mines, rondins et perches :
- catégorie 1 : de 0 à 0,10 m de diamètre : 5 K le m. courant
- catégorie 2 : de 0,11 à 0,30 m " : 20 K "
- catégorie 3 : de 0,30 à 0,50 m " : 50 K "

c) Bois de feu : 0,50 Z. le stère
d) Charbon de bois : 6,00 Z. la tonne
e) Bois de briqueterie : 1,00 Z. pour mille briques

Toutefois les correctifs prévus par l'art. 7 précité s'appliquent pour le bois d'oeuvre et de construction.

Article 6. - La taxe sur la validation des contrats de vente des produits forestiers à l'extérieur du territoire national est de :
- 30 Z./m^3 : grume
- 15 Z./m^3 : placage
- 5 Z./m^3 : sciage
- 5 Z./m^3 : contreplaqué et autres produits finis.

TITRE II : DE LA PECHE
(Art. 7 à 8 Omis)

TITRE III : DES ETABLISSEMENTS DANGEREUX, INSALUBRES OU INCOMMODES
(Art. 9 à 11 Omis)

TITRE IV : DES HOTELS ET RESTAURANTS
(Art.12 à 16 Omis)

TITRE V : DES AGENCES DE VOYAGE
(Art.17 à 18 Omis)

Article 19. - Le Secrétariat Général à l'Environnement, Conservation de la Nature et Tourisme est chargé de l'exécution du présent arrêté qui entre en vigueur à la date de sa signature.

Fait à Kinshasa, le 26 janvier 1983

LE COMMISSAIRE D'ETAT

CIRCULAIRE N° 1640 /SG/DECNT/80

A l'attention de tous les propriétaires d'une Industrie de Transformation de Bois et de tous les promoteurs ou sociétés désireuses de construire une Industrie de Transformation de Bois

DEPARTEMENT DE L'ENVIRONNEMENT, CONSERVATION DE LA NATURE ET TOURISME

Afin de vous éviter des démarches inutiles, il nous apparait opportun d'élucider la notion de garantie d'approvisionnement ainsi que la procédure à suivre pour en bénéficier.

L'exploitant d'une industrie de transformation de bois, a besoin pour assurer l'opération de cette industrie d'un approvisionnement sûr et continu en matière première, pour une période suffisante assurant un rendement financier avantageux justifiant le capital investi.

De la même façon, le promoteur de projets nouveaux a besoin, pour justifier les investissements en capital projeté et leur financement, d'être assuré pour une période donnée d'une disponibilité de la matière première.

Pour satisfaire ces exigences, le Département de l'Environnement, Conservation de la Nature et Tourisme peut accorder pour une période maximum de 20 ans, par convention directe entre la société ou l'industrie et le Département, une garantie d'un certain volume d'approvisionnement annuel d'essences données, en conformité avec la réglementation en vigueur, à l'intérieur d'une forêt de production délimitée.

Cette convention accorde à l'industriel certains avantages et lui confère également certaines obligations dont la principale est de maintenir en opération l'usine pour laquelle elle est conclue au niveau de production prévu.

Comme une convention de garantie d'approvisionnement d'usine de transformation ne peut être conclue qu'avec l'exploitant d'une telle usine et qu'il faut beaucoup de temps pour permettre l'élaboration et la mise en oeuvre d'un projet de construction d'usine, le Conseil Exécutif représenté par le Commissaire d'Etat à l'Environnement, Conservation de la Nature et Tourisme, peut accorder à tout promoteur de projet ayant un dossier technique complet, une lettre d'intention, par laquelle il s'engage à l'avance à signer avec lui une convention de garantie d'approvisionnement d'usine, lorsque son usine sera en construction. Cette lettre d'intention deviendra nulle, si le promoteur du projet ne réalise pas la construction de l'usine avant la fin du délai prévu.

PROCEDURE

Si vous possédez déjà une usine et que vous voulez avoir un approvisionnement régulier en matière première, grâce à une convention de garantie d'approvisionnement, il faudrait formuler une demande comportant un dossier technique complet tel que défini ci-dessous et de manifester le désir d'opérer de façon continue cette usine en conformité avec la réglementation en vigueur.

S'il s'agit d'un projet d'usine nouvelle, il faudra faire une demande conforme, pour démontrer le sérieux du projet.

La demande devra comporter :

1°) Le site de l'usine et un plan général d'implantation;

2°) Le plan de l'usine avec celui des lignes de production, les annexes nécessaires à celle-ci, tels que la centrale d'énergie, le poste de transformation, le brûleur à rebuts etc..

3°) La liste des produits d'usinage ainsi que les taux de transformation ou de rendement, pour chacun des produits, la quantité de matière première nécessaire pour chaque essence;

4°) La liste des équipements importés et leurs coûts;

5°) La liste des investissements en monnaie locale, infrastructure, bâtisse etc..

Pour chaque chantier d'exploitation forestière et pour l'ensemble des chantiers :

6°) Un plan d'exploitation forestière pour les cinq premières années d'exploitation;

7°) La liste du matériel pour :
a) la construction des routes,
b) l'entretien des routes,
c) l'abattage et le tronçonnage,
d) le débardage,
e) le transport routier incluant le chargement et le déchargement,
f) le transport fluvial
g) les investissements en infrastructures permanentes qui seront mis en place durant les cinq premières années (routes principales, ponts, quais etc..),
h) les infrastructures semi-permanentes prévues pour les premiers cinq ans (camps de travailleurs, maisons, garages, entrepôts, etc..),

8°) Les prévisions d'embauche du personnel pour les opérations forestières, les usines et l'ensemble de l'opération ainsi que les experts expatriés prévus;

9°) Une cédule de production,

10°) Une analyse des coûts de production;

11°) Une analyse des marchés pour chaque produit et des prévisions sur les prix moyens de ventes;

12°) Vos prévisions pour les frais fixes, les frais généraux et les frais de vente;

13°) Des bilans et des états de pertes et profits ainsi que le mouvement de trésorerie proforma pour les premiers cinq ans du projet;

14°) Un tableau montrant la structure du capital de la société et la liste des actionnaires et la composition du bureau de la Direction;

15°) Un échéancier précis des travaux de construction de l'usine et du début de l'exploitation;

16°) Vos prévisions pour la protection et le reboisement pour les cinq premières années.

Ce dossier pourra également être présenté à la Commission des Investissements pour obtenir l'approbation du choix du site de l'usine et éventuellement d'autres avantages auxquels vous pouvez avoir droit.

Si vous possédez déjà l'industrie, les informations sur les items 10°, 11°, 12°, 13°, et 15° ne sont pas requises.

Après la réception d'un dossier complet, nous serons en mesure s'il s'agit d'un nouveau projet, compte tenu de l'échéancier prévu pour la construction, d'émettre en votre faveur une lettre d'intention pour la signature avec vous d'une convention de garantie d'approvisionnement d'usine.

Cette lettre d'intention, vous donnera suffisamment d'assurance pour mettre en route votre projet et contiendra les conditions spécifiques qui apparaîtront par la suite dans la convention. Si vous n'arrivez pas à mettre en route votre projet, cette lettre deviendra nulle à l'expiration du délai fixé.

Si vous possédez déjà une usine, une convention de garantie d'approvisionnement répondant aux besoins de votre usine sera immédiatement mise en préparation. Vous serez ensuite convoqué en nos bureaux pour la signature de cette convention. Les frais de notaire sont à la charge du bénéficiaire.

Fait à Kinshasa, le 5 juin 1980

LE COMMISSAIRE D'ETAT

NOTE CIRCULAIRE N° 1986/DECNT/CCE/80
DU 16 DECEMBRE 1980
(Programmes de Reboisement)

NOTE CIRCULAIRE DESTINEE A TOUS LES
EXPLOITANTS FORESTIERS

Département de l'Environnement, Conservation de la Nature et Tourisme;

Bureau du Commissaire d'Etat;

Transmis copies pour informations aux :
- Président Fondateur, Président de la République;
- Premier Commissaire d'Etat;
- Commissaire d'Etat aux Finances et Budget;
- Gouverneurs (tous);
- Coordinateurs (tous);

Au cours de la dernière réunion sur la mercuriale du prix du bois à l'exportation, j'ai attiré votre sérieuse attention sur la stricte application de la politique de reboisement, qui vous a été faite par le Président-Fondateur, Président de la République, lors de son discours du 27 novembre 1977 à la N'SELE.

Je vous rappelle l'alternative sur laquelle nous sommes convenus, de commun accord.

SOIT : l'Entreprise effectue, elle-même, le reboisement, dans les limites de la "zone" de garantie d'approvisionnement, qui lui a été octroyée;

SOIT : l'Entreprise forestière entretenant une pépinière, fournit les moyens logistiques nécessaires aux brigades de reboisement, installés dans la région.

Grâce à ce soutien logistique, (ouverture des pistes, léger labour, transport des plantes), les brigades assurent le reboisement dans et parfois hors de la "zone" d'approvisionnement de l'entreprise.

Chacune de vos entreprises doit établir, avec les directions régionales des brigades de reboisement, un plan triennal ou quinquennal de reboisement.

L'octroi, ou le renouvelement de tous permis de coupe de bois pour l'exercice suivant, est fonction de la mise en application de ce programme de reboisement. La coordination régionale, transmettra ses avis et considérations sur le reboisement, en annexe, au dossier de demande de permis de coupe de bois.

Il vous est également rappelé, que l'obligation de reboisement ne vous dispenserait pas de l'obligation de vous acquitter de la taxe forestière.

LE PRESENT RAPPEL VAUT INSTRUCTION PERMANENTE

Fait à Kinshasa le 11 décembre 1980,

LE COMMISSAIRE D'ETAT

INDEX

1. DEFINITIONS AND FOREST DEVELOPMENT OBJECTIVES

FOREST AND FOREST LANDS

- CAMEROON : Law N° 81/13 - Sec. 3 (Forests).
- C A R : Loi N° 61/273 - Art. 1 (Forests).
- CONGO : Loi N° 004/74 - Art. 2 (Forests).
- EQUATORIAL GUINEA : Decreto-Ley N° 14/1981 - Art. 3(1) (Forest Land).
- GABON : Loi N° 1/82 - Art. 9 (Forests).
- IVORY COAST : Loi N° 65/425 - Art. 1 (Forests); Art. 2 (Protection Zones); Art. 3 (Reforestation Zones). Décret N° 78-231 - Art. 1 (Permanent State Forest Domain); Art. 2 (Rural State Forest Domain).
- LIBERIA : Forest Act 1953 - Sec. II (Forests). Supplementary Act 1957 - PART II Sec. 1 (Forests).
- NIGERIA : For. Ord. Nor. Reg. 1960 - Sec. 2 (Forests).
- ZAIRE : Décret du 11 avril 1949 - Art. 1 (Forests and Forest Lands).

FOREST PRODUCE AND SERVICES

- C A R : Loi N° 61/273 - Art. 1(2) (Forest Produce).
- CONGO : Loi N° 004/74 - Art. 2 (Forest Produce).
- EQUATORIAL GUINEA : Decreto-Ley N° 14/1981 - Art. 2(1) (Forest Produce); Art. 2(2) (Environmental Impact).
- GHANA : For. Ordinance 1927 - Sec. 2 (Forest Produce)
- LIBERIA : Forests Act 1953 - Sec. II (Forest Products). Revenue Law 1977 - Sec. 20.1 (Unprocessed Forest Products; Processed Forest Products).
- NIGERIA : For. Ord. Nor. Reg. 1960 - Sec. 2 (Forest Produce; Minor Forest Produce).
- TANZANIA : For. Ordinance 1982 - Sec. 2 (Forest Produce; Timber; Tree).
- ZAIRE : Décret 11 avril 1949 - Art. 1 (Forest Produce).

OBJECTIVES OF FOREST DEVELOPMENT

- CAMEROON : Law N° 81/13 - Sec. 1, 2 (Objectives of Forestry Regulations).
- CONGO : Loi N° 004/74 - Art. 23 - 26 (Principles of conservation, Management and Economic use of Forest Resources).
- EQUATORIAL GUINEA : Decreto-Ley N° 14/1981 - Introductory Statement; Art. 1.
- GABON : Loi N° 1/82 - Art. 1 - 3 (Objectives of Forest Resources Utilization and Management).
- LIBERIA : Forest Act 1953 - Introductory Statement and Sec. IV. FDA Act 1976 - Sec. 3 (Objectives of Forest Development Authority).

2. CONSTITUTION OF THE FOREST DOMAIN AND CATEGORIES OF FOREST LAND

PERMANENT FOREST DOMAIN

- CAMEROON : Law N° 81/13 - Sec. 13 (State Forests); Sec. 14 (Reservation Decree); Sec. 15 (Minimum Percentage to be classified); Sec. 18 (Local Council Forests). Décret N° 83/169 - Sec. 4 - 8 (Procedure for Constituting State Forests).
- C A R : Loi N° 61/273 - Art. 2(1) (Classified Forests); Art. 3 - 6 (Procedure for Constitution of Classified Forests); Art. 7 (Saving for Maintaining Usage and Ownership Rights).
- CONGO : Loi N° 004/74 - Art. 3 (Classified Forests); Art. 4 - 7 (Applicable Procedure); Art. 8 (Cancellation of Forest Reservation). Décret d'Application - Art. 68 (Marking of Boundaries).

- GABON : Loi N° 1/82 - Art. 10 (Classified Forests); Art. 12 (Procedure); Art. 13 (Minimum Percentage to be Classified).
- GHANA : Forests Ordinance 1927 - Sec. 4 (Forest Reserves); Sec. 5 - 17 (Procedure); Sec. 19 (Transfer of Rights); Sec. 20 (Cancellation of Reservation).
- IVORY COAST : Loi N° 65/425 - Art. 5 (Classified Forests); Art. 6 (Reasons for Reservation). Décret N° 78/231 - Art. 3 - 8 (Constitution of Permanent State Forest Domain); Art. 13 - 16 (Transitory Provisions); Art. 17 (Official Maps).
- LIBERIA : FDA Act 1976 - Sec. VI (Government Forest Reserves); Sec. VII (Native Authority Forest Reserves); Sec. VIII (Communal Forests).
- NIGERIA : For. Ord. Nor. Reg. 1961 - Sec. 4 (Government Forest Reserves); Sec. 5 - 11 (Procedure); Sec. 12 (Order to Constitute Reserve); Sec. 21 (Cancellation of Reservation); Sec. 22 (Communal Forest Reserves); Sec. 23, 24 (Procedure); Sec. 29 (Cancellation of Reservation); Sec. 39 (Marking of Boundaries).
- TANZANIA : For. Ordinance 1982 - Sec. 5(1) (State Forest Reserves); Sec. 5(1 B) (Exchange of Lands of State Forest Reserves); Sec. 5 (1 D) (Amalgation and Fragmentation of Lands); Sec. 5(2) (Demarcation of Boundaries); Sec. 5(A) (Local Authority Forest Reserves); Sec. 5(B) (Mapping of State and Local Authority Forest Reserves); Sec. 6 (Procedure for Constituting Forest Reserves); Sec. 13 (Cancellation of Reservation).
- ZAIRE : Décret 11 avril 1949 - Art. 4 (Classified Forests); Art. 5 (Procedure and Reasons for Forest Reservation); Art. 7 (Cancellation of Reservation).

PROTECTED FOREST DOMAIN

- CAMEROON : Law N° 81/13 - Sec. 21 (Communal Forests).
- C A R : Loi N° 61/273 - Art. 2(2) (Customary Forest Land); Art. 24 (Status of Customary Forest Land).
- CONGO : Loi N° 004/74 - Art. 3 (Protected Forests).
- GABON : Loi N° 1/82 - Art. 10 (Protected Forests).
- GHANA : Decree N° 273 of 1974 Art. 12 (Protected Areas); Art. 13 - 15 (Restrictions and Control).
- IVORY COAST : Loi N° 65/425 - Art. 5 (Protected Forests). Décret N° 78-231 - Art. 9 - 12 (Constitution of Rural State Forest Domain); Art. 17 (Official Maps).
- NIGERIA : For. Ord. Nor. Reg. 1961 - Sec. 4 (Government Protected Forests); Sec. 31, 32 (Comunal Protected Forests); Sec. 39 (Marking of Boundaries).
- TANZANIA : For. Ord. 1982 - Sec. 5(1 A) (Provisional State Forest Reserves).
- ZAIRE : Décret 11 avril 1949 - Art. 4, 6 (Protected Forests).

PRODUCTION AND PROTECTION FORESTS

- CAMEROON : Loi N° 81/13 - Sec. 13. Décret N° 83/169 Sec. 2 (Definitions).
- C A R : Loi N° 61/273 - Art. 3.
- CONGO : Loi N° 004/74 - Art. 3 (1), (2).
- GABON : Loi N° 1/82 - Art. 11.
- IVORY COAST : Loi N° 65/425 - Art. 5

REFORESTATION AREAS

- CAMEROON : Loi N° 81/13 - Sec. 13. Décret N° 83/169 Sec. 2 (Definitions).
- CONGO : Loi N° 004/74 - Art. 10.
- GABON : Loi N° 1/82 - Art. 11
- IVORY COAST : Loi N° 65/425 - Art. 3 (Communal Reforestation).

OTHER CATEGORIES OF FOREST LAND

- CAMEROON : Loi N° 81/13 - Sec. 13 (Integral Nature Reserves, National Parks, Wild Animal and Plant Species Sanctuaries). Décret N° 83/169 - Sec. 2 (Definitions).
- C A R : Loi N° 61/273 - Art. 3 (Fauna and Flora Reserves, Integral Nature Reserves, National Parks).
- CONGO : Loi N° 004/74 - Art. 3(3),(4),(5),(6) (National Parks, Nature Reserves, Integral Nature Reserves, Community Development Forests).
- GABON : Loi N° 1/82 - Art. 11 (National Parks in Forest Areas, Recreation Forests, Flora Sanctuaries, Wildlife Management Areas).

3. CUSTOMARY USAGE RIGHTS AND OWNERSHIP OF FOREST LAND

CUSTOMARY USAGE RIGHTS

- CAMEROON : Loi N° 81/13 - Sec. 39(1) (Protection from Commercial Forest Exploitation). Décret N° 83/169 - Sec. 3 (Practice on State Forest Domain).
- C A R : Loi N° 61/273 - Art. 8 - 11 (in Classified Forests); Art. 25, 28 (on Customary Forest Land); Art. 51 (Protection from Commercial Forest Exploitation).
- CONGO : Loi N° 004/74 - Art. 12, 12-1 (General Practice); Art. 13 (in Protected Forests); Art. 14 - 18 (Restrictions in Reserved Forests); Art. 19 (Abolition in Reforestation Areas).
- EQUATORIAL GUINEA : Decreto-Ley N°14.1981 - Art. 12, 13 (in State Forests).
- GABON : Loi N° 1/82 - Art. 5 (General Acknowledgment); Art. 22(3) (Logging near Villages).
- GHANA : Forests Ord. 1927 - Sec. 9 (Inquiry on and Settlement of Rights).
- IVORY COAST : Loi N° 65/425 - Art. 7; Art. 8 (Restrictions on Permanent State Forest Domain); Art. 9 - 14 (on Protected Forest Land); Art. 18 - 22 (Practice at Commercial Scale).
- NIGERIA : For. Ord. Nor. Reg. 1960 - Sec. 6 - 11 (Inquiry on Existing Rights; Government Forest Reserves); Sec. 12(2) (Admitted Rights); Sec. 13, 14 (Restriction of Rights and Exclusion of Areas); Sec. 15 - 19 (Rights in Government Forest Reserves); Sec. 23 (Inquiry on Existing Rights; Native Authority and Local Government Council Forest Reserves); Sec. 27, 28 (Practice of Usage Rights; Native Authority and Local Government Council Forest Reserves).
- TANZANIA : For. Ord. 1982 - Sec. 6(2-11), 7, 9 (Inquiry and Registration of Usage Rights).
- ZAIRE : Décret 11 Avril 1949 - Art. 8 (in Classified Forests); Art. 9 (in Protected Forests); Art. 10 (Commercial Uses); Art. 11 (Agricultural Uses on Forest Land); Art. 12 - 14 (Small Scale Timber Cutting).

OWNERSHIP / GENERAL

- CAMEROON : Loi N° 81/13 - Sec. 5 (Applicable Legislation); Sec. 9 (Practice of Ownership Rights); Sec. 21 (Forest Produce on Communal Forest Land).
- C A R : Loi N° 61/273 - Art. 2 (Categories of Ownership); Art. 24 (Customary Forest Domain); Art. 29 (Timber Harvesting Rights on Customary Forest Domain).
- EQUATORIAL GUINEA : Decreto-Ley N° 14/1981 - Art. 6 (Categories of Ownership).

- GABON : Loi N° 1/82 - Art. 4 (Applicable Legislation).
- GHANA : For. Ord. 1927 - Sec. 4, 18 (Forest Reserves).
- IVORY COAST : Loi N° 65/425 - Art. 4 (Categories of Ownership).
- TANZANIA : For. Ord. 1982 - Sec. 25 (Trees and Forest Produce Presumed to be Property of the Government, respectively of Local Authorities).

STATE FORESTS

- CAMEROON : Loi N° 81/13 - Sec. 13 (Status and Categories); Sec. 17 (Protection).
- C A R : Loi N° 61/273 - Art. 2(1), 3
- CONGO : Loi N° 004/74 - Art. 1
- EQUATORIAL GUINEA : Decreto-Ley N° 14/1981 - Art. 7, 10.
- GABON : Loi N° 1/82 - Art. 10 (Public and Private State Forest Domain); Art. 11 (Categories of Classified State Forests).
- GHANA : For. Ord. 1927 - Sec. 4 (Government Land).
- IVORY COAST : Loi N° 65/425 - Art. 4; Art. 5 (Categories of State Forest Domain).
- LIBERIA : For. Act 1951 - Sec. VI (Government Forest Reserves).
- NIGERIA : For. Ord. Nor. Reg. 1960 - Sec. 4(1) (Government Forest Reserves); Sec. 4(2) (Government Protected Forests).
- TANZANIA : For. Ord. 1982 - Sec. 5 (State Forest Reserves; Sec. 10 A, 11 A (Management of State Forest Reserves); Sec. 5(1 C) (Local Authority Reserve to become State Forest Reserve).
- ZAIRE : Décret 11 avril 1949 - Art. 2(1); Art. 4, 5 (Categories of State Forests).

COMMUNAL FORESTS

- CAMEROON : Loi N° 81/13 - Sec. 18 (Local Council Forests); Sec. 20 (Rights of Owners); Sec. 22 (Restriction on Uses). Décret N°83/169 - Sec. 16 (Logging and Management).
- C A R : Loi N° 61/273 - Art. 2(2), 3 (Communal and Rural Forests).
- CONGO : Loi N° 004/74 - Art. 3(6) (Forests for Community Development).
- EQUATORIAL GUINEA : Decreto-Ley N° 14/1981 - Art. 8, 14 (Communal Forests); Art. 15 (Timber Harvesting); Art. 16 (Control and Management); Art. 16, 17 (Communal Revenues and Payments for Silvicultural Improvements).
- GHANA : For. Ord. 1927 - Sec. 4 (Tribal and Stool Lands).
- IVORY COAST : Loi N° 65/425 - Art. 14 ; Art. 26 - 28 (Rights and Obligations); Art. 29 - 31 (Land for Communal Reforestation).
- LIBERIA : For. Act 1953 - Sec. VII (Native Authority Forest Reserves); Sec. VIII (Communal Forests). Supplementary Act 1957 - Part III Sec. 10 (Forest Produce from Communal Forests).
- NIGERIA : For. Ord. Nor. Reg. 1960 - Sec. 22 (Native Authority and Local Government Council Forest Reserves); Sec. 20 (Power to Grant Land, Licences and Permits); Sec. 23 - 25 (Procedure for Constituting Forest Reserves); Sec. 26, 29 (Power to Revise, Modify and Repeal Order of Forest Reservation); Sec. 30 (Change of Government Forest Reserves to Communal Forest Reserves); Sec. 31, 32 (Native Authority and Local Government Council Protected Forests); Sec. 33 (Administration and Management of Native Authority and Local Government Council Forest Reserves); Sec. 34 (Communal Forest Areas); Sec. 35, 36, 38 (Constitution and Administration of Communal Forest Areas); Sec. 37 (Power to Make Rules); Sec. 43 (Forest Produce for Public Purposes); Sec. 44 (Fees and Royalties to Communal Forest Owners); Sec. 45 (Acquisition of Land for Public Purposes in Communal Forests).

- TANZANIA : For. Ord. 1982 - Sec. 5 A (Local Authority Forest Reserves); Sec. 10 - 13 (Administration and Management of Local Authority Forest Reserves); Sec. 14 (Forestry Dedication Convenants on Communal Land).
- ZAIRE : Décret 11 avril 1949 - Art. 2(2); Art. 4, 5 (Categories of Communal Forests).

PRIVATE FORESTS

- CAMEROON : Loi N° 81/13 - Sec. 19; Sec. 20 (Rights of Owners). Décret N° 83/169 - Sec. 17 (Restrictions on Logging).
- C A R : Loi N° 61/273 - Art. 2(3); Art. 44 (Rights of Forest Owners); Art. 45 - 48 (Authorization of Forest Clearings).
- EQUATORIAL GUINEA : Decreto-Ley N° 14/1981 - Art. 9; Art. 18 (State Supervision).
- GHANA : For. Ord. 1927 - Sec. 4.
- IVORY COAST : Loi N° 65/425 - Art. 4; Art. 26 - 28 (Rights and Obligations); Art. 29 - 31 (Land for Private Reforestation).
- LIBERIA : Supplementary Act 1957 - Part VII Sec. 11 (Rights and Obligations). FDA Reg.N°5 1979 - Sec. 2, 3 (Assistance to Owners); Sec. 4,5 (Land Titles and Boundaries); Sec. 6, 7 (Management of Land); Sec. 8 - 13 (Salvage Logging).
- TANZANIA : For. Ord. 1982 - Sec. 14 (Forestry Dedication Convenants on Private Land).
- ZAIRE : Décret 11 avril 1949 - Sec. 3; Sec. 37 (Restrictions on Forest Clearings).

PUBLIC WATERWAYS AND COASTLAND

- CAMEROON : Loi N° 81/13 - Sec. 4 (Ownership of Fauna and Flora); Sec. 77 (Restrictions of Uses).
- C A R : Loi N° 61/273 - Art. 49 (Forests on Public Domain).
- EQUATORIAL GUINEA : Decreto-Ley N° 14/1981 - Art. 19 (Status of Rivers).

4. PROTECTION AND MANAGEMENT OF FORESTS AND FOREST LAND

PROTECTION MEASURES

- CAMEROON : Loi N° 81/13 - Sec. 8 (Protection of Forests and Natural Resources); Sec. 40 (Marking of Reserved Trees); Sec. 76 (Protection in Critical Areas); Sec. 77 (Protection along Water Courses). Décret N° 83/169 - Sec. 36 (Logging Coupes).
- C A R : Loi N° 61/273 - Art. 31 (Logging on Customary Forest Land).
- CONGO : Loi N° 004/74 - Art. 21 (Restriction on Uses). Décret d'Application - Art. 8 - 30 (Regulations on Timber Harvesting).
- EQUATORIAL GUINEA : Decreto-Ley N° 14/1981 - Art. 24(1), (2) (Classification of Tree Species and Minimum Cutting Limits).
- GABON : Loi N° 1/82 - Art. 28 (Reserved Trees and Areas); Art. 83 (Import Permits for Plants and Animals).
- IVORY COAST : Loi N° 65/425 - Art. 23 (Protected Tree Species).
- LIBERIA : Supplementary Act 1957 - Part III Sec. 8 (Sacred Trees and Trees close to Villages); Sec. 9 (General Restrictions); Part IV Sec. 2 (Minimum Girth Limits). For. Management Plan - Art. 1 - 6 (Protection, Minimum Diameter Limits, Obligatory Species).
- NIGERIA : For. Ord. Nor. Reg. 1960 - Sec. 40 (Protection against Forest Diceases).
- TANZANIA : For. Ord. 1982 - Sec. 15 A (Regulations on Beekeeping in State Forests); Sec. 17 (National Trees); Sec. 18 A (Measures against Forest Diceases); Sec. 19 B (Minimum Girth Limits).
- ZAIRE : Décret 11 avril 1949 - Art. 34, 35 (Restrictions on Logging and Protected Tree Species).

FOREST INVENTORIES AND MANAGEMENT

- CAMEROON : Loi N° 81/13 - Sec. 16 (Management Plans for State Forests); Sec. 23(1) (Forest Inventories). Décret N° 83/169 - Sec. 8 (Minimum Standards for Inventories); Sec. 9 (Management Plan Standards).
- CONGO : Loi N° 004/74 - Art. 27(1) (Forest Management Plans); Art. 28 (Forest Inventories); Art. 29 (Forest Management Units). Décret d'Application - Art. 69 - 79 (Management and Inventory Regulations).
- EQUATORIAL GUINEA : Decreto-Ley N° 14/1981 - Art. 11 (Management of State Forests); Art. 24(4) (Forest Inventories and Logging Plans).
- GABON : Loi N° 1/82 - Art. 14 (Management Plans for Classified Forests); Art. 23 (Forest Inventories and Logging Plans).
- GHANA : For. Ord. 1927 - Sec. 18 (Management in Forest Reserves).
- IVORY COAST : Décret N° 78-231 - Art. 6 (Management of Permanent State Forest Domain); Art. 10, 11 (Agricultural Development Plans for Rural Forest Domain).
- LIBERIA : Supplementary Act 1957 - PART VII Sec. 10 (Salvage Logging in Development Areas). Timber Concession Agreement - Art. II (Management Plan).
- TANZANIA : For. Ord. 1982 - Sec. 10 (Management and Control in Communal Forest Reserves); Sec. 11 (Appointment of Responsable Forest Officer); Sec. 12 (Supervision of Management in Communal Forest Reserves); Sec. 10 A (Management and Working Plans in State Forest Reserves); Sec. 18 B (Assessment of Natural Forest Resources in Water Catchment Areas)
- ZAIRE : Décret 11 avril 1949 - Art. 15 (Forest Inventories and Allowable Cut).

REGENERATION AND REFORESTATION

- CAMEROON : Loi N° 81/13 - Sec. 7 (Regeneration in State Forests). Décret N° 83/169 - Sec. 1 (Responsibility for Regeneration).
- C A R : Loi N° 61/273 - Art. 14 (Silvicultural Operations).
- GABON : Loi N° 1/82 - Art. 15 (National Reforestation Programme).
- GHANA : For. Ord. 1927 - Sec. 7 (Improvement Measures in Forest Reserves).
- LIBERIA : For. Act 1953 - Art. VIII (Reforestation Obligation for Concessionaires).
- TANZANIA : For. Ord. 1982 - Sec. 14(1), (4) (Forestry Dedication Convenants for Reforestation); Sec. 14(5) (Grants for Tree Planting).
- ZAIRE : Décret 11 avril 1949 - Art. 30 (Reforestation and Regeneration). Note Circulaire N° 1986DECNT/CCE/80 - (Reforestation by Logging Companies).

CONTROL OF FIRES

- CAMEROON : Loi N° 81/13 - Sec. 75 (Restricitons on Use).
- C A R : Loi N° 61/273 - Art. 12 (Restrictions); Art. 13 (Preventive Measures); Art. 26 (Restrictions on Customary Forest Land); Art. 27 (Fighting of Forest Fires).
- CONGO : Loi N° 004/74 - Art. 22 (Restrictions on Use and Fighting of Forest Fires).
- IVORY COAST : Loi N° 65/425 - Art. 32 - 34 (Restrictions).
- TANZANIA : For. Ord. 1982 - Sec. 16 (Fire Protection in State Forests).

5. FORESTS RESOURCES ALLOCATION METHODS

TIMBER ALLOCATION / GENERAL

- CAMEROON : Loi N° 81/13 - Sec. 23(2), (3) (Forest Survey Permits); Sec. 25 (Timber Allocation Methods); Sec. 26 (Timber Allocation in State Forests); Sec. 27 (Timber Allocation on Communal Forest Land); Sec. 35(1) (Application for Licences); Sec. 42 (Use of Drift Timber). Décret N° 83/169 - Sec. 10 - 13 (Sale of Standing Timber in State Forests); Sec. 14 (State Operated Logging); Sec. 15 (Granting of Harvesting Rights in State Forests); Sec. 18 - 22 (Exploration Permits and Forest Inventories on Communal Forest Land); Sec. 23 - 29 (Advisory Technical Commission); Sec. 30, 32, 33 (Procedure for Granting Licences on Communal Forest Land); Sec. 31 (Consultation of Local Population); Sec. 52, 53 (Sale of Standing Timber on Communal Forest Land); Sec. 54 - 65 (Special Permits and Tree Felling Permits); Sec. 74 (Drift Timber).
- CONGO : Loi N° 004/74 - Art. 32, 33, 37 (Methods of Timber Allocation); Art. 34 (Timber Harvesting and Industrial Wood Processing Contracts); Art. 35, 36 (Timber Cutting and Special Permits); Art. 38 (Conditions for Granting Forest Contracts); Art. 39, 40, 43 (Granting Procedure for Forest Contracts); Art. 44 (Granting of Cutting Permits). Décret d'Application - Art. 42 - 56 (Forest Contracts); Art. 57 - 62 (Cutting Permits and Special Permits).
- EQUATORIAL GUINEA : Decreto-Ley N° 14/1981 - Art. 21 (Timber Allocation on State and Communal Forest Land); Art. 22, 23(3) (Forest Concessions); Art. 25 - 29 (Procedure for Auctioning Forest Concessions); Art. 30 (Negotiating Forest Concessions).
- GABON : Loi N° 1/82 - Art. 16 (Permits for Use of Forest Produce); Art. 19 (Timber Allocation in Classified Forests).
- GHANA : Act N° 124 of 1962 - Sec. 16(3), (4), (5) (Granting of Timber Harvesting Rights on Stool Lands).
- IVORY COAST : Loi N° 65/425 - Art. 24 (Timber Allocation on State Forest Domain).
- LIBERIA : Supplementary Act 1957 - PART II Sec. 1, 2, 5 (Authorized Use and Permits); Sec. 3, 4 (Mineral Resources); Sec. 7 (Free Use Permits for Public Works); Sec. 6 (Timber Sale Agreements). FDA Reg. N° 5 1979 - Sec. 8 - 13 (Salvage Logging Contracts). FDA Reg. N° 6 1979 - Sec. 2 - 4 (Application for Forest Exploitation Permits on Non Concession Forest Land); Sec. 5 (Option of Forest Development Authority for Method of Timber Allocation); Sec. 6, 7 (Forest Survey Permits); Sec. 8 - 10 (Forest Exploitation Permits).
- TANZANIA : For. Ord. 1982 - Sec. 11 A (Leasing of Land in State Forest Reserve for Construction Work and Recreational Purposes); Sec. 19(1), (2) (Granting of Licences for all Purposes of Ordinance by Director of Forestry, respectively Local Authorities); Sec. 19(3) (Cancellation and Suspension of Licences); Sec. 19(4) (Refusal of Granting Licences in Case of Repeatedly Committed Offences); Sec. 32 (Saving of Mining Rights and Right of Leasee and Claim Holder to Take Timber); Sec. 33(b) (Power to Vary or Amend Licences Granted under the Provisions of the Previous Ordinance).
- ZAIRE : Décret 11 avril 1949 - Art. 16, 17 (Timber Allocation Procedures in Forest Reserves and on Protected Forest Land); Art. 18 - 25 (Timber Cutting Permits); Art. 26 (Timber Auctioning); Art. 27 (State Operated Logging). Circulaire N° 1640/SG/DCNT/80 - (Raw Material Supply Contracts and Procedure for Application).

CONCESSION AREA

- CAMEROON : Loi N° 81/13 - Sec. 24(1) (Maximum Concession Area); Sec. 24(2) (Acquisition of Majority Shares); Sec. 29 (Special Conditions for National Operators).
- EQUATORIAL GUINEA : Decreto-Ley N° 14/1981 - Art. 23(1) (Location of Concession).
- GABON : Loi N° 1/82 - Art. 24 (Maximum Concession Area); Art. 25 (Capital Participation); Art. 29 (Reduction of Granted Area).
- LIBERIA : Supplementary Act 1957 - Part VII Sec. 2 (Boundaries of Granted Area). Timber Concession Agreement - Art. I (Granted Area, Surrender of All or Part of Area).

CONCESSION DURATION AND RENEWAL

- CAMEROON : Loi N° 81/13 - Sec. 28 (Duration and Renewal of Forest Licences). Décret N° 83/169 - Sec. 44 - 47 (Procedure for Renewal).
- CONGO : Loi N° 004/74 - Art. 38 (Duration of Forest Contracts).
- EQUATORIAL GUINEA : Decreto-Ley N° 14/1981 - Art. 10(2), 23(5), 24(7) (Duration of Concession); Art. 23(9) (Renewal of Concession).
- GHANA : Act N° 124 of 1962 - Sec. 16(10) (Termination of Concession).
- LIBERIA : Supplementary Act 1957 - Part VI Sec. 1 (Expiration of Permits and Agreements). Timber Concession Agreements - Art. I (Duration, Periodic Review, Renewal, Effective Dates); Art. X (Termination of Agreement).

RIGHTS AND OBLIGATIONS OF CONCESSIONAIRE

- CAMEROON : Loi N° 81/13 - Sec. 39(1) (Granted Timber Harvesting Rights); Sec. 3I (Contract Specifications); Sec. 36 (Security Deposit); Sec. 37 (Transfer of Timber Harvesting Rights); Sec. 41 (Validity of Previously Granted Rights). Décret N° 83/169 - Sec. 34 - 43 (Rights and Obligations of Grantee); Sec. 48, 49 (Procedure for Transfer of Licence); Sec. 50, 51 (Release of Concession Area); Sec. 66 - 69 (Control of Timber Harvesting).
- CONGO : Décret d'Application - Art. 80, 81 (Roads and Infrastructure).
- EQUATORIAL GUINEA : Decreto-Ley N° 14/1981 - Art. 23(6) (Obligations of Concessionaire); Art. 23(8) (Right of Access and Construction of Forest Roads); Art. 23(10), 23(7) (Termination and Sanctions). Modela de Concesion - Art. 1 - 10 (Standard Provisions of Concession Contract).
- GABON : Loi N° 1/82 - Art. 21 (Transfer of Permits).
- GHANA : Decree N° 273 of 1974 - Sec. 1 - 11 (Marking of Logs). L.I. N° 23 of 1960 - Sec. 1 - 3 (Certificate for Log Measurement). A.F.R.C.D 47 of 1979 - Sec. 4 (Restrictions on Farm Land); Sec. 2, 3 (Compensation for Damages).
- IVORY COAST : Décret N° 72/125 - Art. 1 - 3 (Transfer of Logging Rights); Art. 4, 5 (Presentation of Logging Plan). Décret N° 78-231 Art. 12 (Salvage Logging).
- LIBERIA : Supplementary Act 1957 - Part IV (Property Marks); Part VII Sec. 1 (Responsibility for Forest Protection); Sec. 3 (Responsibility for Damages); Sec. 5 (Fire Prevention by Concessionaire); Sec. 7 (Recovery of Raw Material); Sec. 8, 9 (Forest Roads and Installations). Timber Concession Agreement - Art. I - III (Rights fo Concessionaire); Art. I (Performance Bond, Minimum Expenditure, Government Inspection); Art. II (Measurement and Marking of Logs); Art. IV (Reports, Records and Communications); Art. VIII (General Obligations of Concessionaire); Art. IX, X (Penalties and Arbitration).

6. FOREST INDUSTRY DEVELOPMENT

FOREST INDUSTRY / GENERAL

- CAMEROON : Décret N° 83/169 - Sec. 27 - 29, 81, 83 (Official Registration of Forest Operators, Processing Units and Timber Exporters); Sec. 75, 76 (Regulations on Capital Participation in the Forest Industry).
- CONGO : Décret d'Application - Art. 1 - 7 (Forestry Professions); Art. 31 - 35 (Transport of Forest Products); Art. 36 - 40 (Commercialization of Timber); Art. 82 (Training Obligation in Forest Contracts).
- EQUATORIAL GUINEA : Decreto-Ley N° 14/1981 - Art. 4, 5 (Classification of Forest Industries).
- GABON : Loi N° 1/82 - Art. 20 (Authorization for Logging and Wood Processing Operators); Art. 22 (Reservation of lrst Logging Zone and other Areas to National Operators); Art. 30 (Official Grading Standards); Art. 81 (Support to Cooperatives of Gabonese Operators).
- GHANA : Decree N° 128 of 1977 - Sec. 7 (Registration of Dealers in Timber Products); Sec. 5 (Power to Demand Information); Sec. 10 (Power to Suspend Manufacturers).
- IVORY COAST : Décret N° 72/606 - Art. 1, 2, 5 (Promotion of Cooperatives for Companies with Temporary Logging Permits); Art. 3, 4 (Authorization of Logging Cooperatives); Art. 6 (Duration of Logging Permits and Granting of Additional Areas in Case of Logging Cooperatives); Art. 7 (Investment Fund for Logging Cooperatives).
- LIBERIA : Timber Concession Agreement - Art. VI (Capitalization, Stock Purchases, Directors, Finance); Art. VII (Employment Conditions; Health and Safety of Employees; Recruitment of Personal; Training).

TIMBER PROCESSING

- CAMEROON : Loi N°81/13 - Sec. 24(1) (Concession Area to be Related to Processing Facilities). Décret N° 83/169 - Sec. 38 - 99 (Timber Processing Units and Minimum Percentage of Processed Raw Material); Sec. 67, 68 (Required Information on Processed Raw Material).
- CONGO : Loi N° 004/74 - Art. 24 (Timber Processing within the Country); Art. 25 (National Participation in Timber Processing); Art. 38, 41, 42 (Industrial Timber Processing Contracts).
- EQUATORIAL GUINEA : Decreto-Ley N° 14/1981 - Art. 24(5) (Processing Obligations).
- GABON : Loi N° 1/82 - Art. 17 (Processing Obligation for Concession Areas); Art. 18(2) (Minimum Percentage for Local Processing).
- IVORY COAST : Décret N° 72/114 - Art. 1, 2 (Regrouping of Logging Permits in Regional Raw Material Supply Areas); Art. 3 (Priority of Raw Material Supply for Established Wood Processing Industries).
- LIBERIA : Timber Concession Agreement - Art. I (Processing of Timber and other Forest Products).

EXPORT OF LOGS AND PROCESSED PRODUCTS

- CAMEROON : Loi N° 81/13 - Sec. 43 (General Restriction on Log Exports); Sec. 44 (Log Export Quota). Décret N° 83/169 - Sec. 70 (Restriction on Log Exports); Sec. 71, 72 (Log Export Quota).
- GHANA : Decree N° 128 fo 1977 - Sec. 1 - 3 (Export of Timber and Timber Products by Ghana Timber Marketing Board); Sec. 4 (Determination of Export and Local Prices of Timber Products by Marketing Board); Sec. 6 (Determination of Percentage of Timber Products Exports by Marketing Board).

- IVORY COAST : Décret N° 82/70 - Art. 1 (Official Registration of Log Exporters); Art. 2 - 5 (Log Supply Quota for all Exported Species); Art. 6, 7 (Advisory Commission on Control of Log Supply Quota and Promotion of Industrial Wood Processing).
- ZAIRE : Arrêté N° 01059 of 1975 - Art. 1 (Abolition of General Log Export Prohibition); Art. 2, 4 (Introduction of General Log Export Quota); Art. 3 (Special Log Export Restriciton on High Value species); Art. 5 (Maximum Log Export Volumes for Operators without Processing); Art. 6 (Limitation of Log Export Volumes for Species to be Promoted); Art. 7 (Minimum Quality Standards for Log Exports).

PROMOTION OF LESSER-USED SPECIES

- CAMEROON : Loi N° 81/13 - Sec. 45 (Promotion of Lesser-Used Species). Décret N° 83/169 - Sec. 73 (Definition of Lesser-Used Species and Promoting Measures).

7. FOREST REVENUES AND FEES

ASSESSMENT AND COLLECTION / GENERAL

- CAMEROON : LOI N° 81/13 - Sec. 11 (Responsible Administrations for Collecting Forestry, Wildlife and Fishery Duties and Taxes); Sec. 31, 33 (Categories of Forest Fees and Taxes); Sec. 32 (Use of Forest Fees and Taxes).
- C A R : Loi N° 61/273 - Art. 58 (Collection of Forest Fees).
- CONGO : Loi N° 004/74 - Art. 26(1) (Assessment of Forest Taxes according to Economic Criteria); Art. 31 (All Timber Harvesting Rights subject to Payment of Forest Taxes); Art. 26(2) (Principle of Special Levies and Development Fund). Loi N° 005/74 - Art. 1 - 7 (Provisions on Assessment and Collection of Forest Fees).
- EQUATORIAL GUINEA : Decreto-Ley N° 14/1981 - Art. 24(8) (Applicable Forest Fees).
- GABON : Loi N° 1/82 - art. 31 (Assessment and Payment of Forest Fees and Taxes); Art. 79 (Determination of Rates of Forest Fees and Taxes by Annual Fiscal Law).
- LIBERIA : Supplementary Act 1957 - Part V (Provisions on Revenues and Receipts). Timber Concession Agreement - Art. V (Fiscal Obligations of Concessionaire - Income Tax; Accounting Principles; Surface Rent; Stumpage Fees; Duties and Exices).
- ZAIRE : Décret 11 avril 1949 - Art. 31 (Payment of Forest Fees); Art. 32 (Payment of Fees to Communal Forest Owners). Ordonnance N° 244/79 - Art. 2(A) (Categories of Forest Fees); Art. 11 (Procedure for Revision of Applicable Rates of Forest Fees).

FOREST FEES AND TAXES

- CAMEROON : Loi N° 81/13 - sec. 23(2), (3) (Fee for Forest Survey Permits); Sec. 31(2) (Territorial Tax); Sec. 31(4) (Selling Price of Forest Produce); Sec. 35(2) (Approval Fee for Licences and Sales of Standing Timber); Sec. 38 (Fee in Case of Transfer of Timber Harvesting Rights); Sec. 42 (Fee on Used Drift Timber); Sec. 136 (Fee for Issuing a Duplicate of a Licence, Permit or Authorization).
- C A R : Loi N° 61/273 - Art. 59 - 61 (Area Tax); Art. 62 (Transfer Tax); Art. 63 (Forest Clearing Fee); Art. 64, 65 (Logging Tax). Loi N° 61/282 - Art. 1 - 8 (Rates of Applicable Forest Taxes and Fees). Ordonnance N° 74/014 - Art. 1 - 3 (Revision of Rates of Logging Tax).

- CONGO : Loi N° 005/74 - Art. 8, 11 - 13 (Assessment and Collection of Roundwood Fee); Art. 9 (Applicable Tax Rates of Roundwood Fee on Exported Logs by Log Quality and Production Region); Art. 10 (Applicable Tax Rates of Roundwood Fee on Logs Processed Locally by Log Quality).
- EQUATORIAL GUINEA : Decreto-Ley N° 14/1981 - Art. 23(2) (Logging Fee).
- GABON : Loi N° 1/82 - Art. 18(3) (Fee on Exported Logs from Industrial Permits).
- GHANA : Act N° 12 of 1960 - Sec. 8 (Stool Land Revenue). Act N° 124 of 1962 - Sec. 16(6) (Payment of Revenues from Timber Harvested on Stool Lands). L.I. N° 1089 of 1976 - Sec. 1 - 6 (Payment of Royalty and Rent); First Schedule (Royalties on Tree Species as of 17.9.1976).
- IVORY COAST : Décret N° 82/70 - Art. 8(a) (Applicable Rates of Export Tax on Processed Forest Products as of 8.2.1982); Art. 8(b) (Applicable Rates of Logging Tax and Groups of Species as of 8.2.1982). Ordonnance N° 82/71 - Art. 1 and Annex (Applicable Rates of Export Tax on Exported Logs by Species as of 13.1.1982).
- LIBERIA : Revenue Law 1977 - Sec. 20.2 (Local Stumpage Fee); Sec. 20.3 (Industrialization Incentive Fee); Sec. 20.4 (Forest Products Fee); Sec. 20.6 (Assessment of Forest Fees). FDA Reg. N° 7 of 1979 - Sec. 1 (Rate of Severance Fee as of 7.12.1979); Sec. 2 (Rates of Industrialization Incentive Fee by Species as of 7.12.1979); Sec. 3 (Rates of Forest Products Fee as of 7.12.1979). FDA REG. N° 10 of 1982 - Sec. 1 (Rates of Industrialization Incentive Fee by Species as of 8.11.1982); Sec. 2, 3 (Rates of Severance and Reforestation Fee); Sec. 4 (Rates of Forest Products Fee as of 8.11.1982).
- NIGERIA : For. Reg. ODSLN of 1981 - Schedule A (Stumpage Rates per Tree by Species as of 24.10.1980); Schedule B (Timber Fee Rates by Species per Hectare, as of 24.10.1980); Schedule C (Tariff of Out-Turn Volume Rates by Species as of 24.10.1980).
- ZAIRE : Ordonnance N° 244/79 - Art. 3 (Logging Permit Fee); Art. 4 (Area Fee for Timber Supply Contracts); Art. 5 - 8 (Fees on Forest Produce); Art. 9 (Log Export Tax). Arrêté N° 0001/CCE/ADRE/83 - Art. 1 - 6 (Applicable Rates of Logging Permit Fee, Area Fee, Fees on Forest Produce, Log Export Tax).

8. SPECIAL FOREST LEVIES AND FOREST DEVELOPMENT FUNDS

FORESTRY DEVELOPMENT, MANAGEMENT AND REFORESTATION

- CAMEROON : Loi N° 81/13 - Sec. 32(2, (3) (Regeneration Fee and Contribution to Forest Inventories); Sec. 32(4) (Contribution to Forestry Development).
- C A R : Loi N° 61/273 - Art. 66 (Payments to National Forestry Office). Ordonnance N° 79/025 - Art. 1 - 5 (Reforestation Tax)
- CONGO : Loi N° 004/74 - Art. 30(1) (Forest Management Tax and Fund); Art. 30(2) (Reforestation Tax and Fund). Loi N° 005/74 - Art. 14 - 17 (Assessment and Applicable Tax Rates of the Forest Management and Reforestation Tax); Art. 26, 27 (Establishment of the Forest Management Fund and the Reforestation Fund).
- EQUATORIAL GUINEA : Decreto-ley N° 14/1981 - Art. 17 (Reinvestment in Forest Improvement).
- GABON : Loi N° 1/82 - Art. 23(3), (4) (Fee for Inventories and Logging Plans executed by Forest Administration).

- GHANA : Act N° 12 of 1960 - Sec. 1 - 6 (Forest Improvement Fund).
- IVORY COAST : Décret N° 81-735 - Art. 7 (Possibility to Establish Forest Development Fund).
- LIBERIA : Revenue Law 1977 - Sec. 20.5 (Reforestation Fee). FDA Reg. N° 7 of 1979 - Sec. 4 (Reforestation Fee).
- NIGERIA : For. Amendment Edict 1969 - Sec. 18 (Establishment and Administration of Forestry Trust Fund).
- ZAIRE : Décret 11 avril 1949 - Art. 30 (Reforestation Levy).

LEVIES FOR OTHER PURPOSES

- CAMEROON : Loi N° 81/13 - Sec. 32(1) (Contribution to Special Council Support Fund); Sec. 32(5) (Contribution to the Execution of Socio-Economic Infrastructures).
- C A R : Ordonnance N° 79/025 - Art. 1 - 5 (Training Tax).
- EQUATORIAL GUINEA : Decreto-Ley N° 14/1981 - Art. 16, 17 (Community Development).
- GHANA : L.I. N° 23 of 1960 - Second Schedule (Fee for Log Measurement Certificates as of 31.8.1976).

9. FOREST ADMINISTRATION

ORGANIZATION AND FUNCTIONING OF FOREST SERVICE

- CAMEROON : Loi N° 81/13 - Sec. 6 (Responsibilities of the Forestry Service); Sec. 10 (Official Marking Hammers).
- CONGO : Loi N° 004/74 - Art. 23, 30(1) (Responsibilities of Forest Administration). Décret d'Application - Art. 63 - 67 (Forest Organization).
- EQUATORIAL GUINEA : Decreto-Ley N° 14/1981 - Art. 31 (Responsibilities of Forest Administration); Art. 32(1) (Organization of Forest Administration).
- GABON : Loi N° 1/82 - Art. 6 (Responsibility and Organization of Forest Administration).
- GHANA : Law N° 42 of 1982 - Sec. 34 (Establishment and Responsibilities of Forestry Commission); Sec. 64 (Dissolvement of Ghana Forestry Commission).
- IVORY COAST : Décret N° 81-735 - Art. 1 (Responsibilities of Ministry of Water and Forests); Art. 2 (Organization of Ministry); Art. 3 - 6 (Territorial Organization of National Forest Administration).
- LIBERIA : Forest Act 1953 - Sec. III, V (Establishment and Functions of Public Forest Administration). FDA Act 1976 - Sec. 2 - 4 (Establishment, Objectives and Powers of the Forest Development Authority); Sec. 5 - 12 (Internal Functioning of FDA).
- TANZANIA : For. Ord. 1982 - Sec. 3 (Responsibility of Public Forest Administration).

STATUS OF FOREST OFFICERS

- CAMEROON : Loi N° 81/13 - Sec. 12 (Responsibility and Status of Forest Officers); Sec. 138(1) (Special Payments to Forest Service Officials). Décret N° 83/169 - Sec. 82 (Regulations on Status of Forest Officers).
- R C A : Loi N° 61/273 - Art. 124 (Payments to Forest Officials).
- GABON : Loi N° 1/82 - Art. 7 (Power of Forest Officials to Collect Forest Fees and Taxes).
- GHANA : For. Ord. 1927 - Sec. 3 (Appointment of Forest Officers). N.R.C.D. 243 of 1974 - Sec. 8/1 (Assistance to Forest Officers).

- LIBERIA : Forests Act 1953 - Sec. IX (Appointment of Forest Officers). Supplementary Act 1957 - Part IX Sec. 1 - 5 (Duties and Responsibilities of Forest Officers). FDA Act 1976 - Sec. 13, 14 (Powers of Forest Officers).
- NIGERIA : For. Ord. Nor. Reg. 1960 - Sec. 3 (Appointment of Forest Officers).
- TANZANIA : For. Ord. 1982 - Sec. 4 (Honorary Forest Officers); Sec. 20, 20 A, 20 B (Powers of Forest Officers and of Director of Forestry); Sec. 19 A (Prohibition for Forest Officers to have Timber Dealings).

OTHER PUBLIC AGENCIES OF THE FOREST SECTOR

- CAMEROON : Décret N° 82-636 - Sec. 1 - 4 (Establishment and Responsibilities of the National Office for Forest Regeneration); Sec. 6 - 21 (Organization and Functioning of Office). Décret N° 81/223 - Sec. 1 - 5 (Establishment and Responsibilities of the National Office for Forestry Development); Sec. 6 - 24 (Organization and Functioning of Forestry Development Office).
- R C A : Ordonnance N° 69/49 - Art. 1 - 3 (Establishment and Responsibilities of the National Forestry Office).
- CONGO : Loi N° 004/74 - Art. 30(2) (National Reforestation Agency).
- GABON : Loi N° 1/82 - Art. 80 (National Reforestation Agency). Loi N° 10/75 - Art. 1 - 6, 13, 14 (Establishment and Educational Programmes of the National Forestry School); Art. 7, 8, 12 (Functioning of School); Art. 10, 11 (Advisory Council).
- IVORY COAST : Décret N° 66/422 - Art. 1 - 3 (Establishment and Responsibilities of State Reforestation Agency); Art. 4 - 10 (Internal Functioning, Budget and Supervisory Board).
- LIBERIA : FDA Act 1976 - Sec. 15 (Advisory Conservation Committees at County Level).
- NIGERIA : For. Amendment Edict 1969 - Sec. 3 A, 3 B (Establishment, Constitution and Functions of Forestry Advisory Commission); Sec. 32, 33 (Transfer of Local Government Forestry Staff and Property); First Schedule (Functioning and Proceedings of the Forestry Advisory Commission).

10. FOREST REGULATIONS (POWER TO MAKE)

- CAMEROON : Loi N° 81/13 - Sec. 7 (Regeneration in State Forests); Sec. 11(3) (Allowances for Forestry, Wildlife and Fisheries Officials); Sec. 12 (Status of Forestry Personal); Sec. 16 (Management.Plans for State Forests and National Parks); Sec. 20 (Use of Local Council and Private Forests); Sec. 21, 22 (Timber Harvesting and Other Uses on Communal Forest Land); Sec. 23(4) (Forest Survey Permits); Sec. 25 (Timber Allocation Methods); Sec. 27 (Timber Harvesting Rights on Communal Forest Land); Sec. 28 (Demarcation and Renewal of Forest Licences); Sec. 34 (Application for Forest Licences and Sales of Standing Timber); Sec. 36 (Security Deposit); Sec. 37 (Transfer of Timber Harvesting Rights); Sec. 39(2) (Other Forest Produce); Sec. 45 (Promotion of Lesser-Used Species); Sec. 77 (Exploitation along Water Courses).
- CONGO : Loi N° 004/74 - Art. 45, 46 (Granting of Timber Harvesting Rights, Wood Processing, Transport and Commercialisation of Forest Products).
- EQUATORIAL GUINEA : Decreto-Ley N° 14/1981 - Art. 32(3) (General Regulations).

- GABON : Loi N° 1/82 - Art. 5 (Customary Usage Rights); Art. 16 (Permits for Forest Produce); Art. 20 (Logging and Wood Processing Operators); Art. 21 (Transfer of Permits); Art. 30 (Grading Standards); Art. 80 (National Reforestation Agency); Art. 81 (Cooperatives of National Operators); Art. 83 (Import of Plant and Animals); Art. 117 (General Regulations).
- GHANA : For. Ord. 1927 - Sec. 33 (General Regulations). Decree N° 273 of 1974 - Sec. 17 (General Regulations). Decree N° 128 of 1977 - Sec. 12 (General Regulations).
- IVORY COAST : Loi N° 65/425 - Art. 25 (Logging Volumes; Raw Material Supply; Wood Processing; Transport; Commercialization and Grading of Logs and Forest Products); Art. 61 (Forest Protection and Management).
- LIBERIA : For. Act 1953 - Sec. IX (General Regulations).
- NIGERIA : For. Ord. Nor. Reg. 1960 - Sec. 46, 46 A, 47 (General Regulations); Sec. 48 (Power of NA or LGC to Make Rules).
- TANZANIA : For. Ord. 1982 - Sec. 30(1), (2) (Power of the Minister to Make Rules); Sec. 30(3) (Power of Local Authorities to Make Rules Subject to Approval); Sec. 31 (Power of the Minister to Grant Exemptions).

11. WILDLIFE RESOURCES

PROTECTION AND MANAGEMENT

- CAMEROON : Loi N° 81/13 - Sec. 1, 2 (Objectives of Wildlife Management); Sec. 6 (Responsibility of Wildlife Service); Sec. 69 - 74 (Wildlife Protection Measures); Sec. 138(1) (Payments to Wildlife Officials).
- GABON : Loi N° 1/82 - Art. 1 - 3 (Objectives of Wildlife Resources Utilization and Management); Art. 5 (Acknowledgment of Customary Rights); Art. 10 (Wildlife Management Areas as Part of the Forest Domain); Art. 32 (Definition); Art. 33 (Wildlife Management Areas and Protected Wildlife Areas); Art. 34 - 40 (Categories of Wildlife Management Areas); Art. 41 (Touristique Development in Wildlife Areas and Hunting Guides); Art. 42 (Responsable Services for Wildlife Management and Touristique Development); Art. 43 (Protected Wildlife Areas); Art. 44 (Management Plans); Art. 45 (Wildlife Management Areas for Local Communities); Art. 46 (Restrictions in Wildlife Management Areas); Art. 47 (List of Protected Animals); Art. 67 (Appointment of Voluntary Personal); Art. 83 (Import Permits for Animals and Trophees).
- EQUATORIAL GUINEA : Decreto-Ley N° 14/1981 - Art. 3(2) (Definition).
- LIBERIA : Supplementary Act 1957 - Part VIII Sec. 1 (Wildlife Regulations); Sec. 13 (Protected Wildlife Areas).

HUNTING PRACTISES AND CONTROL

- CAMEROON : Loi N° 81/13 - Sec. 46 - 62 (Exercise of Hunting Rights); Sec. 63 - 65 (Protection of Persons and Property against Animals); Sec. 66 - 68 (Hunting Arms).
- EQUATORIAL GUINEA : Decreto-Ley N° 14/1981 - Art. 20 (Practice and Control of Hunting).
- GABON : Loi N° 1/82 - Art. 49 - 62 (Hunting Practice and Control); Art. 63 - 66 (Life Game and Hunting Trophees).
- LIBERIA : Supplementary Act 1957 - Part VIII Sec. 1 - 5 (Permits and Licences in Connection with Wildlife Resources); Sec. 6, 7 (Registration of Live Game Animals in Captivity and Confiscation of Unregistered Animals); Sec. 8 (Special Permits for Elephant Hunting); Sec. 9 (Prohibited Acts); Sec. 11 (Permits for Hunting in Government Forest Reserves).

- ZAIRE : Ordonnance N° 244/79 - Art. 2(B), 12 - 25 (Categories of Fees Assessed in Connection with Hunting).

12. NATIONAL PARKS AND NATURE PROTECTION

- CAMEROON : Loi N° 81/13 - Sec. 75 - 77 (Measures for the Protection of Environment).
- CONGO : Loi N° 004/74 - Art. 3(3), (4), (5) (National Parks, Nature Reserves, Integral Nature Reserves); Art. 11 (Objectives and Procedure for Constituting National Parks and Nature Reserves); Art. 20 (Usage Rights in National Parks); Art. 27(2) (Considerations on National Parks and Native Reserves in Forest Management Plans).
- GABON : Loi N° 1/82 - Art. 10 (National Parks as Part of the Forest Domain).
- LIBERIA : For. Act 1953 - Sec. X (Establishment of National Parks). Supplementary Act 1957 - Part VIII Sec. 10 (Prohibition of Hunting and Commercial Fishing in National Parks); Sec. 14 (Advisory Conservation Committees at Provincial and County Level).

13. FISHERY RESOURCES

- CAMEROON : Loi N° 81/13 - Sec. 1, 2 (Objectives of Fishery Regulations); Sec. 6 (Responsibility of Fishery Service); Sec. 78 - 84 (Definitions in Connection with the Use of Fishery Resources); Sec. 85 - 96 (Exercise of Fishing Rights); Sec. 97 - 102 (Management and Conservation of Fishery Resources); Sec. 103 - 105 (Sea Farming and Fish Farming); Sec. 106 - 108 (Installations of Fish Processing Establishments); Sec. 109, 110 (Sanitary Inspection and Control of Fishery Products); Sec. 111 - 113 (Packaging and Transportation of Fishery Products); Sec. 137, 138 (Special Payments to Fishery Service and Fishery Officials).
- GABON : Loi N° 1/82 - Art. 1 - 3 (Objectives of Fishery Resources Development); Art. 5 (Acknowledgment of Customary Usage Rights); Art. 68 (Definition); Art. 69 (Evaluation of Existing Fishery Resources); Art. 70 (Fishing Permits and Authorizations); Art. 71 (Fishing Regulations); Art. 72 (Information to be Submitted); Art. 73 (Investment in Fish Processing); Art. 74 (Reservation of Areas for Small Scale Fishing); Art. 75 (Introduction of New Fishing Techniques); Art. 76 (Aquaculture); Art. 77 (Research and Training); Art. 78 (Sale of Fish and Fish Products).
- ZAIRE : Arrêté N°0001/CCE/ADRE/83 - Art. 7, 8 (Applicable Rates of Fishing Fees).